Lecture Notes in Mathematics

Edited by A. Dold and B. Eckmann

1097

R. M. Dudley
H. Kunita
F. Ledrappier

École d'Été de Probabilités de Saint-Flour XII – 1982

Édité par P. L. Hennequin

Springer-Verlag
Berlin Heidelberg New York Tokyo 1984

Auteurs

R. M. Dudley
Massachusetts Institute of Technology, Department of Mathematics
Cambridge, MA 02139, USA

H. Kunita
Department of Applied Science
Faculty of Engineering, Kyushu University
Fukuoka 812, Japan

F. Ledrappier
Université Pierre et Marie Curie (Paris VI)
Laboratoire de Probabilités, Tour 56
4, Place Jussieu, 75230 Paris Cedex 05, France

Editeur

P. L. Hennequin
Université de Clermont II, Complexe Scientifique des Cézeaux
Département de Mathématiques Appliquées
B.P. 45, 63170 Aubière, France

AMS Subject Classification (1980): 28D05, 28D20, 58F11, 60-02, 60F05, 60G44, 60H05, 60H10, 60J60, 62B05, 62G05

ISBN 3-540-13897-8 Springer-Verlag Berlin Heidelberg New York Tokyo
ISBN 0-387-13897-8 Springer-Verlag New York Heidelberg Berlin Tokyo

CIP-Kurztitelaufnahme der Deutschen Bibliothek
École d'Été de Probabilités: École d'Été de Probabilités de Saint-Fluor. – Berlin; Heidelberg; New York; Tokyo: Springer.
Teilw. mit d. Erscheinungsorten Berlin, Heidelberg, New York.
NE: HST
12. 1982 (1984).
(Lecture notes in mathematics; Vol. 1097)
ISBN 3-540-13897-8 (Berlin ...)
ISBN 0-387-13897-8 (New York ...)
NE: GT

Printing and binding: Beltz Offsetdruck, Hemsbach / Bergstr.
2146/3140-543210

INTRODUCTION

La Douzième Ecole d'Eté de Calcul des Probabilités de Saint-Flour s'est tenue du 22 Août au 8 Septembre 1982 et a rassemblé, outre les conférenciers, une cinquantaine de participants dans les locaux accueillants du Foyer des Planchettes.

Les trois conférenciers, Messieurs Dudley, Kunita et Ledrappier ont entièrement repris la rédaction de leurs cours qui constitue maintenant un texte de référence et ceci justifie le nombre d'années mis pour les publier.

En outre les exposés suivants ont été faits par les participants et ils ont été publiés dans le numéro 76 des Annales Scientifiques de l'Université de Clermont-Ferrand II :

M. HAREL	Convergence pour les processus empiriques éclatés
L.M. LE NY	Forme produit pour des réseaux multiclasses à routages dynamiques
G. LETAC	Mesures sur le cercle et convexes du plan
H.I. PEREIRA	Rate of convergence towards a Frechet type limit distribution
C. SUNYACH	Condition pour qu'une transformation d'un espace uniforme soit une contraction stricte et applications

La frappe du manuscrit a été assurée par les Départements de Kyushu University et de l'Université de Clermont II et nous remercions pour leur soin et leur efficacité les secrétaires qui se sont chargées de ce travail délicat.

Nous exprimons enfin notre gratitude à la Société Springer Verlag qui permet d'accroître l'audience internationale de notre Ecole en accueillant une nouvelle fois ces textes dans la collection Lecture Notes in Mathematics.

P.L. HENNEQUIN
Professeur à l'Université de Clermont II
B.P. n° 45
F-63170 AUBIERE

LISTE DES AUDITEURS

Mr. AZEMA J.	Université de Paris VI
Mr. BADRIKIAN A.	Université de Clermont-Ferrand II
Mr. BALDI P.	Institut de Mathématiques à Pise (Italie)
Mr. BERNARD P.	Université de Clermont-Ferrand II
Mr. BERTHUET R.	Université de Clermont-Ferrand II
Mr. BIRGE L.	Université de Paris X
Mr. BOFILL i SOLIGUER F.	Ecole Technique Supérieure de Tarrasa (Espagne)
Mr. BOUGEROL P.	Université de Paris VII
Mr. CARMONA R.	Université de Californie à Irvine (U.S.A.)
Mme CHALEYAT-MAUREL M.	Université de Paris VI
Mle CHEVET S.	Université de Clermont-Ferrand II
Mme COCOZZA C.	Université de Paris VI
Mr. CRAUEL H.	Université de Bremen (R.F.A.)
Mr. DERRIENNIC Y.	Université de Brest
Mr. DE SAM LAZARO J.	Université de Rouen
Mme ELIE L.	Université de Paris VII
Mr. FOURT G.	Université de Clermont-Ferrand II
Mr. GAREL B.	Université de Savoie à Chambéry
Mr. McGILL P.	Université Ulster à Coleraine (Irlande du Nord)
Mr. GOLDBERG J.	I.N.S.A. de Villeurbanne
Mr. HAREL M.	I.U.T. de Limoges
Mr. HE SHENG W.	Université de Strasbourg I
Mr. HENNEQUIN P.L.	Université de Clermont-Ferrand II
Mme IGLESIA PEREIRA H.	Faculté des Sciences de Lisbonne (Portugal)
Mr. ITMI M.	Faculté des Sciences de Rouen
Mr. LE JAN Y.	Université de Paris VI
Mr. LE PAGE E.	Université de Rennes I
Mr. LETAC G.	Université de Toulouse III
Mr. LIN Cheng	Faculté des Sciences de Rouen
Mr. LOOTGIETER J.C.	Université de Paris VI
Mr. MARTIAS C.	C.N.E.T. à Issy-les-Moulineaux
Mr. MASSART P.	Université de Paris XI
Mr. MASSE J.C.	Université Laval, Québec (Canada)
Mr. MOGHA G.	Université de Saint-Etienne
Mr. NDUMU NGU M.	Faculté des Sciences de Yaoundé (Cameroun)
Mr. NUALART D.	Université de Barcelone (Espagne)
Mr. PICHARD J.F.	Université de Haute-Normandie
Mr. PRATELLI M.	Institut de Statistiques à Padova (Italie)

Mr. REVUZ D.	Université de Paris VII
Mr. ROUSSIGNOL M.	Université de Paris VI
Mr. ROUX D.	Université de Clermont-Ferrand II
Mr. SALISBURY	Université de Vancouver (Canada)
Mme SANZ-SOLE M.	Université de Barcelone (Espagne)
Mr. UPPMAN A.	Université de Rouen
Mle WEINRYB S.	Ecole Polytechnique de Palaiseau
Mr. ZHENG X.	Université de Tientsin (Chine)
Mr. ZHENG W.	Université de Strasbourg I

TABLE DES MATIERES

R.M. DUDLEY : "A COURSE ON EMPIRICAL PROCESSES"

H. KUNITA : "STOCHASTIC DIFFERENTIAL EQUATIONS AND STOCHASTIC FLOW OF DIFFEOMORPHISMS"

F. LEDRAPPIER - "QUELQUES PROPRIETES DES EXPOSANTS CARACTERISTIQUES"

A COURSE ON EMPIRICAL PROCESSES

PAR R.M. DUDLEY

Preface

In this course I have tried to organize and present currently known limit theorems for empirical processes based on variables which are independent and identically distributed with some law P, in rather general (at least, multidimensional) sample spaces. For each limit theorem (Glivenko-Cantelli law of large numbers, Donsker central limit theorem, or law of the iterated logarithm) the object is to find the weakest possible conditions on a class of sets in the sample space, or functions on the space, such that convergence holds simultaneously and uniformly over the class. In this scheme, cumulative empirical distribution functions appear in the quite special case where the class of sets is the class of intervals $(-\infty,x]$ in the line or a Euclidean space. The limit theorems have been extended to much more general classes of sets and functions, especially in the past five years.

Bringing together the material brought up some questions, some of which could be answered, so that there are a few new results in the course. Recent improvements in technique (due to V. I. Kolčinskii and David Pollard) also allow shorter proofs of some older results. I will next briefly survey the contents of the course and mention what is new. (Readers new to the subject may prefer to begin with Chapter 1.)

The main emphasis will be on central limit theorems (Donsker classes). Laws of large numbers are treated in Sections 6.1, 11.1 and 11.2. Some laws of the iterated logarithm and corresponding strong invariance principles will be only briefly mentioned (Sections 1.1, 4.2); such results are proved in a forthcoming paper with Walter Philipp, "Invariance principles for sums of Banach space valued random elements and empirical processes".

Since the supremum norm for the space of all bounded functions on an infinite set is non-separable, there are difficulties of measurability (too many Borel sets) and even of choosing the space in which the empirical process and limiting Gaussian process take their values. The above-mentioned work with W. Philipp develops a new approach to the foundations of empirical processes via P. Major's formulation of Donsker's invariance principle (Section 1.1 below). For a possibly non-measurable function f on a probability space let f^* be the essential infimum of the set of measurable functions g with $g > f$. Let $(S, ||.||)$ be a Banach space, generally non-separable. It turns out that a useful set of inequalities for random variables in separable Banach spaces carry over to $||.||^*$, with only minor modifications in their proofs (Chap. 3). Thus the foundations given in Chapters 1-4 can avoid most of the measurability problems in previous versions of the theory. The new approach avoids the need to define convergence of laws, laws themselves, or sigma-algebras in non-separable metric or Banach spaces.

In connection with the Vapnik-Červonenkis property, which restricts only the intersections of sets in a class with finite sets, some further measurability conditions are really needed, as shown by an example (10.3.3) of Mark Durst and the author.

Limit theorems for empirical measures uniformly over classes $\mathcal{F}$ of functions are not only special cases of limit theorems in a Banach space S, but are actually equivalent to such theorems (Sec. 4.1), where $\mathcal{F}$ is (any norming subset of) the unit ball in the dual of S. For example, the classical strong law of large numbers of E. Mourier in separable Banach spaces, and the Blum-DeHardt Glivenko-Cantelli theorem, both follow from an easy extension of DeHardt's theorem on empirical measures to possibly unbounded functions (Sec. 6.1). Among central limit theorems in separable Banach spaces which do not restrict the geometry of the space, perhaps the best up to now is that of Naresh Jain and Michael Marcus. The Jain-Marcus theorem follows directly from a new central limit theorem of Pollard for empirical processes (Sec. 11.3).

There are at least three kinds of hypotheses sufficient for a central limit theorem uniformly over a class $\mathcal{F}$. If $\mathcal{F}$ is countable, special conditions are available (Chapter 5). In Chap. 6, for each $\varepsilon > 0$, $\mathcal{F}$ is covered by finitely many "brackets" $[f_i, f_j] = \{f: f_i \leq f \leq f_j\}$ where $||f_j - f_i||_p < \varepsilon$ in some $L^p(P)$ norm. If the number $N_p(\varepsilon)$ of brackets needed is always finite, we have a law of large numbers (Sec. 6.1). If $\mathcal{F}$ is uniformly bounded and $\int_0^1 (\log N_1(x^2))^{1/2}\, dx < \infty$ we have a central limit theorem (Sec. 6.2). The hypothesis on $N_1(x^2)$ is sharp (I. S. Borisov's theorem, Sec. 6.3).

In Chap. 7, $N_p(\varepsilon)$ is estimated for concrete classes of functions (satisfying uniform differentiability and Hölder conditions) and sets (regions with differentiable boundaries, convex sets and lower layers).

Chapter 8 is based on my paper "Empirical and Poisson processes on classes of sets or functions too large for central limit theorems". It is shown that the law of the iterated logarithm fails (by a factor of $(\log n)^{1/2}$) for the class of lower layers in the plane. The same holds for the convex sets in 3-space and the class of sets in $\mathbb{R}^d$ with uniformly C^{d-1} smooth boundaries.

Chapter 9 brings in the Vapnik-Červonenkis combinatorial condition on a class $\mathcal{C}$ of sets: for some n, no set with n elements has all its 2^n subsets cut out by sets in $\mathcal{C}$. In the purely combinatorial part of the theory, there are notable results of Vapnik and Červonenkis themselves, N. Sauer, P. Assouad, R. Wenocur and others, yet questions remain open in sets with as few as 6 elements (but many pairs of collections of subsets!).

Section 10.1 notes (a folk theorem?) that the empirical measure is a sufficient statistic for an unknown P. The rest of Chap. 10 develops measurability structures (admissibility, Suslin property) to be used in Chap. 11.

Section 11.1 introduces a kind of entropy, due independently to Kolčinskii and Pollard, which provides a common generalization for a) Vapnik-Červonenkis conditions and b) metric entropy of functions in the supremum norm. Section 11.2 proves the Vapnik-Červonenkis-Steele necessary and sufficient condition on a class $\mathcal{C}$ of sets (with suitable measurability structure) for the weak <u>or</u> strong law of large numbers (Glivenko-Cantelli theorem) uniformly over $\mathcal{C}$, for a given law P. (The recent extension by Vapnik and Červonenkis to uniformly bounded classes of functions is also mentioned.) The proof that the weak law implies the strong law in this case, due to Mike Steele, uses Kingman's subadditive ergodic theorem. Also, a new simpler proof by Kolčinskii and Pollard's methods is given for the Vapnik-Červonenkis weak law (which Steele just referred to in his proof).

Section 11.3 proves Pollard's central limit theorem, but with his separability replaced by the "admissible Suslin" condition from Chap. 10. Here $\mathcal{F}$ is of the form $\{Fg: g \in \mathcal{G}\}$ where $\int F^2 dP < \infty$ and $\mathcal{G}$ is a uniformly bounded class of functions satisfying an integral condition for Pollard's entropy, which does not depend on P. Thus if $F = 1$ one finds classes of functions satisfying the uniform central limit theorem for all P, hence useful in nonparametric statistics where P is unknown (Sec. 12.1).

If $\mathcal{G}$ is a class of indicators of sets, finiteness of Pollard's entropy is equivalent to the Vapnik-Červonenkis property (Sec. 11.1). Also, if the central limit theorem holds uniformly over a class $\mathcal{C}$ of sets for all laws P defined on $\mathcal{C}$, then $\mathcal{C}$ must have the Vapnik-Červonenkis property, a result of Mark Durst and the author (Sec. 11.4). In this sense, Vapnik-Červonenkis classes (with mild measurability conditions, which are satisfied for classes in reasonable applications) are exactly the right classes of sets for non-parametric statistics.

I am very grateful to all those who sent me preprints or reprints of their work on empirical processes, and to those whose comments led to corrections and improvements in the course: to Patrice Assouad, Erich Berger, Yves Derrienic, Joseph Fu, Lucien LeCam, Pascal Massart, Jim Munkres, Walter Philipp, He Sheng Wu, and especially to Lucien Birgé and Tom Salisbury, my heartiest thanks. I would much appreciate hearing from readers of any further errors or improvements.

Many thanks, as well, to Irene Fontaine-Gilmour for excellent typing.

Chapter 1 Introduction

1.1 Invariance principles

The classical results of Donsker (1951, 1952) have recently been put into simpler yet stronger forms as follows. Let λ be Lebesgue measure on $[0,1]$ and let $N(m,\sigma^2)$ denote the normal law with mean m and variance σ^2. Call a probability space nonatomic if there is a random variable on it with law λ. "Independent and identically distributed with law P" will be abbreviated "i.i.d. P".

Theorem 1.1.1. Let P be any probability law on $\mathbb{R}^1$ with mean 0 and variance 1. Then on any nonatomic probability space there exist random variables $X_1, X_2, \ldots$ i.i.d. P, and $Y_1, Y_2, \ldots$ i.i.d. $N(0,1)$, such that

$$(1.1.2)\quad \lim_{n\to\infty} n^{-1/2}\max_{k\le n}\left|\sum_{j=1}^k X_j - Y_j\right| = 0 \text{ in probability.}$$

Remark. The sequence $\{X_j\}_{j\ge1}$ is not, of course, independent of the sequence $\{Y_j\}_{j\ge1}$.

Let $S_n := X_1 + \ldots + X_n$ (throughout, ":=" means "equals by definition"). Let $S_0 := 0$, $f_n(k/n) := S_k/n^{1/2}$, $k = 0, 1, \ldots, n$, and let f_n be linear on each interval $[k/n, (k+1)/n]$. Donsker (1951) proved that the law of f_n, in the separable Banach space $C[0,1]$ of continuous real functions on $[0,1]$ with supremum norm, converges to that of standard Brownian motion. Theorem 1.1.1 is easily seen to imply Donsker's theorem, yet Theorem 1.1.1 itself avoids mentioning any infinite-dimensional sample space such as $C[0,1]$. Further, Theorem 1.1.1 replaces convergence in law by a stronger form of convergence, in probability.

The formulation of Theorem 1.1.1 emerged from proofs of Donsker's "invariance principle" in the books of Breiman (1968, Theorem 13.8, p. 279) and Freedman (1971, p. 83, (130)) and was brought out explicitly by Major (1976, p. 222).

The term "invariance principle", by the way, refers to the fact that the limit in distribution of any functional such as $\max_{k\le n} S_k/n^{1/2}$ is the same, or <u>invariant</u>, for all X_j i.i.d. with mean 0 and variance 1.

Theorem 1.1.1 extends in a natural way to finite-dimensional Euclidean spaces, and an alternate result implies laws of the iterated logarithm, as follows. On $\mathbb{R}^d$

let $N(m,C)$ be the normal law with mean m and covariance matrix C.

<u>Theorem 1.1.3</u> Let P be a law on $\mathbb{R}^d$ with mean 0 and covariance matrix C ($\int x_i x_j \, dP = C_{ij}$, $i,j = 1, \ldots, d$, $x = \langle x_1, \ldots, x_d \rangle \in \mathbb{R}^d$). Then on any nonatomic probability space there exist random vectors $X_1, X_2, \ldots,$ i.i.d. P and $Y_1, Y_2, \ldots,$ i.i.d. $N(0,C)$, such that (1.1.2) holds (where $|\cdot|$ is any norm on $\mathbb{R}^d$) for convergence in L^2. Or, Y_j can be chosen i.i.d. $N(0,C)$ such that (instead of (1.1.2)) we have

(1.1.4) $$\lim_{n\to\infty} \left(\sum_{j=1}^n X_j - Y_j\right)/(nLLn)^{1/2} = 0 \quad \text{a.s.}$$

where $Lx := \log \max(x,e)$, $LLx := L(Lx)$.

Theorem 1.1.3 is due to Walter Philipp: for the (1.1.4) case see Philipp (1979, Cor.1; cf. corrections in Dudley and Philipp, 1982, sec.5). For the (1.1.2) case see Philipp (1980, Theorem 1, with correction). For $k = 1$, Theorem 1.1.3 for the (1.1.4) case can be proved by the Skorohod imbedding (Strassen, 1964, Theorem 2).

Next we turn to the other main theorem of Donsker, (1952), on empirical distribution functions. Let $x_1, x_2, \ldots,$ be i.i.d. λ. Then the empirical distribution function F_n is defined by

$$F_n(t) := n^{-1}\sum_{j=1}^n 1_{[0,t]}(x_j).$$

Let

$$X_j(t) := 1_{[0,t]}(x_j) - t, \quad 0 \leqslant t \leqslant 1.$$

Let $Y(t)$ be a Brownian bridge process, that is a Gaussian process for $0 \leqslant t \leqslant 1$ with mean 0 and covariance $EY(s)Y(t) = s(1-t)$, $0 \leqslant s \leqslant t \leqslant 1$. We can take $Y(t) = W(t) - tW(1)$, where W is a standard Brownian motion (Wiener process), and (thus) $t \longrightarrow Y(t)(\omega)$ is continuous in t for (almost) all ω. One then has

<u>Theorem 1.1.5</u> On any nonatomic probability space there exist random variables $x_1, x_2, \ldots,$ i.i.d.(λ) and i.i.d. Brownian bridges $Y_1(t), Y_2(t), \ldots,$ such that almost surely as $n \longrightarrow \infty$

$$\sup_{0\leqslant t\leqslant 1}\left|\sum_{j=1}^n 1_{[0,t]}(x_j) - t - Y_j(t)\right| = \mathcal{O}((\log n)^2).$$

Theorem 1.1.5 is due to Komlós, Major and Tusnády (1975, Theorem 4). It improves on a theorem of Kiefer (1972) who had $n^{\varepsilon+1/3}$, $\varepsilon > 0$, in place of $(\log n)^2$; here any sequence $o(n^{1/2})$ is enough to imply the result of Donsker (1952).

Further, Theorem 1.1.5 avoids the measurability problems which affect Donsker's formulation (e.g., Billingsley, 1968, p. 153; Dudley, 1966). Note also that the speed of convergence in Theorem 1.1.5 makes unnecessary the division into cases as in (1.1.2) and (1.1.4) in Theorem (1.1.3).

A price that will be paid for avoiding some problems of measurability and choice of function space is to assign joint distributions to some random variables X_j and Y_j that do not naturally have any joint law. Methods for such assignment will be presented beginning in the next section.

Note Kolmogorov (1931, 1933) showed that if $a(\cdot)$ and $b(\cdot)$ are two C^1 functions with $a(0) < 0 < b(0)$ and $a(t) < b(t)$ for all t, then (for the above piecewise-linear f_n) $\lim_{n\to\infty} \Pr\{a(t) < f_n(t) < b(t), 0 < t \leq 1\}$ converges to the corresponding probability when $f_n(t)$ is replaced by Brownian motion. This result overlaps substantially with Donsker's invariance principle. Kolmogorov's papers were overlooked to some degree in the 1950's. On some later developments see Borovkov and Korolyuk (1965) and Nagaev (1970). Thanks to L. LeCam for pointing out Kolmogorov's papers.

I. S. Borisov (1981a) has announced improvements to Theorem 1.1.5.

All the references are collected at the end of the text. On several topics in empirical processes, up to 1978, Gaenssler and Stute (1979) give many further references. Other, forthcoming, extensive treatments are those of Gaenssler (1983) and Pollard (1982b).

1.2 Construction of joint distributions. This section presents some lemmas useful in the construction of random variables for which joint laws may or may not be given.

A Polish space is a topological space metrizable as a complete, separable metric space. A law is a probability measure; if defined on a Polish space, the σ-algebra is that of Borel sets unless otherwise specified. Let us recall:

1.2.1. Lemma (disintegration). Let X and Y be Polish spaces and Q any law on $X \times Y$. Then there exists a map $x \to Q_x$ from X into the set of all laws on Y, such that $Q = \int Q_x \, dq$ where q is the marginal of Q on X, i.e., for every bounded measurable real function f on $X \times Y$,

$$\int f(x,y)\,dQ(x,y) = \int_X \int_Y f(x,y)\,dQ_x(y)\,dq(x).$$

Proof. For every uncountable Polish space H there is a 1-1 Borel-measurable isomorphism of H onto $[0,1]$ (Parthasarathy, 1967, p. 14). Thus every Polish space is Borel-isomorphic to some compact subset of $[0,1]$ (either the whole interval, a finite set, or the sequence $\{0\} \cup \{1/n\}_{n\geq 1}$). So we may assume $X = Y = [0,1[$ with usual metric. Let

$f_n(x,t) := Q([(k-1)/2^n, k/2^n[\times[0,t[)/q([(k-1)/2^n, k/2^n[)$, $k-1 \leq 2^n x < k$, $k = 1, \ldots, 2^n$, $0 \leq t \leq 1$, or $f_n(x,t) := 0$ if the denominator is 0. Then $0 \leq f_n \leq 1$. For each t, $f_n(x,t)$ is a martingale sequence with respect to $q(dx)$ and converges a.s. to some $F(x,t)$. By first restricting t to be rational we may assume $F(x,\cdot)$ to be non-decreasing with $F(x,0) \equiv 0$, $F(x,1) \equiv 1$. Then $F(x,\cdot)$ is a (left continuous) probability distribution function defining a law Q_x. Now the result holds for $f(x,y) = 1_{[a,b[}(x)1_{[c,d[}(y)$ where a, b, c and d are dyadic rationals (of the form $k/2^n$). The smallest collection $\mathcal{F}$ of functions containing all finite linear combinations of such f, and such that bounded pointwise limits of sequences in $\mathcal{F}$ are in $\mathcal{F}$, gives all bounded measurable f as stated. □

Next we have:

1.2.2 Lemma For Polish spaces X, Y and Z, let P be a law on $X \times Y$ and Q on $Y \times Z$ where P and Q have the same marginal on Y. Then some law R on $X \times Y \times Z$ has marginals P on $X \times Y$ and Q on $Y \times Z$.

Proof. From Lemma 1.2.1 let $P = \int P_y\,dp$, $Q = \int Q_y\,dp$ for some laws P_y on X and Q_y on Z and the common marginal p on Y. For each $y \in Y$ let R_y be the product law $P_y \times Q_y$ on $X \times Z$. Then (as in 1.2.1) let $R = \int R_y\,dp$ on $X \times Y \times Z$. It is easily checked that for any product measurable bounded real function f on $X \times Y \times Z$, $y \longrightarrow \iint f(x,y,z)\,dP_y(x)\,dQ_y(z)$ is completion measurable, so that R is a well-defined law on $X \times Y \times Z$ which has the desired marginals. □

1.2.3 Lemma Let Q be a law on $S \times T$ for Polish spaces S and T. Let q be the marginal of Q on S. Let $(\Omega, \mathcal{S}, P)$ be a probability space and V a random variable on Ω with values in S and law $\mathcal{L}(V) = q$. Suppose there is a real random variable U on Ω independent of V with continuous distribution function F_U. Then there is a random variable $W: \Omega \to T$ such that the joint law $\mathcal{L}(\langle V,W\rangle) = Q$.

Proof. As in the proof of 1.2.1 we may assume $S = T = [0,1]$ with usual metric. Taking $F_U(U)$, we may assume U is uniformly distributed on $[0,1]$. By Lemma 1.2.1 take a disintegration $Q = \int Q_x dq(x)$. Let F_x be the distribution function of Q_x and

$$F_x^{-1}(t) := \inf\{u: F_x(u) \geq t\}, \quad 0 < t < 1.$$

Then $F_x^{-1}(t) \leq z$ if and only if $F_x(z) \geq t$, $0 < t < 1$, $-\infty < z < \infty$, and $x \to F_x(z)$ is measurable. Thus $x \to F_x^{-1}(t)$ is measurable for each t, $0 < t < 1$. For each x, F_x^{-1} is left continuous and non-decreasing in t. Hence $\langle x,t\rangle \to F_x^{-1}(t)$ is jointly measurable. Thus for $W(\omega) := F_{V(\omega)}^{-1}(U(\omega))$, $\omega \to W(\omega)$ is measurable. For each x, $\lambda \circ (F_x^{-1})^{-1} = Q_x$. Thus for any bounded Borel function g, Fubini's theorem gives

$$\begin{aligned}
\int g dQ &= \int_0^1 \int_0^1 g(x,y) dQ_x(y) dq(x) \\
&= \int_0^1 \int_0^1 g(x,F_x^{-1}(y)) dy dq(x) \\
&= \int_0^1 \int_0^1 g(x,F_x^{-1}(y)) d(q \times \lambda)(x,y) \\
&= E(g(V,F_V^{-1}(U))) = Eg(V,W),
\end{aligned}$$

so $\mathcal{L}(V,W) = Q$. □

Notes In Lemma 1.2.1, $X \times Y$ can be replaced by a Polish space Z with a Borel measurable map g into X and $q = Q \circ g^{-1}$, Q_x on Z; completion of q may then be needed (Bourbaki, 1959, Chap.6, 58-59).

Lemma 1.2.2 and its proof are due to Vorob'ev (1962) for finite sets X, Y, Z, and to Berkes and Philipp (1977, Lemma A1) for separable Banach spaces. Vorob'ev (1962) and Shortt (1982) treat more complicated systems of marginals. Shortt (1982) allows suitable spaces more general than Polish spaces, finds for what X the result holds for all separable Y, Z, etc.

Lemma 1.2.3 can be found essentially in the proof of Skorohod (1976, Theorem 1), as Erich Berger kindly pointed out. See also Eršov (1975).

Chapter 2 Empirical processes, Gaussian limits, and inequalities.

2.1 Gaussian limit processes. Theorem 1.1.5 now has a variety of extensions and related results, where "$\mathcal{O}((\log n)^2)$ a.s." is replaced by various modes and rates of convergence, while the sets $[0,t]$ or functions $1_{[0,t]}$ are replaced by suitable classes of sets or functions. In fact, the development of such classes, in more general (usually multidimensional) spaces (but without rates) will be the main object of the rest of the course.

Let $(A,\mathcal{A},P)$ be any probability space and $\mathcal{L}^2 := \mathcal{L}^2(A,\mathcal{A},P) := \{f: A \to \mathbb{R}$, $\mathcal{A}$-measurable, with $\int f^2 dP < \infty\}$. (Here functions equal P-a.s. are not identified into equivalence classes; if they were, the space would be called L^2.) On $\mathcal{L}^2$ we have the usual pseudo-metric which I call $e_P(f,g) := (\int(f-g)^2 dP)^{1/2}$.

Let $x_1, x_2, \ldots,$ be i.i.d. P, $x(j) := x_j$. For technical reasons, I take x_j to be the coordinates for a countable product $(A^\infty, \mathcal{A}^\infty, P^\infty)$ of copies of $(A,\mathcal{A},P)$.

Let $P(f) := \int f dP$, $\delta_x(f) := f(x)$ and $P_n := n^{-1}(\delta_{x(1)} + \ldots + \delta_{x(n)})$. Here P_n is the nth empirical measure, itself random. Let $\nu_n := n^{1/2}(P_n - P)$.

Let W_P be the isonormal Gaussian process indexed by $\mathcal{L}^2(A,\mathcal{A},P)$ (e.g. Dudley, 1973), i.e., the $W_P(f)$ $f \in \mathcal{L}^2$, are jointly Gaussian with mean 0 and covariance

$$E\, W_P(f) W_P(g) = P(fg), \quad f, g \in \mathcal{L}^2.$$

Let G_P be another Gaussian process, also indexed by $\mathcal{L}^2(A,\mathcal{A},P)$, with mean 0 and covariance

(2.1.1) $$E\, G_P(f) G_P(g) = P(fg) - P(f)P(g), \; f, g \in \mathcal{L}^2.$$

Here G_P can be defined by

(2.1.2) $$G_P(f) = W_P(f) - P(f) W_P(1).$$

Then $t \to W_\lambda(1_{[0,t]})$ is a Brownian motion and $t \to G_\lambda(1_{[0,t]})$ a Brownian bridge, $0 \leq t \leq 1$.

For definiteness and to provide a suitable base for later developments I take as an underlying probability space

$$(\Omega, \mathcal{S}, \Pr) := (A^\infty, \mathcal{A}^\infty, P^\infty) \times ([0,1], \mathcal{B}, \lambda)$$

where $\mathcal{B}$ is the Borel σ-algebra. Let $\Pr^*(B) := \inf\{\Pr(C): C \supset B\}$, $B \subset \Omega$.

For any constant function f on A, $f(x) - P(f) \equiv 0$, $\nu_n(f) \equiv 0$ and $G_P(f) = 0$ a.s. For each $f \in \mathcal{L}^2$ and $g := f - P(f)$ we have $\nu_n(g) \equiv \nu_n(f)$ and $G_P(f) \equiv G_P(g)$. Thus additive constant functions are "ignored" by the ν_n and G_P. To factor out the constants let $||\cdot||_{2,0}$ be the seminorm on $\mathcal{L}^2(A,\mathcal{A},P)$ given by

$$||f||^2_{2,0} := \sigma^2_P(f) := P(f^2) - P(f)^2, \quad f \in \mathcal{L}^2.$$

Let $\mathcal{L}^2_0$ denote the vector space $\mathcal{L}^2$ but with the inner product $(f,g)_{2,0} := P((f - P(f))(g - P(g)))$. Let ρ_P be the corresponding pseudo-metric; $\rho_P(f,g) := \sigma_P(f-g) \geqslant 0$. Then G_P is a linear isometry from $\mathcal{L}^2_0$ to a space of Gaussian variables (it becomes an "isonormal" process with respect to the inner product given by its covariance (2.1.1)).

When restricted to any finite subset of $\mathcal{L}^2(A,\mathcal{A},P)$, the normalized empirical measure ν_n converges in law to G_P as $n \to \infty$ by the finite-dimensional central limit theorem. To extend the convergence to suitable infinite subsets $\mathcal{F} \subset \mathcal{L}^2(A,\mathcal{A},P)$, while keeping the convergence uniform over $\mathcal{F}$ (as in Theorem 1.1.5 for $\mathcal{F} := \{1_{[0,t]}: t \in \mathbb{R}\}$), it is convenient if the limit process G_P is itself well-behaved on $\mathcal{F}$ in the following sense.

__Definition.__ A class $\mathcal{F} \subset \mathcal{L}^2(A,\mathcal{A},P)$ will be called a __G_PBUC__ class if and only if the process $G_P(f)(\omega)$ can be chosen so that for all ω, the sample functions $f \to G_P(f)(\omega)$, restricted to $f \in \mathcal{F}$, are __b__ounded and __u__niformly __c__ontinuous for ρ_P.

A class $\mathcal{F} \subset \mathcal{L}^2(A,\mathcal{A},P)$ will be called a __functional Donsker class__ (for P) if and only if it is a G_PBUC class and there are processes $Y_j(f,\omega)$, $f \in \mathcal{F}$, $\omega \in \Omega$, where Y_j are independent copies of G_P, with $f \to Y_j(f,\omega)$ bounded and ρ_P-uniformly continuous on $\mathcal{F}$ for each j, such that for every $\varepsilon > 0$

$$(2.1.3) \qquad \Pr^*\{n^{-1/2}\max_{m\leqslant n}\sup_{f\in\mathcal{F}}|\textstyle\sum_{j=1}^m f(x_j) - P(f) - Y_j(f)| > \varepsilon\} \to 0, \quad n \to \infty.$$

Say that $\mathcal{F}$ satisfies a __strong invariance principle__ if Y_j can be chosen so that, instead of (2.1.3), there are measurable functions U_n on Ω such that as $n \to \infty$, almost surely $U_n \to 0$ and

$$(2.1.4) \qquad \sup_{f\in\mathcal{F}}|\textstyle\sum_{j=1}^n f(x_j) - P(f) - Y_j(f)| \leqslant (nLLn)^{1/2}U_n.$$

To apply the above definitions to a class of sets $\mathcal{C} \subset \mathcal{A}$ let $\mathcal{F} = \{1_B: B \in \mathcal{C}\}$.

<u>Remarks</u>. The almost sure boundedness and ρ_P-uniform continuity of sample functions of (a suitable version of) G_P on $\mathcal{F}$ do not follow from (2.1.3) nor (2.1.4). Indeed, suppose the functions in $\mathcal{F}$ have a Gaussian joint distribution with mean 0 for P. Then one can set $Y_j(f) := f(x_j)$ and the left sides of both (2.1.3) and (2.1.4) will be identically 0, although $\mathcal{F}$ may be an infinite-dimensional linear subspace of $\mathcal{L}^2$.

Since in (2.1.3) and (2.1.4) we are concerned with convergence uniformly over $\mathcal{F}$, it is natural to work in the space $\ell^\infty(\mathcal{F})$ of all bounded real functions on $\mathcal{F}$ with supremum norm, and to assume that G_P has its sample functions at least bounded on $\mathcal{F}$.

There exist sets $\mathcal{F}$ jointly Gaussian with respect to P on which G_P is a.s. bounded and even uniformly continuous with respect to e_P but not for ρ_P. For example let $\{f_n\}$ be a sequence of functions in $\mathcal{L}^2$ which are i.i.d. N(0,1) and let $\mathcal{F} = \{n + f_n/(L(n+1))^{1/2}\}_{n\geq 1}$ (Sonis, 1966; Dudley, 1967, Props. 6.7, 6.9). For such an $\mathcal{F}$, letting $Y_j(f) := f(x_j) - P(f)$, we again have the left sides of (2.1.4) and (2.1.3) identically 0. So to avoid such pathological cases, the G_PBUC property (for ρ_P) was included in the definitions.

The word "functional" in "functional Donsker class" refers not only to the fact that $\mathcal{F}$ is a class of functions, but more to the $\max_{m\leq n}$ in (2.1.3), so that we also consider functions of the argument m (which classically, though not here, have been interpolated to give functions of a continuous variable).

2.2 <u>Inequalities</u>. To prove that certain classes $\mathcal{F}$ are functional Donsker classes or satisfy strong invariance principles, some inequalities for a fixed f will be useful. Let

$$S_n(f) := \sum_{j=1}^n (f(x_j) - P(f)).$$

Several inequalities will follow from the Bernštein-Chebyshev moment-generating functions inequality: for any real random variable X and $t \in R$,

$$\text{(2.2.1)} \qquad \Pr(X \geq t) \leq \inf_{u\geq 0} e^{-tu} E e^{uX}.$$

First we have :

2.2.2 <u>Bernštein's inequality</u>. If $|f(x) - P(f)| \leq M < \infty$ for P-almost all x and $\sigma := \sigma_P(f)$, then for $0 < K < \infty$,

(2.2.3) $\quad \Pr\{|S_n(f)| \geq Kn^{1/2}\} \leq 2\exp(-K^2/(2\sigma^2 + 2Mn^{-1/2}K/3))$.

<u>Proof</u>. For any $u \geq 0$,

$E\exp(uS_n(f)) = (E\exp(uS_1(f)))^n = (1 + \sigma^2u^2F/2)^n < \exp(n\sigma^2u^2F/2)$

where $F := F(u) := 2\sigma^{-2}\sum_{r=2}^{\infty} u^{r-2}ES_1(f)^r/r!$.

For $r \geq 2$, $|S_1(f)|^r \leq S_1(f)^2M^{r-2}$, so $F \leq 2\sum_{r=2}^{\infty}(Mu)^{r-2}/r! \leq \sum_{r=2}^{\infty}(Mu/3)^{r-2}$

$\leq 1/(1 - Mu/3)$ if $0 < u < 3/M$.

Set $v := Kn^{1/2}$ and $u := v/(n\sigma^2 + Mv/3)$, so that $v = n\sigma^2u/(1 - Mu/3)$. Then $0 < u < 3/M$ and $n\sigma^2u^2F/2 \leq n\sigma^2u^2/(2 - 2Mu/3) = vu/2$ so by (2.2.1)

$\Pr(S_n(f) \geq v) \leq e^{-uv/2} = \exp(-v^2/(2n\sigma^2 + 2Mv/3)) \leq \exp(-K^2/(2\sigma^2 + 2MKn^{-1/2}/3))$. □

Next, let $Q(\{1\}) = Q(\{-1\}) = 1/2$ and let s_j be i.i.d. Q ("Rademacher" variables).

Then we have:

2.2.4 <u>Proposition</u> (Hoeffding). For any $t \geq 0$ and $a_j \in \mathbb{R}$

$$\Pr\{\sum_{j=1}^{n} a_js_j > t\} \leq \exp(-t^2/(2\sum_{j=1}^{n} a_j^2)).$$

<u>Proof</u>. Since $1/(2n)! \leq 2^{-n}/n!$, $n = 0, 1, \ldots$, we have $\cosh x \leq \exp(x^2/2)$ for all x. Now $\inf_u \exp(-ut + \sum_j a_j^2u^2/2)$ is attained at $u = t/\sum a_j^2$, giving the result. □

Other inequalities, applicable to $S_n(f)$ when f is the indicator of a set, concern binomial probabilities

$$B(k,n,p) := \sum_{0 \leq j \leq k} \binom{n}{j} p^jq^{n-j}, \quad 0 \leq p := 1 - q \leq 1,$$

$$E(k,n,p) := \sum_{k \leq j \leq n} \binom{n}{j} p^jq^{n-j},$$

where k is not necessarily an integer.

2.2.5 <u>Chernoff-Okamoto inequalities</u>. We have

(2.2.6) $\quad E(k,n,p) \leq (np/k)^k(nq/(n-k))^{n-k}$ if $k \geq np$,

(2.2.7) $\quad B(k,n,p) \leq \exp(-(np-k)^2/(2npq))$ if $k \leq np \leq n/2$.

<u>Proof</u>. We have the moment generating function

$$\sum_{k=0}^{n} e^{uk}\binom{n}{k}p^kq^{n-k} = (pe^u + q)^n$$

and apply (2.2.1) where $e^u = tq/((n-t)p)$ to get (2.2.6) and the same bound for $B(k,n,p)$ if $k \leq np$ by symmetry.

For (2.2.7), let $f(x) := x\cdot\log(x/p) + (1-x)\log((1-x)/q)$, $0 \leqslant x \leqslant 1$, where $0\cdot\log 0 := 0$. Then (2.2.6) implies $B(k,n,p) \leqslant e^{-nf(k/n)}$, $k \leqslant np$. We have $f(p) = f'(p) = 0$, $f''(x) = 1/(x(1-x)) \geqslant 0$. Let $g(x) := (x-p)^2/(2pq)$. Then $g(p) = g'(p) = 0$, $g''(x) = 1/(pq) \leqslant f''(x)$ for $0 \leqslant x \leqslant p \leqslant 1/2$, so in that range $g(x) \leqslant f(x)$ and $e^{-nf(k/n)} \leqslant e^{-ng(k/n)}$, hence the result. □

Setting, in (2.2.6), $x := nq/(n-k) \leqslant e^{x-1}$ gives

(2.2.8) $$E(k,n,p) \leqslant (np/k)^k e^{k-np} \quad \text{if } k \geqslant np.$$

Here is a more special inequality that will be useful later :

2.2.9 Proposition. If $k \leqslant n/2$ then

$$2^n B(k,n,1/2) \leqslant (ne/k)^k.$$

Proof. By (2.2.6) and symmetry,

$$B(k,n,1/2) \leqslant (n/2)^n k^{-k}(n-k)^{k-n}.$$

Letting $y := n/(n-k) \leqslant e^{y-1}$ then gives the result. □

For any real x let $x^+ := \max(x,0)$. A Poisson random variable z with parameter m satisfies $\Pr(z = k) = e^{-m}m^k/k!$ $k = 0, 1, \ldots$.

2.2.10 Lemma. For any Poisson variable z with parameter $m \geqslant 1$,

$$E(z-m)^+ \geqslant m^{1/2}/8.$$

Proof. Let $j := [m]$ (greatest integer $\leqslant m$). Then Stirling's formula with an error bound (Feller, vol.I, Sec.II.9, p.54) and a telescoping sum give

$$\begin{aligned} E(z-m)^+ &= \textstyle\sum_{k>m} e^{-m}m^k(k-m)/k! \\ &\geqslant me^{-m}m^j(e/j)^j(2\pi j)^{-1/2}e^{-1/(12j)} \\ &\geqslant (m^{j+1}/j^{j+1/2})e^{-13/12}(2\pi)^{-1/2} \\ &\geqslant m^{1/2}/8. \end{aligned}$$

□

Notes. S. Bernštein (1927, pp. 159-165) published his inequality. The above proof is based on Bennett (1962, p.34) with some incorrect, but unnecessary steps (his (3), (4) ..) removed as suggested by Giné (1974). For related and stronger inequalities under weaker conditions (non-identically distributed, unbounded variables) see also Bernštein (1924, 1927), Hoeffding (1963), and Uspensky (1937, p.205).

Hoeffding (1963, Theorem 2) implies Prop. 2.2.4. Chernoff (1952, (5.11)) proved (2.2.6). Okamoto (1958, Lemma 2(b')) proved (2.2.7). For the last three inequalities I do not know references except my own papers (1978, Lemma (2.7) for (2.2.8); 1983(?), Lemma 3.3 for (2.2.10)).

Chapter 3. Measurability and sums in general Banach spaces.

3.1 Measurable cover functions and norms. In the classical case of empirical distribution functions for the uniform law on $[0,1]$ as in Theorem 1.1.5, we have a collection of functions $\mathcal{F} = \{1_{[0,t]}: 0 \leq t \leq 1\}$. For a bounded real function g on $\mathcal{F}$ let

$$||g||_{\mathcal{F}} := \sup_{f\in\mathcal{F}}|g(f)|.$$

Then if $x \neq y$ in $[0,1]$, $||\delta_x - \delta_y||_{\mathcal{F}} = 1$. Thus the set D of all δ_x, $0 \leq x \leq 1$, is discrete and non-separable for the norm $||.||_{\mathcal{F}}$. When x has uniform law λ, to make all closed subsets of D for $||.||_{\mathcal{F}}$ measurable is to extend λ to all subsets of $[0,1]$, which is impossible under the axiom of choice and continuum hypothesis (or some weaker assumptions: Banach and Kuratowski, 1929; Ulam, 1930).

For such reasons one has to deal with some non-measurable sets and functions. It will be useful to approximate a possibly non-measurable function f from above by a measurable function, as follows.

Let $(\Omega,\mathcal{S},\Pr)$ be a probability space. For a possibly non-measurable set $A \subset \Omega$ the notion of measurable cover is rather familiar: it is a set $B \in \mathcal{S}$ such that $A \subset B$ and

$$\Pr(B) = \Pr^*(A) := \inf\{\Pr(C): C \supset A, C\in\mathcal{S}\}.$$

If B and C are two measurable covers of A then $\Pr\{1_B = 1_C\} = 1$.

Let $\bar{\mathbb{R}} := [-\infty,\infty]$ and let $\mathcal{L}^o := \mathcal{L}^o(\Omega,\mathcal{S},\Pr, \bar{\mathbb{R}})$ denote the set of all $\mathcal{S}$-measurable functions from Ω into $\bar{\mathbb{R}}$. Then $\mathcal{L}^o$ is a lattice: for any $f, g \in \mathcal{L}^o$, $f\vee g := \max(f,g) \in \mathcal{L}^o$ and $f\wedge g := \min(f,g) \in \mathcal{L}^o$. But this $\mathcal{L}^o$ is not a vector space because infinite values are allowed (even on sets of positive probability). On it I put the pseudo-metric

$$d(f,g) := \inf\{\varepsilon > 0: \Pr\{|\tan^{-1}f-\tan^{-1}g| > \varepsilon\} < \varepsilon\}.$$

Then $d(f,g) = 0$ if and only if $f = g$ a.s.

For any subset $\mathcal{J} \subset \mathcal{L}^o(\Omega,\mathcal{S},\Pr,\bar{\mathbb{R}})$, a function $f \in \mathcal{L}^o$ is called an essential infimum of $\mathcal{J}$, $f :=$ ess.inf $\mathcal{J}$, if and only if for all $j \in \mathcal{J}$, $f \leq j$ a.s., and for any $g \in \mathcal{L}^o$ such that $g \leq j$ a.s. for all $j \in \mathcal{J}$, we have $g \leq f$ a.s. Clearly an essential infimum, if it exists, is unique up to a.s. equality.

3.1.1. Theorem. For any probability space $(\Omega,\mathcal{S},\mathrm{Pr})$ and $\mathcal{J} \subset \mathcal{L}^0(\Omega,\mathcal{S},\mathrm{Pr},\bar{\mathbb{R}})$, an essential infimum of $\mathcal{J}$ exists. If for some (in general non-measurable) function $f: \Omega \to \mathbb{R}$ we have $\mathcal{J} = \{j \in \mathcal{L}^0;\ j \geq f \text{ everywhere}\}$ then we can choose $f^* := \text{ess.inf}\ \mathcal{J}$ so that $f^* \geq f$ everywhere.

Proof. Take $j_m \in \mathcal{J}$ such that $\mathrm{Pr}(\tan^{-1} j_m) \downarrow \inf_{j \in \mathcal{J}} \mathrm{Pr}(\tan^{-1} j)$. Then $\min(j_1, \ldots, j_m) \downarrow j$ for some $j \in \mathcal{L}^0(\Omega,\mathcal{S},\mathrm{Pr},\bar{\mathbb{R}})$ and $j = \text{ess.inf}\ \mathcal{J}$. If $\mathcal{J} = \{h \in \mathcal{L}^0 : h \geq f\}$ then $j \geq f$ everywhere. □

Here f^* has been called a measurable cover function. If f is real valued and bounded above by some finite constant (or measurable real function) then f^* is also real valued. But whenever there exist non-measurable $A_n \downarrow \emptyset$ with $\mathrm{Pr}^*(A_n) \equiv 1$, let $f := n$ on $A_n \setminus A_{n+1}$. Then f is real valued but $f^* = +\infty$ a.s.

Some inequalities for f^* will be useful. The next two Lemmas are straightforward.

3.1.2 Lemma. For any two functions $f,g : \Omega \to]-\infty,\infty]$, we have $(f+g)^* \leq f^* + g^*$ a.s. We also have $(f-g)^* \geq f^* - g^*$ a.s. whenever both sides are defined a.s.

3.1.3 Lemma. Let S be a vector space with a seminorm $||\cdot||$. Then for any two functions X, Y from Ω into S,

$$||X+Y||^* \leq (||X|| + ||Y||)^* \leq ||X||^* + ||Y||^* \quad \text{a.s.}$$

and $||cX||^* = |c|\ ||X||^*$ a.s. for each real c.

Next we consider products.

3.1.4 Lemma. Let $(A_j,\mathcal{a}_j,P_j)$, $j = 1, \ldots, n$, be any n probability spaces. Let f_j be functions: $A_j \to [-\infty,\infty]$. Suppose either

a) $f_j \geq 0$, $j = 1, \ldots, n$, or

b) $f_1 \equiv 1$ and $n = 2$.

Then on the Cartesian product $\Pi_{j=1}^n (A_j,\mathcal{a}_j,P_j)$, with $x := (x_1, \ldots, x_n)$, if $f(x) = \Pi_{j=1}^n f_j(x_j)$, then $f^*(x) = \Pi_{j=1}^n f_j^*(x_j)$ a.s., where we set $0\cdot\infty = 0$.

Proof. Clearly $f^*(x) \leq \Pi_{j=1}^n f_j^*(x_j)$ a.s., with $1^* \equiv 1$. For the converse inequality we may assume $n = 2$, by induction in case a). Suppose $f^*(x) < f_1^*(x_1) f_2^*(x_2)$ with positive probability. Then for some rational $r < s$,

$f^*(x) < r < s < f_1^*(x_1)f_2^*(x_2)$ with positive probability. If $f_1 \equiv 1 \equiv f_1^*$ then (for some x_1) this contradicts the choice of f_2^*. So assume $f_1 \geq 0$ and $f_2 \geq 0$. Then for $x_1 \in B$ with $P_1(B) > 0$,

$$0 \leq f_1(x_1)f_2^*(x_2) \leq f^*(x_1,x_2) < r < s < f_1^*(x_1)f_2^*(x_2)$$

for some x_2. Then $f_1(x_1) < \frac{r}{s} f_1^*(x_1)$ for all $x_1 \in B$, contradicting the choice of f_1^*. □

3.1.5. <u>Lemma</u>. Let $(\Omega,\mathcal{S},Pr) = \Pi_{i=1}^3(\Omega_i,\mathcal{S}_i,P_i)$ with coordinate projections Π_i: $\Omega \to \Omega_i$, $i = 1,2,3$. Then for any bounded real function $f \geq 0$ on $\Omega_1 \times \Omega_3$, with $g(\omega_1,\omega_2,\omega_3) := f(\omega_1,\omega_3)$,

$$E(g^*|(\Pi_1,\Pi_2)^{-1}(\mathcal{S}_1 \times \mathcal{S}_2)) = E(g^*|\Pi_1^{-1}(\mathcal{S}_1)) \quad \text{a.s. (Pr)}.$$

<u>Proof.</u> By Lemma 3.1.4 b), g^* equals Pr-a.s. a measurable function not depending on ω_2, thus independent of $\Pi_2^{-1}(\mathcal{S}_2)$. □

3.1.6. <u>Lemma</u>. Let $X: \Omega \to \mathbb{R}$. Then for any $t \in \mathbb{R}$,

a) $Pr^*(X > t) = Pr(X^* > t)$;

b) for any $\varepsilon > 0$, $Pr^*(X \geq t) \leq Pr(X^* \geq t) \leq Pr^*(X \geq t - \varepsilon)$.

<u>Proof</u>. Clearly $\{X > t\} \subset \{X^* > t\}$ and $\{X \geq t\} \subset \{X^* \geq t\}$, so we have "$\leq$" in a) and the first inequality in b).

Take a measurable cover $A \supset \{X > t\}$, so $Pr^*(X > t) = Pr(A)$. Then $X^* \leq t$ a.s. outside A, so $Pr(X^* > t) \leq Pr(A)$, proving a). Thus for $0 < \delta \leq \varepsilon$, $Pr(X^* > t - \delta) \leq Pr^*(X > t - \varepsilon)$. Letting $\delta \downarrow 0$ proves b). □

3.1.7. <u>Corollary</u> For any $X_n: \Omega \to \mathbb{R}$, $|X_n|^* \to 0$ (i.e. by definition, $X_n \to 0$ in probability) if and only if for every $\varepsilon > 0$, $Pr^*(|X_n| > \varepsilon) \to 0$ as $n \to \infty$.

The next fact is useful in conjuction with Lemma 3.1.2 when f and g are "independent" like the f_j in Lemma 3.1.4.

3.1.8 <u>Lemma</u>. Let $(X,\mathcal{S},Pr) = (X_1,\mathcal{S}_1,P_1) \times (X_2,\mathcal{S}_2,P_2)$. Then for any functions $f_i: X_i \to \mathbb{R}$, $B := \{x_1 \in X_1: f_1^*(x_1) = +\infty\}$, and $f(x_1,x_2) := f_1(x_1) + f_2(x_2)$, we have $f^*(x_1,x_2) = +\infty$ a.s. for $x_1 \in B$.

Proof. Let $C = \{x: f^*(x) < +\infty\}$ and suppose $\Pr(C \cap (B \times X_2)) > 0$. Then by Fubini's theorem there is a y such that $P_1(x_1 \in B: \langle x_1, y\rangle \in C) > 0$. On this event, $f_1(x_1) \leq f^*(x_1,y) - f_2(y) < +\infty$, contradicting the choice of $f_1^* = +\infty$ on B. □

Notes. Eames and May (1967) defined f^*, and studied it when f is bounded. Results 3.1.2 through 3.1.8 are more or less as in Dudley and Philipp (1982, Sec.2). Theorem 3.1.1 and its proof are as in Vulikh (1967, pp. 78-79) for $\Omega = [a,b]$ with Lebesgue measure (the proof needs no change).

3.2 Independent random elements and partial sums. We need a notion of independence for functions which may not be measurable. Let $(A_j, \mathcal{A}_j, P_j)$, $j = 1,2,\ldots$, be probability spaces, and form a product $\Pi_{j=1}^n (A_j, \mathcal{A}_j, P_j) = (B, \mathcal{B}, P)$ with points $x := \{x_j\}_{j=1}^n$. If X_j are functions on B of the form $X_j = h_j(x_j)$, $j = 1,\ldots,n$, where each h_j is a function on X_j (not necessarily measurable) then we call X_j independent random elements. If the h_j are measurable this implies independence in the usual sense.

In the rest of this section, $X_j = h(x_j)$, $j = 1,\ldots,n$, are independent random elements with values in a real vector space S on which there is a seminorm $||\cdot||$, and $S_k := \sum_{j=1}^k X_j$, $k = 1,2,\ldots$, with $S_0 := 0$.

The next fact extends Kuelbs (1977, Lemma 2.1).

3.2.1. Lemma. Suppose given X_j with $\tau_n \geq \sum_{j=1}^n E||X_j||^{*2}$ and

(3.2.2) $$||X_j|| \leq M,\ j = 1,\ldots,n.$$

Then for $0 \leq \gamma \leq 1/(2M)$

(3.2.3) $$E\exp(\gamma||S_n||^*) \leq \exp(3\gamma^2\tau_n + \gamma E||S_n||^*).$$

Hence for any $K > 0$ and $0 \leq \gamma \leq 1/(2M)$

(3.2.4) $$\Pr\{||S_n||^* \geq K\} \leq \exp(3\gamma^2\tau_n - \gamma(K - E||S_n||^*)).$$

3.2.5. Remarks. If $K < E||S_n||^*$ then the infimum of the right side of (3.2.4) is attained at $\gamma = 0$ and the bound is trivial.

If $0 \leq K - E||S_n||^* \leq 3\tau_n/M$ then the infimum is attained at $\gamma = (K - E||S_n||^*)/(6\tau_n)$ and equals $\exp(-(K - E||S_n||^*)^2/(12\tau_n))$.

If $K - E||S_n||^* > 3\tau_n/M$ then the infimum (again, for $0 \leq \gamma \leq 1/(2M)$) is at $1/(2M)$ and equals

$$\exp(3\tau_n/(4M^2) - (K-E||S_n||^*)/(2M)) \leq \exp(-(K-E||S_n||^*)/(4M)).$$

<u>Proof.</u> First, (3.2.2) is equivalent to $||X_j||^* \leq M$, $1 \leq j \leq n$. Clearly, (3.2.4) follows from (3.2.3). To prove (3.2.3), set $Y_k := S_n - X_k$. For any $\eta \in \mathcal{L}^1(A, \mathcal{A}, P)$, let $E_k\eta := E(\eta|x_1,\ldots,x_{k-1})$, $k \leq n$; $E_{n+1}\eta := \eta$, $E_1\eta := E\eta$. Then $E_{k+1}||Y_k||^* = E_k||Y_k||^*$ by Lemma 3.1.5 with $\omega_1 = \langle x_1,\ldots,x_{k-1}\rangle$, $\omega_2 = x_k$ and $\omega_3 = \langle x_{k+1},\ldots,x_n\rangle$. Let $\eta_k := E_{k+1}||S_n||^* - E_k||S_n||^*$. Then

$$\begin{aligned} \mathrm{E}\exp(\gamma||S_n||^*) &= E(E_n(\exp[\gamma E_n||S_n||^* + \gamma\eta_n])) \\ &= E(\exp(\gamma E_n||S_n||^*)E_n\exp(\gamma\eta_n)). \end{aligned} \tag{3.2.6}$$

By boundedness and Lemmas 3.1.2 and 3.1.3,

$$\begin{aligned} E_{k+1}||S_n||^* - E_k||S_n||^* &\leq E_{k+1}||Y_k||^* + E_{k+1}||X_k||^* - E_k||Y_k||^* + E_k||X_k||^* \\ &= ||X_k||^* + E||X_k||^*, \quad k = 1,\ldots,n, \end{aligned}$$

and likewise

$$E_{k+1}||S_n||^* - E_k||S_n||^* \geq -||X_k||^* - E||X_k||^*.$$

Thus $E_k|\eta_k|^j \leq 2^jE||X_k||^{*j}$, $j \geq 2$. Clearly $E_k\eta_k = 0$. Thus for $0 \leq \gamma \leq 1/(2M)$,

$$\begin{aligned} E_k(\exp(\gamma\eta_k)) &= 1 + \textstyle\sum_{j=2}^{\infty} \gamma^j E_k\eta_k^j/j! \\ &\leq 1 + \frac{4}{2!}\gamma^2E||X_k||^{*2}\left[1 + \frac{2\gamma M}{3} + \frac{2^2\gamma^2M^2}{4\cdot 3} + \ldots\right] \\ &\leq 1 + 2\gamma^2E||X_k||^{*2}/(1 - 2\gamma M/3) \\ &\leq \exp(2\gamma^2E||X_k||^{*2}(1 + \gamma M)) \end{aligned}$$

since $1 + x \leq e^x$, $x \in \mathbb{R}$. Combining with (3.2.6) gives

$$\mathrm{E}\exp(\gamma||S_n||^*) \leq \exp(2\gamma^2E||X_n||^{*2}(1+\gamma M))\mathrm{E}\exp(\gamma E_n||S_n||^*).$$

Now as in (3.2.6),

$$\mathrm{E}\exp(\gamma E_n||S_n||^*) = E(\exp(\gamma E_{n-1}||S_n||^*)E_{n-1}\exp(\gamma\eta_{n-1})),$$

etc., so the argument can be iterated n times to give (3.2.3). □

Next, Ottaviani's inequality (Breiman, 1968, Lemma 3.2.1, p.45) also extends to starred norms.

3.2.7 <u>Lemma</u>. Suppose X_j are such that, for some $\alpha > 0$,

$$\max_{j\leq n}\Pr(||S_n - S_j||^* > \alpha) \quad := \quad c < 1.$$

Then

$$\Pr(\max_{j\leq n}||S_j||^* \geq 2\alpha) \quad \leq \quad (1-c)^{-1}\Pr(||S_n||^* \geq \alpha).$$

Proof. Let $j^*(\omega)$ be the least $j \leq n$ such that $||S_j||^* \geq 2\alpha$ (or, say, $n+1$ if there is no such j).

Then

$$\begin{aligned}
\Pr(||S_n||^* \geq \alpha) &\geq \Pr(||S_n||^* \geq \alpha,\ \max_{j\leq n}||S_j||^* \geq 2\alpha) \\
&= \textstyle\sum_{j=1}^{n} \Pr(||S_n||^* \geq \alpha,\ j^* = j) \\
&\geq \textstyle\sum_{j=1}^{n} \Pr(||S_n - S_j||^* \leq \alpha,\ j^* = j)
\end{aligned}$$

since $||S_j||^* \leq ||S_n||^* + ||S_j - S_n||^*$ (Lemma 3.1.3).

Now $||S_n - S_j||^*$ is a measurable function of $\langle x_{j+1},\ldots,x_n\rangle := y_2$, by Lemma 3.1.4 with $n = 2$, $y_1 := \langle x_1,\ldots,x_j\rangle$. Thus $||S_n - S_j||^*$ is independent of the event $\{j^* = j\}$ in the usual sense, so the last sum equals

$$\begin{aligned}
&\textstyle\sum_{j=1}^{n} \Pr(||S_n - S_j||^* \leq \alpha)\Pr(j^* = j) \\
&\geq (1-c)\textstyle\sum_{j=1}^{n} \Pr(j^* = j) \quad = \quad (1-c)\Pr(\max_{j\leq n}||S_j||^* \geq 2\alpha). \qquad \square
\end{aligned}$$

The next fact is proved by a method similar to that of Kahane (1968, p.16):

3.2.8 <u>Lemma</u>. Let $K > 0$ and suppose X_j are such that

(3.2.9) $\qquad \Pr(||S_j - S_i||^* \geq K) \leq 1/2, \quad 0 \leq i < j \leq n.$

Then for any $t > K$ and $s \geq 0$

$$\Pr\{||S_n||^* \geq 4t+s\} \quad \leq \quad 4\Pr\{||S_n||^* \geq t\}^2 + \Pr\{\max_{m\leq n}||X_m||^* \geq s\}.$$

Proof. Let $T(\omega) := \min\{j \geq 1:\ ||S_j(\omega)||^* \geq 2t\}$, or $+\infty$ if there is no such j.

Then

$$\Pr\{||S_n||^* \geq 4t+s\} \quad = \quad \textstyle\sum_{m=1}^{n} \Pr\{T = m,\ ||S_n||^* \geq 4t+s\}$$

(3.2.10) $$\leq \left(\textstyle\sum_{m=1}^{n}\Pr\{T = m,\ ||S_n||^* \geq 4t+s,\ ||X_m||^* < s\}\right) + \Pr\{\max_{m\leq n}||X_m||^* > s\}.$$

By Lemma 3.1.3, $||S_n||^* \le ||S_{m-1}||^* + ||X_m||^* + ||S_n - S_m||^*$, so the last sum $\sum$ in (3.2.10) satisfies

$$\begin{aligned}\textstyle\sum &\le \textstyle\sum_{m=1}^{n} \Pr(T=m,\ ||S_n-S_m||^* \ge 2t) \\ &= \textstyle\sum_{m=1}^{n} \Pr(T=m)\Pr(||S_n-S_m||^* \ge 2t) \\ &\le \Pr(\max_{0\le m\le n}||S_n-S_m||^* \ge 2t)\textstyle\sum_{m=1}^{n} \Pr(T=m) \\ &\le 2\Pr(||S_n||^* \ge t)\Pr(\max_{m\le n}||S_m||^* \ge 2t) \\ &\le 4\Pr(||S_n||^* \ge t)^2\end{aligned}$$

using Lemma 3.1.4 as in the proof of 3.2.7 for independence, then using Lemma 3.2.7 itself twice, the first time with order of summation reversed. The Lemma now follows from (3.2.10). □

The random elements X_j will be called <u>symmetric</u> if we can write $x_j = \langle y_j, z_j\rangle$ where for each j, y_j and z_j are independent and have the same distribution in some space B_j ($A_j = B_j \times B_j$ with $P_j = Q_j \times Q_j$ for some Q_j) and $X_j = \psi_j(y_j) - \psi_j(z_j)$ for some function ψ_j from B_j into S.

3.2.11 <u>Lemma</u>. If X_j are symmetric then for any $r \ge 0$

$$\Pr(\max_{j\le n}||S_j||^* > r) \le 2\Pr(||S_n||^* > r).$$

<u>Proof</u> (adapted from Kahane, 1968, p.12, Lemma 1).

Let $M_k(\omega) := \max_{j\le k}||S_j||^*$. Let C_k be the disjoint events $\{M_{k-1} \le r < ||S_k||^*\}$, $k = 1,2,\ldots$, where $M_0 := 0$. By Lemma 3.1.3,

$$2||S_m||^* \le ||S_n||^* + ||2S_m - S_n||^*, \quad 1 \le m \le n,$$

so if $||S_m||^* > r$, then either $||S_n||^* > r$ or $||2S_m-S_n||^* > r$ or both. The transformation which interchanges y_j and z_j for $j > m$ preserves probabilities and interchanges S_n and $2S_m - S_n$, so interchanges $||S_n||^*$ and $||2S_m-S_n||^*$, while preserving all x_j, $j \le m$. Thus

$$\Pr(C_m \cap \{||S_n||^* > r\}) = \Pr(C_m \cap \{||2S_m-S_n||^* > r\}) \ge \tfrac{1}{2}\Pr(C_m).$$

Thus

$$\Pr(M_n > r) = \textstyle\sum_{m=1}^{n} \Pr(C_m) \le 2\Pr(||S_n||^* > r). \quad \square$$

The next and last proof in this section, also related to an argument of Kahane (1968, p.16), appears more explicitly in Hoffmann-Jørgensen (1974, p.164) and Jain and Marcus (1975b, Lemma 3.4).

3.2.12 Lemma. Let X_j be symmetric and $s, t, > 0$. Then

$$\Pr\{\|S_n\|^* \geq 2t + s\} \leq 4(\Pr\{\|S_n\|^* \geq t\})^2 + \Pr\{\max_{m\leq n}\|X_m\|^* \geq s\}.$$

Proof. Let $T(\omega) := \inf\{k \geq 1: \|S_k\|^* \geq t\}$, or $+\infty$ if there is no such k. Let $N_k := \max_{m\leq k}\|X_m\|^*$. Since $\|S_n\|^* \geq 2t + s$ implies $T \leq n$, we have

$$(3.2.13)\quad \Pr\{\|S_n\|^* \geq 2t + s\} = \sum_{j=1}^n \Pr\{T = j, \|S_n\|^* \geq 2t + s\}.$$

If $T = j \leq n$ and $\|S_n\|^* \geq 2t + s$, then $\|S_{j-1}\|^* < t$ and

$\|S_n\|^* \leq \|S_{j-1}\|^* + \|X_j\|^* + \|S_n-S_j\|^*$ a.s. by Lemma 3.1.3, so

$\|S_n-S_j\|^* \geq t + s - N_n(\omega)$. Thus

$$\begin{aligned}\Pr\{T = j, \|S_n\|^* \geq 2t + s\} &\leq \Pr\{T = j, \|S_n-S_j\|^* \geq t + s - N_n\}\\ &\leq \Pr\{T = j, \|S_n-S_j\|^* \geq t\} + \Pr\{T = j, N_n \geq s\}\\ &= \Pr(T = j)\Pr(\|S_n-S_j\|^* \geq t) + \Pr(T = j, N_n \geq s)\end{aligned}$$

where for the last step we again use 3.1.4 as in 3.2.7. Now we sum on j and use (3.2.13) to get

$$\Pr(\|S_n\|^* \geq 2t + s) \leq \Pr(N_n \geq s) + \sum_{j=1}^n \Pr(T = j)\Pr(\|S_n-S_j\|^* \geq t).$$

By the Lévy inequality (3.2.11) we have

$$\Pr(\|S_n-S_j\|^* \geq t) \leq 2\Pr(\|S_n\|^* \geq t)$$

and

$$\sum_{j=1}^n \Pr(T = j) = \Pr(\max_{k\leq n} \|S_k\|^* \geq t) \leq 2\Pr(\|S_n\|^* \geq t).$$

Combining the last three facts gives the result. □

Note The Lemmas in this section all appeared in Dudley and Philipp (1982) which for some, gave only literature references rather than the above complete proofs.

3.3 A central limit theorem implies measurability in separable Banach spaces.

As noted at the beginning of Section 3.1 above, if F_n is an empirical distribution function, $\Pr(F_n \in A)$ need not be defined if A is complete and discrete, hence Borel, for the (non-separable) supremum norm. Thus the variables

$X_j := \delta_{x(j)} - P$ need not be Borel measurable in general. But in separable Banach spaces one usually assumes that variables are Borel measurable. It will be shown here that in a separable normed space, a weak form of central limit theorem (or a stronger invariance principle such as (1.1.2)) can hold only if X_1 is measurable. After proving this in $\mathbb{R}^1$ it will be extended easily to separable normed spaces.

Define the inner measure $Pr_*(B) := \sup\{Pr(C): C \subset B\}$ and let

$$f_* := -((-f)^*) = \text{ess.sup}\{g: g \leq f, g \text{ measurable}\}.$$

3.3.1 Theorem. Let $(A, \mathcal{A}, P)$ be a probability space, x_n coordinates on $(A^\infty, \mathcal{A}^\infty, P^\infty)$, $n = 1, 2, \ldots$. Let $h: A \to \mathbb{R}$ (where h is not assumed measurable), $X_i := h(x_i)$. Let $S_n := X_1 + \ldots + X_n$. If for all t,

$$\lim_{n\to\infty} Pr^*(S_n/n^{1/2} \leq t) = \lim_{n\to\infty} Pr_*(S_n/n^{1/2} \leq t) = N(0,1)(]-\infty,t]),$$

then h is measurable for the completion of P, so that X_i are measurable, $EX_i = 0$ and $EX_i^2 = 1$.

Proof. If h is measurable then $EX_i = 0$ and $EX_i^2 = 1$ (Gnedenko and Kolmogorov, 1968, sec.35, Theor.4, p.181). Suppose h is non-measurable for P, so that the X_n are non-measurable.

Let $B = \{x \in A: h^*(x) = +\infty\}$. Suppose $P(B) > 0$. Then for all j, $Pr(X_j^* = +\infty) = P(B)$ by Lemma 3.1.4, case b).

By Lemma 3.1.8, $S_n^* = +\infty$ a.s. on the set where $X_j^* = +\infty$ for any $j \leq n$. Thus

$Pr((S_n/n^{1/2})^* = +\infty) \geq Pr(X_j^* = +\infty \text{ for some } j \leq n) = 1 - (1-P(B))^n \to 1$ as $n \to \infty$.

By Lemma 3.1.6 this contradicts the hypothesis. Thus $P(B) = Pr(X_j^* = +\infty) = 0$.

Let $B_j := B(j) := \{X_j \geq X_j^* - 2^{-j}\}$.

Then $Pr^*(B_j) = 1$, $j = 1,2, \ldots$. Let $C_n := \bigcap_{j=1}^n B_j$. Apply Lemma 3.1.4 with $P_j = P$ and $f_j = 1_{B(j)}$ to obtain $Pr^*(C_n) = 1$. On C_n,

$$S_n \leq X_1^* + \ldots + X_n^* \leq S_n + 1.$$

Thus $\mathcal{L}((X_1^* + \ldots + X_n^*)/n^{1/2}) \to N(0,1)$.

Hence $EX_1^* = 0$. Likewise $EX_{1*} = 0$. Thus $X_{1*} = X_1 = X_1^*$ a.s., i.e. X_1 is completion measurable. □

3.3.2 Corollary. Suppose $(S,|\cdot|)$ is a separable normed space and $X_n = h(x_n)$ where x_n are independent, identically distributed random variables with values in some measurable space $(A,\mathcal{A})$ and h is any function from A into S (not assumed measurable).

Suppose Y_n are i.i.d. Gaussian variables in S with mean 0 and

$$(3.3.3) \qquad \lim_{n\to\infty} n^{-1/2}\left|\sum_{j\leq n} X_j - Y_j\right|^* = 0$$

in probability. Then the X_j are completion measurable for the Borel σ-algebra on S.

Proof. Let $F = S'$, the dual Banach space. Then (3.3.3) implies that for each $f \in F$, the central limit theorem as in Theorem 3.3.1 holds for the $f(h(x_j))$, with some limit law $N(0,\sigma_f^2)$, $\sigma_f^2 := \sigma^2(f(Y_1))$. Thus by that Theorem, each $f \circ h$, $f \in F$, is measurable for the completion of $\mathcal{L}(x_1)$ on $\mathcal{A}$. So each $f(X_j)$ is completion measurable, $f \in F$. Since S is separable, by the Hahn-Banach Theorem there is a countable norming subset $\{f_n\} \subset S'$, i.e. $|s| = \sup_n |f_n(s)|$ for all $s \in S$, and the Borel σ-algebra of S is generated by the f_n. Thus the X_j are completion measurable. □

3.3.4 Corollary. If the hypotheses of Cor. 3.3.2 hold except that S is not necessarily separable, and H is a bounded linear operator from S into a separable normed space, then the $H(X_j)$ are completion measurable.

Note. The results of this section are essentially from Dudley and Philipp, 1982, Section 8.

Chapter 4 Asymptotic equicontinuity and limit theorems.

4.1 A characterization of functional Donsker classes.

Let $(S,||\cdot||)$ be a Banach space (in general non-separable). A subset $\mathcal{F}$ of the unit ball $\{f \in S': ||f||' \leq 1\}$ of the dual space $(S',||\cdot||')$ is called a norming subset if and only if $||s|| = \sup_{f\in\mathcal{F}} |f(s)|$ for all $s \in S$. Clearly the whole unit ball is always a norming subset (by the Hahn-Banach theorem).

Conversely, given any set $\mathcal{F}$, let $S = \ell^\infty(\mathcal{F})$ be the Banach space of all bounded real functions on $\mathcal{F}$, with sup norm

$$||s|| = ||s||_{\mathcal{F}} := \sup_{f\in\mathcal{F}} |s(f)|, \quad s \in S.$$

Then the natural map $f \to (s \to s(f))$ takes $\mathcal{F}$ one-to-one onto a norming subset of S'.

So, the forms of limiting behavior for empirical measures defined in Sec.2.1 (functional Donsker class, strong invariance principle) can be viewed as limit behaviour for a norm $||\cdot||_{\mathcal{F}}$ on a Banach space S. Conversely, limit theorems in a general Banach space S can be viewed as limit theorems for empirical measures on S, uniformly over a class $\mathcal{F}$ of functions, such as the unit ball of S', since for $f \in S'$ and $x(1), \ldots, x(n) \in S$,

$$(\delta_{x(1)} + \ldots + \delta_{x(n)})(f) = f(x(1) + \ldots + x(n)).$$

With definitions as in Sec. 2.1, here is a characterization of the functional Donsker property:

4.1.1. Theorem. Given any probability space $(A,\mathcal{Q},P)$ and $\mathcal{F} \subset \mathcal{L}^2(A,\mathcal{Q},P)$, $\mathcal{F}$ is a functional Donsker class if and only if both

a) $\mathcal{F}$ is totally bounded in $\mathcal{L}^2_0(A,\mathcal{Q},P)$;

b) for every $\varepsilon > 0$ there is a $\delta > 0$ and an N such that for $n \geq N$

$$Pr^*\{\sup[|\textstyle\int f-g\, d\nu_n|: f, g \in \mathcal{F}, \rho_P(f,g) < \delta] > \varepsilon\} \leq \varepsilon.$$

Proof. The "only if" direction is easy and will be proved first. Let $\mathcal{F}$ be a functional Donsker class. Let x_j and Y_j be as in (2.1.3). Given $\varepsilon > 0$, take n_o large enough so that for $n \geq n_o$,

$$Pr^*\{n^{-1/2} \sup_{f\in\mathcal{F}} |\textstyle\sum_{j=1}^n f(x_j) - P(f) - Y_j(f)| > \varepsilon/3\} < \varepsilon/2.$$

Since by definition $\mathcal{F}$ is a G_PBUC class, and G_P is an isonormal process on $\mathcal{L}_0^2$ with metric ρ_P, a) follows (e.g. Dudley, 1967, Prop.3.4, p.295). Thus there is a $\delta > 0$ such that for a suitable version of G_P

$$\Pr\{\sup\{|G_P(f) - G_P(g)|: f,g \in \mathcal{F}, \rho_P(f,g) < \delta\} > \varepsilon/3\} < \varepsilon/2.$$

By assumption, each $(Y_1 + \ldots + Y_n)/n^{1/2}$ is such a version of G_P. Combining gives b), proving "only if".

To prove "if", assume a) and b). Let $U := UC(\mathcal{F})$ denote the space of all real valued functions on $\mathcal{F}$ uniformly continuous for ρ_P. Then U is a separable subspace of $S := \ell^\infty(\mathcal{F})$.

For any finite subset $\mathcal{G}$ of $\mathcal{F}$, by the finite-dimensional central limit theorem, we can let $n \to \infty$ in b) and so replace $\int f - g \, d\nu_n$ by $G_P(f) - G_P(g)$, $f, g \in \mathcal{G}$. Since δ does not depend on $\mathcal{G}$ we can let $\mathcal{G}$ increase up to a countable dense set $\mathcal{H}$ in $\mathcal{F}$ for ρ_P. For each $k = 1,2, \ldots$, we obtain $\delta_k > 0$ and N_k such that b) holds in the form

(4.1.2) $\quad \Pr^*\{\sup[|\int f-g \, d\nu_n|: f,g \in \mathcal{F}, \rho_P(f,g) < \delta_k] > 2^{-k}\} < 2^{-k}, n \geq N_k.$

We may assume N_k increases with k. If follows that

(4.1.3) $\quad \Pr\{\sup\{|G_P(g)-G_P(h)|: g,h \in \mathcal{H}, \rho_P(g,h) < \delta_k\} > 2^{-k}\} \leq 2^{-k}.$

Thus almost surely for k large enough

$$\sup\{|G_P(g) - G_P(h)|: g, h \in \mathcal{H}, \rho_P(g,h) < \delta_k\} \leq 2^{-k}.$$

Then almost surely, for every $f \in \mathcal{F}$ the limit

$$\lim\{G_P(g): g \in \mathcal{H}, \rho_P(g,f) \to 0\}$$

exists and defines a version of $G_P(f)$ (defined as 0 on the set of probability 0 where convergence fails). Thus $\mathcal{F}$ is a G_PBUC class and the above version of G_P has a law μ defined on the Borel sets of U.

For $k = 1,2, \ldots$, let $\mathcal{F}_k$ be a finite subset of $\mathcal{F}$, by a), such that

$$\sup_{f \in \mathcal{F}} \inf\{\rho_P(f,g): g \in \mathcal{F}_k\} < \delta_k.$$

Let T_k denote the finite-dimensional space of all real functions on $\mathcal{F}_k$, also with supremum norm $||\cdot||_k$. Let $\mathcal{F}_k := \{g_1, g_2, \ldots g_{m(k)}\}$. For each $f \in \mathcal{F}$ let $f_k = g_j$ for the least j such that $\rho_P(f,g_j) < \delta_k$. For any $\varphi \in S$ let

$\varphi_k(f) := \varphi(f_k)$, $f \in \mathcal{F}$. Then $\varphi_k \in S$. Let $\Lambda_k(\varphi) := \varphi_k$, $X_{kj} := X_j - \Lambda_k X_j$. The union of the (finite-dimensional) ranges of the Λ_k, $k = 1,2, \ldots$, is included in a complete separable subspace T of S with $U \subset T$. Note that $||\Lambda_k\varphi|| \le ||\varphi||$ for all k and all $\varphi \in S$. Then by (4.1.2) we have for $n \ge N_k$

(4.1.4) $$\Pr^*\{n^{-1/2}||\textstyle\sum_{j=1}^n X_{kj}|| > 2^{-k}\} \le 2^{-k}.$$

Then by Lemma 3.1.6

$$\Pr\{n^{-1/2}||\textstyle\sum_{j=1}^n X_{kj}||^* > 2^{-k}\} \le 2^{-k}, \quad n \ge N_k.$$

Then if $n \ge 2N_k$ and $m = 0,1, \ldots, n$, we have by Lemma 3.1.3 a.s.

$$n^{-1/2}||\textstyle\sum_{j=m+1}^n X_{kj}||^* \le n^{-1/2}||\sum_{j=1}^n X_{kj}||^* + m^{-1/2}||\sum_{j=1}^m X_{kj}||^*,$$

and since either $n - m \ge N_k$ or $m \ge N_k$,

$$\Pr\{n^{-1/2}||\textstyle\sum_{j=m+1}^n X_{kj}||^* > 2^{1-k}\} \le 2^{1-k}.$$

Thus for $k \ge 2$, Ottaviani's inequality (3.2.7) with $C = 1/2$ gives

(4.1.5) $$\Pr\{n^{-1/2} \max_{m \le n} ||\textstyle\sum_{j=1}^m X_{kj}||^* > 2^{2-k}\} \le 2^{2-k}, \quad n \ge 2N_k.$$

Let P_k be the law on T_k of $f \to f(x_1) - \int f\,dP$, $f \in \mathcal{F}_k$. Then by Theorem 1.1.3 there exist V_{kj} i.i.d. P_k, W_{kj} i.i.d. with a Gaussian law Q_k, and some $n_k := n(k) \ge 2N_k$ such that for all $n \ge n_k$, $k \ge 1$,

(4.1.6) $$\Pr\{n^{-1/2}\max_{m \le n}||\textstyle\sum_{j \le m} V_{kj} - W_{kj}||_k > 2^{-k}\} \le 2^{-k}.$$

Also, Q_k must be the law of the restriction of G_P to $\mathcal{F}_k$. We may assume that the sequences $\{\langle V_{kj}, W_{kj}\rangle\}_{j \ge 1}$ are independent of each other for different k. We also take $1 := n_0 < n_1 < n_2 < \ldots$.

For each $k = 0, 1, \ldots$, if $n_k \le j < n_{k+1}$, write $k := k(j)$ and set $V_j := V_{kj}$, $W_j := W_{kj}$. Then $\{V_j, j \ge 1\}$ and $\{W_j, j \ge 1\}$ are sequences of independent random variables. Each sequence has its values in a countable product of Polish spaces which itself is Polish. Let $\{Y_j, j \ge 1\}$ be i.i.d. μ on U. Then W_j has the law of the restriction $Y_j \restriction \mathcal{F}_k$ of Y_j to $\mathcal{F}_k$, $n_k \le j < n_{k+1}$. Let T_{kj} be a copy of T_k and U_j of U for each j, $T_{(j)} := T_{k(j)j}$. Then, apply Lemma 1.2.2 where $\mathcal{F}_{(j)} := \mathcal{F}_{k(j)}$,

$$X = Y = \Pi_{j=1}^\infty T_{(j)}, \quad Z = \Pi_{j=1}^\infty U_j,$$

$P = \mathcal{L}(\{<V_j,W_j>\}_{j\geq 1})$, and $Q = \mathcal{L}(\{Y_j \restriction \mathcal{F}_{(j)}, Y_j\}_{j\geq 1})$.

Thus we may assume $W_j = Y_j \restriction \mathcal{F}_{(j)}$ for all $j \geq 1$.

Now for $<\{X_j\}_{j\geq 1}, t> \in A^\infty \times [0,1] := \Omega$, let $S_1 := X$ above and $V(\omega) := \{f \to f(x_j) - \int f dP,\ f \in \mathcal{F}_{(j)}\}_{j\geq 1} \in S_1$. Let $T_1 := Y \times Z$ above and $Q := \mathcal{L}(\{<V_j,W_j,Y_j>\}_{j\geq 1})$ on $S_1 \times T_1$.

Then by Lemma 1.2.3 above we can take V_j and Y_j to be defined on Ω with $V(\omega)_j = V_j$ for all j. It remains to prove (2.1.3) for the given Y_j, with $X_j(f) := f(x_j) - \int f\, dP$, $f \in \mathcal{F}$.

Given $\varepsilon > 0$, take k large enough so that $2^{7-k} < \varepsilon$. We have $X_j \restriction \mathcal{F}_{(j)} = V_j$, $j \geq 1$. Let $M_k \geq n_k$ be large enough so that for all $n \geq M_k$

(4.1.7) $$\Pr^*\{n^{-1/2} \textstyle\sum_{j=1}^{n(k)} ||\Lambda_k X_j|| + ||Y_j|| > 2^{-k}\} < 2^{-k}.$$

Fix $n \geq M_k$. Choose r such that $n_r \leq n \leq n_{r+1}$.

Then

$$I_n := \max_{m\leq n} ||\textstyle\sum_{j\leq m} X_j - Y_j||$$

(4.1.8)
$$\leq \max_{m\leq n(k)} \{||\textstyle\sum_{j\leq m} X_j - Y_j||\}$$
$$+ \textstyle\sum_{i=k}^{r-1} \max_{n(i)\leq m<n(i+1)} ||\textstyle\sum_{j=n(i)}^{m} X_j - Y_j||$$
$$+ \max_{n(r)\leq m\leq n} ||\textstyle\sum_{n(r)\leq j\leq m} X_j - Y_j||.$$

Then a.s.

$$L_k := \max_{m\leq n(k)} ||\textstyle\sum_{j\leq m} X_j - Y_j||^*$$
$$\leq \max_{m\leq n(k)} (||\textstyle\sum_{j\leq m} X_{kj}||^* + \textstyle\sum_{j\leq m} ||\Lambda_k X_j|| + ||Y_j||).$$

Thus by (4.1.5) and (4.1.7),

(4.1.9) $$\Pr(L_k \geq 2^{3-k}) \leq 2^{3-k}.$$

Next, since restriction to $\mathcal{F}_i$ is a linear isometry from $\Lambda_i S$ to $(T_i, ||.||_i)$, we have for $k \leq i \leq r$

$$A_i := \max_{n(i)\leq m<n(i+1)} ||\textstyle\sum_{j=n(i)}^{m} X_j - Y_j||^*$$
$$\leq \max_{n(i)\leq m<n(i+1)} (||\textstyle\sum_{j=n(i)}^{m} X_{ij}||^*$$
$$+ ||\textstyle\sum_{j=n(i)}^{m} V_j - W_j||_i + ||\textstyle\sum_{j=n(i)}^{m} Y_j - \Lambda_i Y_j||).$$

Since $n \geq n_r \geq 2N_i$, we have by (4.1.5) and stationarity, if $k \leq i < r$,

$$\Pr\{\max_{n(i)\leq m<n(i+1)} ||\sum_{j=n(i)}^{m} X_{ij}||^* > 2^{2-i} n^{1/2}\} \leq 2^{2-i}.$$

Likewise by (4.1.6), if $k \leq i < r$,

$$\Pr\{n^{-1/2} \max_{n(i)\leq m<n(i+1)} ||\sum_{j=n(i)}^{m} V_j - W_j||_i > 2^{-i}\} \leq 2^{-i}.$$

Next, $(n_{i+1} - n_i)^{-1/2} \sum_{j=n(i)}^{n(i+1)-1} Y_j$ is equal in law to a version of G_P (or Y_1) with uniformly continuous sample functions, and for $i < r$, $n \geq n_{i+1} - n_i$. By Lévy's inequality (Lemma 3.2.11 above) we have

$$\Pr\{n^{-1/2} \max_{n(i)\leq m<n(i+1)} ||\sum_{j=n(i)}^{m} Y_j - \Lambda_i Y_j|| > 2^{-i}\}$$

$$\leq 2\Pr\{n^{-1/2} ||\sum_{j=n(i)}^{n(i+1)-1} Y_j - \Lambda_i Y_j|| > 2^{-i}\}$$

$$\leq 2\Pr\{||Y_1 - \Lambda_i Y_1|| > 2^{-i}\} \leq 2^{1-i}$$

by (4.1.3). Collecting terms we have

$$\Pr(A_i > 2^{3-i} n^{1/2}) \leq 2^{3-i}, \quad k \leq i < r.$$

For $i = r$, replacing n_{i+1} by n throughout yields the same result. From this, (4.1.8) and (4.1.9) we obtain since $2^{5-k} < \varepsilon$

$$\Pr(\max_{m\leq n} ||\sum_{j\leq m} X_j - Y_j||^* \geq n^{1/2}\varepsilon)$$

$$\leq 2^{3-k} + \sum_{i=k}^{r} 2^{3-i} < \varepsilon,$$

proving (2.1.3) as desired and hence Theorem 4.1.1. □

In Theorem 4.1.1 a) and b), if the metric ρ_P is replaced by the usual $\mathcal{L}^2$ metric e_P, the conditions are no longer necessary (let $\mathcal{F}$ be the set of all constant functions) but remain sufficient :

4.1.10 <u>Theorem</u>. Given any probability space $(A, \mathcal{A}, P)$ and $\mathcal{F} \subset \mathcal{L}^2(A, \mathcal{A}, P)$, $\mathcal{F}$ is a functional Donsker class if both

a) $\mathcal{F}$ is totally bounded for e_P;

b) for every $\varepsilon > 0$ there is $\delta > 0$ and an N such that for $n \geq N$

$$\Pr^*\{\sup[|\int f-g\, d\nu_n|: f, g \in \mathcal{F}, e_P(f,g) < \delta] > \varepsilon\} \leq \varepsilon.$$

<u>Proof</u>. Letting $n \to \infty$ in b), we can take G_P to have sample functions almost surely uniformly continuous for e_P.

As in (2.1.2) one can write $W_P(f) = G_P(f) + XP(f)$ for all $f \in \mathcal{L}^2$, where X is a standard normal variable independent of G_P, $X = W_P(1)$. Clearly $f \to XP(f)$ is uniformly continuous on $\mathcal{F}$ for e_P, and bounded on $\mathcal{F}$ by a). Thus W_P has sample functions a.s. bounded and uniformly continuous on $\mathcal{F}$ for e_P. Hence the closure of $\mathcal{F}$ in $L^2(A, \mathcal{Q}, P)$ is a compact "GC-set" (Dudley, 1967; Feldman, 1971) and the closed, convex, symmetric hull K of $\mathcal{F}$ is also a compact GC-set (Feldman, 1971). For an orthonormal basis $\{e_n\}_{n=1}^{\infty}$ of the linear span of K and 1 with $e_1 = 1$, the series $\sum_{n=1}^{\infty} (\cdot, e_n) W_P(e_n)$ converges uniformly on K almost surely (Dudley, 1967; Feldman, 1971). Thus so does the series

$$\sum_{n=2}^{\infty} (\cdot, e_n) W_P(e_n) = \sum_{n=2}^{\infty} (\cdot, e_n) G_P(e_n) = G_P(\cdot).$$

Each finite partial sum of the series is continuous for ρ_P. Hence $G_P(\cdot)$ is a.s. continuous for ρ_P on K, hence uniformly continuous, and so a fortiori for ρ_P on $\mathcal{F}$.

Thus $\mathcal{F}$ is "G_PBUC" not only for e_P but also for ρ_P, as used in the definition here. Now, in the proof that a) and b) are sufficient in Theorem 4.1.1, one can replace ρ_P by e_P throughout. The second half of the proof needs no changes and shows that $\mathcal{F}$ is a functional Donsker class. □

Notes. A first result in the direction of Theorem 4.1.1 was Dudley (1966, Prop.2). A theorem similar to 4.1.1 was proved under a measurability assumption on the class $\mathcal{F}$, if $\mathcal{F}$ is a collection of indicators of sets, in Dudley (1978). Dudley (1981a) stated an extension to more general classes of functions, but with an error ($\mathcal{L}^2$ instead of $\mathcal{L}_0^2$ in a)). The present form corrects the error, completely removes the measurability assumption and strengthens a central limit theorem to an invariance principle, as in Dudley and Philipp (1982).

4.2 Strong invariance principles.

For any class $\mathcal{F}$ of functions on a set X define the envelope function by

$$F_{\mathcal{F}}(x) := \sup\{|f(x)| : f \in \mathcal{F}\} := ||\delta_x||_{\mathcal{F}}, \qquad x \in X.$$

Then we have (with the definitions (1.1.4) and (2.1.4)):

Theorem 4.2.1. Let $\mathcal{F}$ be any functional Donsker class such that

$$\int F_{\mathcal{F}}(x)^{*2} / \; LLF_{\mathcal{F}}^{*}(x) \quad dP(x) \;<\; \infty.$$

Then $\mathcal{F}$ satisfies a strong invariance principle.

4.2.2 Corollary. If $\mathcal{F} = \{1_A : A \in \mathcal{C}\}$ for a collection $\mathcal{C}$ of sets, $\mathcal{C} \subset \mathcal{A}$, and if $\mathcal{F}$ is a functional Donsker class then $\mathcal{F}$ also satisfies a strong invariance principle.

The proof of Theorem 4.2.1 is rather long and technical. As given in Dudley and Philipp (1982), it is based on corresponding results in separable Banach spaces due to Heinkel (1979) and Goodman, Kuelbs and Zinn (1981), using Kuelbs and Zinn (1979). Substantial further work seems to be needed for the extension to the non-separable case.

If the moment condition is strengthened a little to $\int F_{\mathcal{F}}(x)^{*2}\, dP(x) < \infty$, then the proof can be simplified substantially. Here the corresponding result in the separable case, obtaining a law of the iterated logarithm (LIL), is due to Pisier (1975). An LIL for Donsker classes of sets, under a measurability condition, was proved by Kuelbs and Dudley (1980). For uniformly bounded classes of functions, Kolčinskii (1981b) has such a result.

Theorem 4.2.1 will not be proved in this course.

Chapter 5 Sequences of sets and functions

Given a countable set $\mathcal{F} \subset \mathcal{L}^2(A, \mathcal{A}, P)$, say $\mathcal{F} = \{f_m\}_{m=1}^{\infty}$, some conditions will be given for $\mathcal{F}$ to be a functional Donsker class.

5.1 Sequences of sets

Recall that a collection $\mathcal{C}$ of measurable sets is called a functional Donsker class (Sec. 2.1) if and only if $\mathcal{F} := \{1_B : B \in \mathcal{C}\}$ is one.

5.1.1 Theorem. Let $\mathcal{C} = \{C_m\}_{m=1}^{\infty}$ be a sequence of measurable sets. If for some $r < \infty$,

(5.1.2) $$\sum_{m=1}^{\infty} (P(C_m)(1-P(C_m))^r < \infty ,$$

then $\mathcal{C}$ is a functional Donsker class.

Proof. Since $\min(P(C_m), 1-P(C_m)) \to 0$ as $m \to \infty$, $\mathcal{C}$ is totally bounded for e_P and ρ_P as in 4.1.10a). It remains to check the asymptotic equicontinuity condition 4.1.10b). The integers $\{m = 1, 2, \ldots\}$ can be decomposed into two subsequences, along one of which $\frac{1}{2} \geq P(C_m) \to 0$ while along the other $\frac{1}{2} < P(C_m) \to 1$. Then

$$\inf\{P(C_m) - \tfrac{1}{2} : P(C_m) > \tfrac{1}{2}\} > 0$$

and by choosing δ smaller than this infimum we need only verify 4.1.1b) for the subsequences separately.

We may assume $P(C_i \,\Delta\, C_j) > 0$ for $i \neq j$ and $0 < P(C_i) < 1$ for all i since

$$\Pr\{\nu_n(C_i) = \nu_n(C_j) \text{ whenever } P(C_i \,\Delta\, C_j) = 0\}$$
$$= 1 = \Pr\{\nu_n(C_i) = 0 \text{ whenever } P(C_i) = 0 \text{ or } 1\}$$

using countability.

By symmetry, we may now assume that

$$\tfrac{1}{2} \geq p_m := 1 - q_m := P(C_m) \downarrow 0, \; m \to \infty,$$
$$p_m > 0 \text{ for all } m, \text{ and } \textstyle\sum_m p_m^r < \infty.$$

Given $0 < \varepsilon \leq 1$, suppose we can find M and N such that for all $n \geq N$

(5.1.3) $$\Pr\{\sup_{m \geq M} |\nu_n(C_m)| > \varepsilon\} < \varepsilon.$$

Then for J large enough, $p_m < p_M/2$ for $m \geq J$. Let

$$\gamma := \min\{P(C_i \,\Delta\, C_j) : i < j \leq J\},$$
$$\delta' := \min(\gamma, p_M)/2.$$

Then

$$\sup\{|\nu_n(C_i) - \nu_n(C_j)|: P(C_i \,\Delta\, C_j) < \delta'\}$$
$$\leqslant \sup\{|\nu_n(C_i) - \nu_n(C_j)|: i, j \geqslant M\}$$
$$\leqslant 2 \sup\{|\nu_n(C_j)|: j \geqslant M\} \leqslant 2\varepsilon$$

with probability $\geqslant 1 - \varepsilon$, implying 4.1.10b).

To prove (5.1.3), recalling the binomial probabilities defined before 2.2.5, it will suffice to find M and N such that for $n \geqslant N$

(5.1.4) $$\sum_{m\geqslant M} E(np_m + \varepsilon n^{1/2}, n, p_m) < \varepsilon/2$$

and

(5.1.5) $$\sum_{m\geqslant M} B(np_m - \varepsilon n^{1/2}, n, p_m) < \varepsilon/2.$$

For (5.1.5), the Chernoff-Okamoto inequality (2.2.7) gives

$$B(np_m - \varepsilon n^{1/2}, n, p_m) \leqslant \exp(-\varepsilon^2/(2p_m q_m)).$$

For some K, $1 \leqslant K < \infty$, $p_m \leqslant Km^{-1/r}$ for all m. We may assume $r \geqslant 1$. Choose M large enough so that

$$\sum_{m\geqslant M} \exp(-m^{1/r}\varepsilon^2/(2K)) < \varepsilon/2$$

to give (5.1.5) for all n.

The other side, (5.1.4), is more difficult. Berňstein's inequality 2.2.2 gives, if

$$2pn^{1/2} \geqslant \varepsilon \quad \text{and} \quad p \leqslant 1/2,$$
$$E(np + \varepsilon n^{1/2}, n, p) \leqslant \exp(-\varepsilon^2/(2pq + \varepsilon n^{-1/2}))$$
$$\leqslant \exp(-\varepsilon^2/(6pq)).$$

Then

$$\sum_{m\geqslant M}\{E(np_m + \varepsilon n^{1/2}, n, p_m): 2p_m n^{1/2} \geqslant \varepsilon\}$$

(5.1.6) $$\leqslant \sum_{m\geqslant M} \{\exp(-\varepsilon^2/(6p_m)): 2p_m n^{1/2} \geqslant \varepsilon\}$$
$$\leqslant \sum_{m\geqslant M} \exp(-\varepsilon^2 m^{1/r}/(6K)) < \varepsilon/4$$

for all n if M is large enough.

It remains to treat the sum, say S_2, of (5.1.4) restricted to $2p_m n^{1/2} < \varepsilon$. Let $s(n) := n^{1/2}$ and $k := np + \varepsilon s(n)$, so that $\varepsilon = (k-np)/s(n)$. Then for $k \geqslant np$, (2.2.8) implies

$$E(k,n,p) \leq (np/np + \varepsilon s(n))^{np+\varepsilon s(n)} e^{\varepsilon s(n)}.$$

Let $p = p_m$, $0 < x := n^{1/2}p/\varepsilon$. Then

$$\varepsilon n^{1/2} - (np + \varepsilon n^{1/2})\ln(1 + \varepsilon/(pn^{1/2})) = \varepsilon n^{1/2}(1-f(x))$$

where $f(x) := (1 + x)\ln(1 + x^{-1})$ decreases for $x > 0$, so $f(x) \geq f(1/2) > 3/2$. Thus $1 < 2f(x)/3$, $\varepsilon n^{1/2}(1 - f(x)) \leq -\varepsilon n^{1/2} f(x)/3$, and

$$S_2 \leq \sum_m \{\exp(-\varepsilon n^{1/2} + np_m)(\ln(1 + \varepsilon/(n^{1/2}p_m))/3): 2p_m n^{1/2} < \varepsilon\}$$
$$\leq \sum_m \{(p_m n^{1/2}/\varepsilon)^{\varepsilon s(n)/3}: 2p_m n^{1/2} < \varepsilon\}.$$

Since $p_m \leq Km^{-1/r}$ we have $S_2 \leq S_3 + S_4$ where

$$S_3 := \sum_m \{(p_m n^{1/2}/\varepsilon)^{\varepsilon s(n)/3}: 2Km^{-1/r}n^{1/2} < \varepsilon\}$$
$$\leq (Kn^{1/2}/\varepsilon)^{\varepsilon s(n)/3} \sum_{m>G} m^{-\varepsilon s(n)/(3r)}$$

where $G := (2Kn^{1/2}/\varepsilon)^r \geq 2$. Then for any $n_1 > (3r/\varepsilon)^2$ and $n \geq n_1$,

$$S_3 \leq (Kn^{1/2}/\varepsilon)^{\varepsilon s(n)/3} \int_{G-1}^{\infty} x^{-\varepsilon s(n)/(3r)} dx$$
$$\leq (Kn^{1/2}/\varepsilon)^{\varepsilon s(n)/3}(\varepsilon n^{1/2}(3r)^{-1}-1)^{-1}(G-1)^{1-\varepsilon s(n)/(3r)}.$$

For K, ε and r fixed and $n \to \infty$, the logarithm of the latter expression is asymptotic to $-\varepsilon n^{1/2}(\ln 2)/3 \to -\infty$, so $S_3 \to 0$. Thus $S_3 \leq \varepsilon/8$ for $n \geq n_2$ for some n_2.

Finally,

$$S_4 := \sum_m \{(p_m n^{1/2}/\varepsilon)^{\varepsilon s(n)/3}: 2p_m n^{1/2} \leq \varepsilon \leq 2Km^{-1/r}n^{1/2}\}$$
$$\leq (2Kn^{1/2}/\varepsilon)^r\, 2^{-\varepsilon s(n)/3} \to 0$$

as $n \to \infty$, so $S_4 < \varepsilon/8$ for $n \geq n_3$ for some n_3. Thus for $n \geq \max(n_1,n_2,n_3)$, $S_2 < \varepsilon/4$. This and (5.1.6) give (5.1.4). □

5.1.7. <u>Theorem</u>. If the measurable sets $\{C_m\}_{m>1}$ are independent for P, then they form a functional Donsker class for P if and only if for some $r < \infty$,

$$\sum_m (p_m(1-p_m))^r < \infty$$

where

$$p_m := P(C_m).$$

Proof. "If" follows from the last theorem. To prove "only if", first suppose $\limsup_{m\to\infty} p_m := p < 1$ and for all n, $\sum_m p_m^n = +\infty$. Then for each n, $\Pr\{P_n(C_m) = 1$ for infinitely many $m\} = 1$ by the Borel-Cantelli lemma. Now $P_n(C_m) = 1$ implies $\nu_n(C_m) = n^{1/2}(1-p_m) \geq n^{1/2}(1-p)/2$ for m large. But Theorem 4.1.1a) and b) imply that

$$\lim_{M\to\infty} \sup_n \Pr\{\sup_{f\in\mathcal{F}} |\nu_n(f)| > M\} = 0,$$

so $\{C_m\}_{m\geq 1}$ is not a functional Donsker class. A symmetrical argument applies if $\liminf_{m\to\infty} p_m > 0$ and for all n, $\sum_m (1-p_m)^n = +\infty$.

If $\{C_m\}$ is a functional Donsker class, let us write it as a union of two subsequences $\{C_{r(j)}\}_{j\geq 1} \cup \{C_{t(k)}\}_{k\geq 1}$ with $P(C_{r(j)}) \leq 1/2 < P(C_{t(k)})$ for all j and k. (One of the subsequences may be finite or empty.) Then by the above, for some $n < \infty$

$$\sum_j P(C_{r(j)})^n < \infty \quad \text{and} \quad \sum_k (1-P(C_{t(k)}))^n < \infty,$$

so that $\sum_m (p_m(1-p_m))^n < \infty$. □

Note. The results of this section are from Dudley (1978, Sec.2). The proofs here are slightly improved.

5.2 Sequences of functions.

For a probability space $(A,\mathcal{A},P)$ and $f \in \mathcal{L}^2(A,\mathcal{A},P)$ let $\sigma_P^2(f) := \int f^2 dP - (\int f\, dP)^2$ (the variance of f). Here is a sufficient condition for the functional Donsker property of a sequence $\{f_m\}$ which is easy to prove, yet turns out to be optimal of its kind:

5.2.1 Theorem. If $\{f_m\}_{m\geq 1} \subset \mathcal{L}^2(A,\mathcal{A},P)$ and $\sum_{m=1}^{\infty} \sigma_P^2(f_m) < \infty$, then $\{f_m\}_{m\geq 1}$ is a functional Donsker class for P.

Proof. Since $f_m \to 0$ in $\mathcal{L}_0^2$, clearly $\{f_m\}$ is totally bounded there.

For any $0 < \varepsilon < 1$, $n > 1$ and $m \geq 1$, by Chebyshev's inequality

$$\sum_{j\geq m} \Pr(|\nu_n(f_j)| \geq \varepsilon/2) \leq 4\sum_{j\geq m} \sigma_P^2(f_j)/\varepsilon^2 < \varepsilon$$

for $m \geq m_o$ for some $m_o < \infty$. We have a.s. for all n, $\nu_n(f_j) = 0$ for all j such that $\sigma_P^2(f_j) = 0$ and $\nu_n(f_j) = \nu_n(f_k)$ for all j and k with

$\sigma_p^2(f_j-f_k) = 0$. Let

$$\alpha := \inf\{\sigma_P^2(f_j-f_k): \sigma_P^2(f_j-f_k) > 0,\ j < m_o,\ \sigma_P^2(f_j) > 0\}.$$

Then $\alpha > 0$. Let $\delta = \min(\alpha,1)$. Then $\Pr\{|\nu_n(f_j) - \nu_n(f_k)| > \varepsilon$ for some j, k with $\sigma_P^2(f_j-f_k) < \delta\} < \varepsilon$, implying 4.1.1b) and hence Theorem 5.2.1. □

The following shows 5.2.1 to be sharp in one sense (although 5.2.1 does not imply Theorem 5.1.1):

5.2.2 Proposition. Let $a_m > 0$ satisfy $\sum a_m = +\infty$. Then there is a sequence $\{f_m\}$ in an $\mathcal{L}^2(A,\mathcal{A},P)$ with $\sigma_P^2(f_m) \leqslant a_m$ where $\{f_m\}$ is not a functional Donsker class.

Proof. We may assume $a_m \downarrow 0$. There exist $b_m \downarrow 0$ such that $\sum_m a_m b_m = +\infty$. In $A = [0,1]$ with Lebesgue measure P let $C_m := C(m)$ be independent sets with $P(C_m) = a_m b_m$. Let $f_m := b_m^{-1/2} 1_{C(m)}$. Then $\sigma_P^2(f_m) \leqslant \int f_m^2\, dP = a_m$. For each n, almost surely $P_n(C_m) \geqslant 1/n$ for infinitely many m. Then $\nu_n(C_m) \geqslant n^{-1/2}/2$ for infinitely many m and $\sup_m \nu_n(f_m) = +\infty$ a.s. Then by Theorem 4.1.1 $\{f_m\}$ is not a functional Donsker class for P. □

From the above proof we have:

5.2.3 Proposition. Whenever $\{C(m)\}_{m\geqslant 1}$ are independent for P, $\limsup_{m\to\infty} P(C(m)) < 1$, $\sum_m P(C(m)) = +\infty$ and $c_m \to +\infty$, then $\{c_m 1_{C(m)}\}_{m\geqslant 1}$ is not a functional Donsker class.

Note. Theorem 5.2.1 and Proposition 5.2.3 are in Dudley (1981a, Sec.2). I have not previously published 5.2.2.

Chapter 6 Metric entropy, with inclusion and bracketing.

6.0 Metric entropy and capacity. Let (S,d) be a metric space and $B \subset S$. In case B is totally bounded, there are several, more or less equivalent, ways to measure "how totally bounded" it is. They are defined as follows. Let $\varepsilon > 0$.

A set $\{x_1, \ldots, x_n\} \subset S$ (not necessarily included in B!) is called an ε-net for B if and only if for all $x \in B$, $d(x,x_i) \leq \varepsilon$ for some i. Let $\mathcal{N}(\varepsilon,B,S,d)$ denote the minimal number of points in an ε-net (in S) for B.

The diameter of a set $C \subset S$ is defined by

$$\operatorname{diam} C \ := \ \sup\{d(x,y): x, y \in C\}.$$

Let $N(\varepsilon,B,d)$ denote the smallest n such that $B \subset \bigcup_{i=1}^n B_i$ for some B_i with $\operatorname{diam} B_i \leq 2\varepsilon$, $i = 1, \ldots, n$.

Let $D(\varepsilon,B,d)$ denote the largest n such that for some $x_1, \ldots, x_n \in B$, $d(x_i,x_j) > \varepsilon$ whenever $i \neq j$. The points x_i may be called ε-distinguishable.

Note that an ε-distinguishable set $\{x_1, \ldots, x_n\}$ with maximal n is also an ε-net for B.

The following chain of inequalities is straightforward to prove:

6.0.1. Theorem. For any $\varepsilon > 0$, and set B in a metric space S,

$$D(2\varepsilon,B,d) \leq N(\varepsilon,B,d) \leq \mathcal{N}(\varepsilon,B,S,d) \leq \mathcal{N}(\varepsilon,B,B,d) \leq D(\varepsilon,B,d).$$

It follows that as $\varepsilon \downarrow 0$, when all these functions go to ∞ unless B is a finite set, they all have the same asymptotic behavior to within a factor of 2 in ε. Which of the four functions is used as a "basic" measure of the size of B is thus essentially a matter of convenience.

The points x_i, whether ε-distinguishable or (hence) an ε-net, are usually useful, and $N(\varepsilon,B,d)$ is usually bounded above just by $\mathcal{N}(\varepsilon,B,B,d)$. Thus $N(\varepsilon,B,d)$ may be a little less handy.

Now $\mathcal{N}(\varepsilon,B,S,d)$ has the additional complication that it depends not only on B but on the larger space S. On the other hand $\mathcal{N}(\varepsilon,B,B,d)$ does not necessarily increase as the set B increases (let B be the surface of a sphere of radius ε in a Euclidean space S).

For these reasons I am inclined to adopt $D(\varepsilon,B,d)$ as basic: it is intrinsic to B, not depending on $S \supset B$; it increases with B; and it directly gives a sequence $\{x_1, \ldots, x_n\}$ of points which are both ε-distinguishable and an ε-net.

In several expositions stress has been placed on the logarithms (to the base 2) of the above quantities, where $\log_2 N(\varepsilon,B,d)$ has been called the ε-entropy of B, and $\log_2 D(\varepsilon,B,d)$ its ε-capacity. For a number of interesting compact sets in function spaces $N(\varepsilon,B,d)$ grows like $\exp(\varepsilon^{-r})$ for some power r, suggesting that logarithms may result in simpler functions like ε^{-r}. But for a general theory it seems to me that the doubling of notations, to name the logarithms of the basic quantities, may not be worthwhile.

Note. The defined quantities and Theorem 6.0.1 all appeared in Kolmogorov and Tikhomirov (1959, sec.1, Theorem IV). Lorentz (1966) prefers "metric entropy" to "ε-entropy"; the latter has been used by others with different meanings.

6.1 Definitions and the Blum-DeHardt law of large numbers.

Definitions. Given a measurable space $(A,\mathcal{A})$, let $\mathcal{L}^0(A,\mathcal{A})$ denote the set of all real-valued $\mathcal{A}$-measurable functions on A. Given $f,g \in \mathcal{L}^0(A,\mathcal{A})$ let $[f,g] := \{h \in \mathcal{L}^0(A,\mathcal{A}) : f \leq h \leq g\}$ (empty unless $f \leq g$). Given a probability space $(A,\mathcal{A},P)$, $1 \leq q \leq \infty$, $\mathcal{F} \subset \mathcal{L}^q(A,\mathcal{A},P)$ with usual seminorm $||\cdot||_q$, and $\varepsilon > 0$, let $N_{[]}^{(q)}(\varepsilon,\mathcal{F},P)$ denote the smallest m such that for some $f_1, \ldots, f_m$ in $\mathcal{L}^q(A,\mathcal{A},P)$,

$$\mathcal{F} \subset \bigcup_{i,j} \{[f_i,f_j] : ||f_j - f_i||_q \leq \varepsilon\}. \tag{6.1.1}$$

Here a set $[f_i,f_j]$ is called a bracket and $\log N_{[]}^{(q)}(\varepsilon,\mathcal{F},P)$ will be called a metric entropy with bracketing.

Note that the f_j are not required to be in $\mathcal{F}$. If $\mathcal{F} \subset \mathcal{L}^r$ then for $q \leq r \leq \infty$, $\mathcal{F} \subset \mathcal{L}^q$ and

$$N_{[]}^{(q)}(\varepsilon,\mathcal{F},P) \leq N_{[]}^{(r)}(\varepsilon,\mathcal{F},P) \quad \text{for all } \varepsilon > 0. \tag{6.1.2}$$

For $r = \infty$, more is true. Let $d_{sup}(f,g) := \sup_x |(f-g)(x)|$, $||f||_{sup} := d_{sup}(f,0)$. It is easily seen using brackets $[f_j - \varepsilon, f_j + \varepsilon]$ that for any law P,

$$N_{[]}^{(\infty)}(2\varepsilon,\mathcal{F},P) \leq 2D(\varepsilon,\mathcal{F},d_{sup}). \tag{6.1.3}$$

Thus, for example, a set $\mathcal{F}$ of continuous functions, totally bounded in the usual supremum norm with given bounds $D(\varepsilon,\mathcal{F},\infty)$ will have the same bounds, multiplied at most by 2, on all $N_{[\,]}^{(q)}(2\varepsilon,\mathcal{F},P)$, $1 \le q \le \infty$.

If $\mathcal{F}$ consists of indicator functions of measurable sets then in finding brackets $[f_i,f_j]$ to cover $\mathcal{F}$, it is no loss to assume $0 \le f_k \le 1$ for all k. Next, if $C(i) := \{x: f_i(x) > 0\}$, $D(j) := \{x: f_j(x) = 1\}$, and $f_i \le 1_C \le f_j$ then

$$f_i \le 1_{C(i)} \le 1_C \le 1_{D(j)} \le f_j.$$

If $\mathcal{C}$ is a collection of measurable sets and $\varepsilon > 0$ let

$N_I(\varepsilon,\mathcal{C},P) := \inf\{m:$ for some $C_1, \ldots, C_m \in \mathcal{A}$, for all $C \in \mathcal{C}$ there are i,j with $C_i \subset C \subset C_j$ and $P(C_j \setminus C_i) \le \varepsilon\}$.

Then it follows that

(6.1.4) $\qquad N_I(\varepsilon,\mathcal{C},P) \le 2N_{[\,]}^{(1)}(\varepsilon,\mathcal{F},P)$ where $\mathcal{F} = \{1_C: C \in \mathcal{C}\}$.

We have the following law of large numbers, or generalized Glivenko-Cantelli theorem:

6.1.5 <u>Theorem</u>. Suppose $\mathcal{F} \subset \mathcal{L}^1(A,\mathcal{A},P)$ and for all $\varepsilon > 0$, $N_{[\,]}^{(1)}(\varepsilon,\mathcal{F},P) < \infty$. Then $\lim_{n\to\infty}\|P_n-P\|_{\mathcal{F}}^* = 0$ a.s.

<u>Proof.</u> Given $\varepsilon > 0$ take $f_1,\ldots,f_m \in \mathcal{L}^1$, $m < \infty$, to satisfy (6.1.1) for q = 1. By the ordinary strong law of large numbers there is an N such that

$$\Pr\{\sup_{n\ge N} \max_{j\le m} |(P_n-P)(f_j)| > \varepsilon\} < \varepsilon.$$

For each $f \in \mathcal{F}$ let $f_i \le f \le f_k$ with $P(f_k-f_i) \le \varepsilon$. Then if $\max_{j\le m}|(P_n-P)(f_j)| \le \varepsilon$ we have

$$\begin{aligned}|(P_n-P)(f)| &\le |(P_n-P)(f_i)| + |(P_n-P)(f-f_i)|\\ &\le \varepsilon + (P_n+P)(f_k-f_i)\\ &\le 3\varepsilon + (P_n-P)(f_k-f_i) \le 5\varepsilon.\end{aligned}$$

Thus

$$\Pr{}^*\{\sup_{n\ge N}\|P_n-P\|_{\mathcal{F}} > 5\varepsilon\} < \varepsilon$$

and the conclusion follows from Lemma 3.1.6. □

6.1.6 <u>Remarks</u>.

A) The sufficient condition in Theorem 6.1.5 is not necessary. For example let $C_m := C(m)$ be independent sets with $P(C_m) = 1/m$. Then by Theorems 5.1.1 and 4.2.1,

$$\lim_{n\to\infty} \sup_m |(P_n-P)(C_m)| = 0 \text{ a.s.}$$

(This also follows, more directly, using Bernstein's inequality (2.2.2) and $\sum_{m,n=1}^{\infty} E(\varepsilon n,n,\frac{1}{m}) \leqslant 2\sum_{m,n=1}^{\infty} \exp(-\varepsilon^2 mn/4) + \exp(-\varepsilon^2(mn^3)^{1/2}/2) < \infty$.)

Given $f_1, \ldots, f_k \in \mathcal{L}^1$, suppose

$$\{C_m\}_{m\geqslant 1} \subset \bigcup\{[f_i,f_j]: P(f_j-f_i) < 1/2\}.$$

We may assume $f_i \geqslant 0$ for all i. For each i and j with $P(f_j-f_i) < 1/2$, we have

$$\sum\{P(C_m): f_i \leqslant 1_{C(m)} \leqslant f_j\} < +\infty$$

since if $P(f_i) > 0$, $f_i \leqslant 1_{C(m)}$ for only finitely many m, while if for a subsequence $C_{m(r)}$ we have $\sum_r P(C_{m(r)}) = +\infty$ and $C := \bigcup_r C_{m(r)}$, then $P(C) = 1$, so $f_i \leqslant 1_C \leqslant f_j$ implies $P(f_j) = 1$ and $P(f_i) > 0$. Thus

$$N_{[]}^{(1)}(\varepsilon,\{C_m\},P) = +\infty \quad \text{for every} \quad \varepsilon < 1.$$

B) On the other hand let $\mathcal{C}$ be the collection of all finite subsets of $[0,1]$ with Lebesgue law P. Then $||P_n-P||_{\mathcal{C}} \equiv 1 \not\to 0$ although $1_A = 0$ a.s. for all $A \in \mathcal{C}$. This shows that in (6.1.5) $N_I < \infty$ cannot be replaced by $N(\varepsilon,\mathcal{F},d_p) \equiv 1$ for any $\mathcal{L}^p$ distance d_p.

One way to apply Theorem 6.1.5 is via the following:

6.1.7 Proposition. Let $(S,||.||)$ be a separable Banach space and P a law on the Borel sets of S such that $\int ||x||dP(x) < \infty$. Let $\mathcal{F}$ be the unit ball of the dual space S'. Then for every $\varepsilon > 0$, $N_{[]}^{(1)}(\varepsilon,\mathcal{F},P) < \infty$.

Proof. By Ulam's theorem (e.g. Parthasarathy, 1967, p.29) take a compact $K \subset S$ such that $\int_{S\setminus K} ||x||dP(x) < \varepsilon/3$. The elements of $\mathcal{F}$, restricted to K, form a uniformly bounded, equicontinuous family, hence totally bounded for the sup norm $||.||_K$ on K by the Arzelà-Ascoli theorem. Take $f_1, \ldots, f_m \in \mathcal{F}$, $m < \infty$, such that for all $f \in \mathcal{F}$, $||f-f_j||_K < \varepsilon/3$ for some j.

Let $g_j := f_j - \varepsilon/3$ on K, $g_j(x) := -||x||$, $x \notin K$; $g_{j+m} := f_j + \varepsilon/3$ on K, $g_{j+m}(x) := ||x||$, $x \notin K$, all for $j = 1, \ldots, m$.

Then for any $f \in \mathcal{F}$, if $||f-f_j||_K < \varepsilon/3$, then $g_j \leqslant f \leqslant g_{j+m}$ and $P(g_{j+m}-g_j) < \varepsilon$, so $N_{[]}^{(1)}(\varepsilon,\mathcal{F},P) \leqslant 2m < \infty$. □

6.1.8 Corollary. (Mourier's strong law of large numbers).

Let $(S,||\cdot||)$ be a separable Banach space, P a law on S such that $\int ||x|| dP(x) < \infty$, and $X_1, X_2, \ldots$, i.i.d. P. Let $S_n := X_1 + \ldots + X_n$. Then S_n/n converges a.s. in $(S,||\cdot||)$ to some $x_0 \in S$.

Proof. By 6.1.5 and 6.1.7, S_n/n is a Cauchy sequence a.s. for $||\cdot||$, hence converges a.s. to some random variable $Y \in S$. For each $f \in \mathcal{F}$, $f(Y) = P(f)$ a.s., i.e. $Y \in f^{-1}(\{P(f)\})$ a.s. Let $\{f_m\}_{m\geq 1} \subset \mathcal{F}$ be a countable total set: if $f_m(x) = 0$ for all m, then $x = 0$. Such f_m exist by the Hahn-Banach theorem. Let $D := \bigcap_m f_m^{-1}(\{P(f_m)\})$. Then $Y \in D$ a.s., so D is non-empty. But if $y, z \in D$ then $||y-z|| = \sup_m |f_m(y-z)| = 0$, so $D = \{x_0\}$ for some x_0. □

Direct proof. Given $\varepsilon > 0$, there is a function g from S into a finite subset of itself such that $P(||x-g(x)||) < \varepsilon$. The strong law of large numbers holds for the finite-dimensional variables $g(X_j)$, with some limit x_ε, and

$$n^{-1}||\textstyle\sum_{j=1}^n X_j - g(X_j)|| \leq n^{-1}\sum_{j=1}^n ||X_j - g(X_j)||$$

$$\longrightarrow E||X_1 - g(X_1)|| < \varepsilon \quad \text{as} \quad n \to \infty$$

by the one-dimensional strong law. Thus

$$\limsup_{n\to\infty} ||x_\varepsilon - S_n/n|| < \varepsilon \quad \text{a.s.}$$

Letting $\varepsilon \downarrow 0$ through some sequence, we get S_n/n converging a.s. to some x_0. □

6.1.9 Corollary. If $\mathcal{F} \subset \mathcal{L}^1(A,\mathcal{A},P)$ and $\{\delta_x : x \in A\}$ is separable for $||\cdot||_{\mathcal{F}}$, then $||P_n - P||_{\mathcal{F}} = ||P_n - P||^*_{\mathcal{F}} \to 0$ a.s.

Proof. This follows from 6.1.5 and 6.1.7 or 6.1.8. □

Remarks. The proof of Mourier's strong law 6.1.8 via 6.1.5 and 6.1.7 is not presented as being shorter than the direct proof. Rather, the point is that 6.1.5 is stronger than 6.1.9. For any class $\mathcal{F}$ to which 6.1.9. applies, 6.1.5 will also apply (though, perhaps, less obviously). On the other hand if $\mathcal{F} = \{1_{[0,t]} : 0 < t < 1\}$ and P is Lebesgue measure on $[0,1]$ then 6.1.5 applies but 6.1.9 does not. In this sense 6.1.5 is strictly stronger than 6.1.9.

Notes. Theorem 6.1.5 is due to Blum (1955, Lemma 1) for families of (indicators of) sets and to De Hardt (1971, Lemma 1) for uniformly bounded families of functions. Mourier (1951, 1953 pp. 195-196) proved Theorem 6.1.8. I do not know a reference for Remark 6.1.6 or Prop. 6.1.7. In the proof of 6.1.8, (Bochner or Pettis) integrals of Banach-valued functions were not assumed, so they had to be, in part, reconstructed.

For a survey of laws of large numbers for empirical measures up to 1978, see Gaenssler and Stute (1979).

6.2 A central limit theorem for sets or bounded functions.

6.2.1 Theorem. If $\mathcal{F}$ is a uniformly bounded class of measurable functions with

$$\int_0^1 (\log N_{[]}^{(1)}(x^2, \mathcal{F}, P))^{1/2} dx < \infty$$

then $\mathcal{F}$ is a functional Donsker class.

Proof. We may assume $0 \leqslant f \leqslant 1$ for all $f \in \mathcal{F}$. Since $\mathcal{F}$ is totally bounded for e_P, to apply Theorem 4.1.10, it is enough to verify 4.1.10(b).

Let $0 < \varepsilon < 1$ and $N_I(y) := N_{[]}^{(1)}(y, \mathcal{F}, P)$. Since $N_I(x) \geqslant N_I(y)$ for $0 < x < y$, we have $x \cdot \log N_I(x) \to 0$ as $x \downarrow 0$. Take $\gamma > 0$ such that

(6.2.2) $$N_I(x) \leqslant \exp(\varepsilon^2/(600x)), \quad 0 \leqslant x < \gamma.$$

Take $\alpha > 0$ small enough so that

(6.2.3) $$\exp(-\varepsilon^2/(1800\alpha)) < \varepsilon/4.$$

The hypothesis on N_I is equivalent to

$$\int_0^1 (\log N_I(y))^{1/2} y^{-1/2} dy < \infty$$

and also to

$$\sum_{i \geqslant 1} (2^{-i} \log N_I(2^{-i}))^{1/2} < \infty.$$

Take $u \geqslant 2$ large enough so that

(6.2.4) $$\sum_{i \geqslant u} (2^{-i} \log N_I(2^{-i}))^{1/2} < \varepsilon/96,$$

and such that

(6.2.5) $$\sum_{i \geqslant 0} \exp(-2^{i+u} \varepsilon^2/(20{,}000(i+1)^4)) < \varepsilon/32.$$

Let $r \geqslant u$ be large enough so that $\delta_0 := 2^{-r} \leqslant \min(\alpha, \gamma)$. For $k = 1, 2, \ldots,$ let $\delta_k := \delta(k) := \delta_0/2^k = 2^{-k-r}$. Let $m(k) := N_I(\delta_k)$ and $b_k := (2^{-k} \log m(k))^{1/2}$.

Take functions $f_{k1}, \ldots, f_{km(k)}$ as in the definition of $N_{[\,]}^{(1)}(\delta_k, \mathcal{F}, P)$ so that for each $f \in \mathcal{F}$ and $k = 0,1,2,\ldots,$ there are $r(k) := r(k,f)$ and $s(k) := s(k,f)$ with $f_{kr(k)} \leq f \leq f_{ks(k)}$ and $P(f_{ks(k)} - f_{kr(k)}) \leq \delta_k$. We may assume $0 \leq f_{kj} \leq 1$ for all k,j since $0 \leq f \leq 1$, $f \in \mathcal{F}$.

Let $g_k := g_k(f) := f_{ks(k)} - f_{k+1,s(k+1)}$.

Then

(6.2.6) $$P(g_k^2) \leq P(|g_k|) \leq 2\delta_k.$$

Let $n_0 := n_0(\varepsilon) := \varepsilon^2/(256\delta_0^2)$. By (6.2.3), $\delta_0 \leq \alpha < \varepsilon^2/1800$, so $n_0 > 12{,}000/\varepsilon^2 \longrightarrow \infty$ as $\varepsilon \downarrow 0$. For each $n > n_0$ there is a unique $k = k(n)$ such that

(6.2.7) $$1/2 < 8\delta_k n^{1/2}/\varepsilon \leq 1.$$

Then for each n, $k = k(n)$, each $f \in \mathcal{F}$, $r = r(k,f)$ and $s = s(k,f)$, we have

(6.2.8) $$\nu_n(f_{kr}) - \frac{\varepsilon}{8} \leq \nu_n(f_{kr}) - \delta_k n^{1/2} \leq \nu_n(f) \leq \nu_n(f_{ks}) + \frac{\varepsilon}{8}$$

and

(6.2.9) $$|\nu_n(f_{ks(k)} - f_{0s(0)})| \leq \textstyle\sum_{0 \leq i < k} |\nu_n(g_i)|.$$

Let $\mathcal{G}_i$ be the set of all functions $g_i(f)$, $f \in \mathcal{F}$. Then the number of functions in $\mathcal{G}_i$ is bounded by

(6.2.10) $$\operatorname{card}(\mathcal{G}_i) \leq m(i)m(i+1).$$

We have $\log m(i) = 2^i b_i^2$. Let
$$d_i := \max((i+1)^{-2}\varepsilon/32,\ 6b_{i+1}2^{-r/2}).$$

Then by (6.2.4),

(6.2.11) $$\textstyle\sum_{i \geq 0} d_i < \varepsilon/8.$$

For each $i < k = k(n)$, by (6.2.7), $n^{1/2}\delta_i > n^{1/2}\delta_k > \varepsilon/16$. Thus by (6.2.11), $d_i \leq 2n^{1/2}\delta_i$. Bernstein's inequality (2.2.2) gives for each $g \in \mathcal{G}_i$, by (6.2.6),

$$\Pr\{|\nu_n(g)| > d_i\} \leq 2\exp(-d_i^2/(4\delta_i + d_i n^{-1/2}))$$
$$\leq 2\exp(-d_i^2/(6\delta_i)).$$

Let $M_i := 2m(i)m(i+1) \leq 2m(i+1)^2 = 2\exp(2^{i+2}b_{i+1}^2)$. Then by (6.2.10),

$$P_{in} := \Pr\{|\nu_n(g)| > d_i \text{ from some } g \in \mathcal{G}_i\}$$
$$\leq M_i \exp(-d_i^2/(6\delta_i))$$
$$= M_i \exp(-2^i d_i^2/(6\delta_0))$$
$$\leq 2\exp(2^i(4b_{i+1}^2 - d_i^2/(6\delta_0))).$$

Now by definition of d_i, $4b_{i+1}^2 \leqslant d_i^2/(9\delta_0)$, and

$$P_{in} \leqslant 2\exp(-2^i d_i^2/(18\delta_0))$$
$$\leqslant 2\exp(-2^{i+r}\varepsilon^2/(18(32)^2(i+1)^4)).$$

Thus by (6.2.5),

(6.2.12) $$\sum_{0\leqslant i<k} P_{in} < \varepsilon/8.$$

Now with $k := k(n)$, let

$V_n := \max\{|\nu_n(f_{ku}-f_{kt})|: f_{kt}\leqslant f_{ku}, P(f_{ku}-f_{kt}) < \delta_k, t,u, = 1, \ldots, m(k)\}$.

Let $Q_n := \Pr(V_n \geqslant \varepsilon/8)$. Then by Bernstein's inequality (2.2.2), and (6.2.7),

$$Q_n \leqslant m(k)^2\exp(-\varepsilon^2 64^{-1}/(2\delta_k + \varepsilon 8^{-1}n^{-1/2}))$$
$$\leqslant m(k)^2\exp(-\varepsilon^2/(128\delta_k + 128\delta_k))$$
$$\leqslant \exp(2^k(2b_k^2 - \varepsilon^2/(256\delta_0))).$$

Now for $j := k + r$, by (6.2.2) and choice of r

$$2b_k^2 = 2^{1-k}\log N_I(2^{-j})$$
$$= 2^{r+1}(2^{-j}\log N_I(2^{-j}))$$
$$\leqslant 2^r\varepsilon^2/300.$$

Thus by (6.2.3) and choice of r

$$Q_n \leqslant \exp(-2^{k+r}\varepsilon^2/1800) < \varepsilon/4.$$

If $V_n < \varepsilon/8$ then by (6.2.8), $|\nu_n(f-f_{k,s(k,f)})| \leqslant \varepsilon/4$ for all $f \in \mathcal{F}$. Then by (6.2.9) through (6.2.12),

$$\Pr^*\{\sup_{f\in\mathcal{F}}|\nu_n(f-f_{0s(0,f)})| > \varepsilon/2\} < \varepsilon/2.$$

We also have, again by Bernstein's inequality,

$$P_0 := \Pr\{\sup\{|\nu_n(f_{0i}-f_{0j})|: P(|f_{0i}-f_{0j}|)<3\delta_0\} > \varepsilon/4\}$$
$$\leqslant m(0)^2\exp(-\varepsilon^2 16^{-1}/(6\delta_0 + \varepsilon 4^{-1}n^{-1/2})).$$

For $n > n_0$, by its definition above (6.2.7), $n^{-1/2}\varepsilon/4 < 4\delta_0$, so

$$P_0 < m(0)^2\exp(-\varepsilon^2/(160\delta_0)).$$

Now by (6.2.2) and choice of δ_0, we have $m(0)^2 \leqslant \exp(2\varepsilon^2/(600\delta_0))$, so

$$P_0 \leq \exp(-\varepsilon^2/(400\delta_0)) < \varepsilon/4$$

by (6.2.3) since $\delta_0 \leq \alpha$. Thus for $n > n_0$

$$\Pr^*\{\sup\{|\nu_n(f-h)|: f,h \in \mathcal{F},\ P(|f-h|) < \delta_0\} > 3\varepsilon\} < 3\varepsilon,$$

proving 4.1.10(b) in this case. □

Note. The result of this section (6.2.1) is as in Dudley (1978, Theorem 5.1), extended straightforwardly from sets to bounded functions, and with some numbers corrected. See Sec. 11.3 below for further Donsker class results.

6.3 The power set of a countable set: BORISOV-DURST theorem.

Let P be a law on the set $\mathbb{N}$ of nonnegative integers. The next theorem gives a criterion for the functional Donsker property of the collection $2^{\mathbb{N}}$ of all subsets of $\mathbb{N}$, for P, in terms of the numbers $P(\{m\}) := p_m$, $m \geq 0$. We also find that the sufficient condition given in Theorem 6.2.1 is necessary for $2^{\mathbb{N}}$.

6.3.1 Theorem. The following are equivalent:

a) $2^{\mathbb{N}}$ is a functional Donsker class for P;

b) $\sum_m p_m^{1/2} < \infty$;

c) $\int_0^1 (\log N_I(x^2, 2^{\mathbb{N}}, P))^{1/2} dx < \infty$.

Proof. We have c) $\Longrightarrow$ a) by Theorem 6.2.1. Next, to prove a) $\Longrightarrow$ b), suppose $\sum p_m^{1/2} = +\infty$. The random variables $W(m) := W_P(1_{\{m\}})$ (for the isonormal W_P on $L^2(P)$ as defined in Sec. 2.1) are independent and Gaussian with mean 0 and variances p_m. The sum of truncated means $\sum_m E|W(m)|1_{\{|W(m)|\leq 1\}} = +\infty$ since its mth term is asymptotic to $(2p_m/\pi)^{1/2}$ if $p_m \to 0$; if $p_m \not\to 0$ the series clearly diverges. Now $\sum_m p_m|W_P(\mathbb{N})| < \infty$ a.s., so $\sum_m |G_P(\{m\})| = +\infty$ a.s. Hence $\sup_{A\subset \mathbb{N}} G_P(1_A) = +\infty$ a.s. and $2^{\mathbb{N}}$ is not a G_PBUC class, so a fortiori not a Donsker class. Thus a) $\Longrightarrow$ b).

Next, to prove b) $\Longrightarrow$ c). Equivalently, let us prove

$$\sum_{k=1}^{\infty} 2^{-k}(\log N_I(4^{-k}, 2^{\mathbb{N}}, P))^{1/2} < \infty.$$

We can assume $p_m \geq p_r > 0$ for $m \leq r$. Let r_j be the number of values of m such that $4^{-j-1} < p_m^{1/2} \leq 4^{-j}$, $j = 0,1,2,\ldots$, and $C_j := r_j/4^j$. Then

$\sum_j c_j < \infty$. For $k \geq k_0$ large enough there is a unique $j(k)$ such that

(6.3.2) $$\sum_{j>j(k)} c_j/4^j \leq 4^{-k} < \sum_{j\geq j(k)} c_j/4^j.$$

Let $m(k) := m_k := \sum_{j=1}^{j(k)} r_j$. Then

$$\sum_{m>m(k)} p_m \leq \sum_{j>j(k)} r_j/4^{2j} \leq 4^{-k}.$$

Let A_i run over all subsets of $\{1, \ldots, m(k)\}$ where $i = 1, \ldots, 2^{m(k)}$. Let $B_i := A_i \cup \{m \in \mathbb{N}: m > m(k)\}$. Then for any $C \subset \mathbb{N}$, let $A_i = C \cap \{1, \ldots, m(k)\}$. Then $A_i \subset C \subset B_i$ and $P(B_i \setminus A_i) \leq 4^{-k}$. So $N_I(4^{-k}, 2^{\mathbb{N}}, P) \leq 2^{m(k)+1}$. Thus it will be enough to prove

(6.3.3) $$\sum_k m_k^{1/2}/2^k < \infty,$$

with $\sum_k$ restricted to $k \geq k_0$. We have

$$\sum_k m_k^{1/2}/2^k = \sum_k (\sum_{j=0}^{j(k)} 4^j c_j)^{1/2}/2^k$$

$$\leq \sum_k \sum_{j=0}^{j(k)} 2^{j-k} c_j^{1/2} = \sum_{j=0}^{\infty} c_j^{1/2} \sum_{k:\ j\leq j(k)} 2^{j-k}.$$

To prove this converges, since $\sum_j c_j < \infty$, it is enough by Cauchy's inequality to prove $\sum_j (\sum_{k:\ j\leq j(k)} 2^{j-k})^2 < \infty$.

Let $k(j)$ be the smallest k such that $j(k) \geq j$. Then

$$\sum_{k:\ j\leq j(k)} 2^{j-k} \leq 2^{j+1-k(j)}.$$

It is now to be proved that $\sum_j 4^{j-k(j)} < \infty$. Setting $j(k_0-1) := 0$ we have

$$\sum_{j\geq 1} 4^{j-k(j)} \leq \sum_k \sum_{j:\ j(k-1)<j\leq j(k)} 4^{j-k} \leq \sum_k 4^{1+j(k)-k}.$$

Now for each k, let $\kappa(k)$ be the smallest κ such that $j(\kappa) = j(k)$. Then from (6.3.2) for κ, letting $\mathcal{K}$ denote the range of $\kappa(\cdot)$,
$4^{-\kappa(k)} < \sum_{j\geq j(k)} c_j/4^j$, so

$$\sum_k 4^{j(k)-k} \leq \sum_k 4^{j(k)-k+\kappa(k)} \sum_{j\geq j(k)} c_j/4^j$$

$$= \sum_j c_j 4^{-j} \sum_{\kappa\in\mathcal{K},\ j(\kappa)\leq j} 4^{j(\kappa)+\kappa} \sum_{k:\ \kappa(k)=\kappa} 4^{-k}$$

$$< 2\sum_j c_j 4^{-j} \sum_{\kappa\in\mathcal{K},\ j(\kappa)\leq j} 4^{j(\kappa)}.$$

Since $j(\cdot)$ is one-to-one on $\mathcal{K}$, the sum is at most $4\sum_j c_j < \infty$. □

<u>Notes</u>. The equivalence of a) and b) (to show that b) $\Longrightarrow$ a) is also not difficult) was proved by Durst and Dudley (1981). The more difficult implication b) $\Longrightarrow$ c) was discovered and announced by I. S. Borisov (1981b). At this writing I have not seen his proof.

6.3.3. <u>Remark</u>. If $\mathcal{C} = 2^N$ then it is easy to see that $\sup_{A\in\mathcal{C}}|(P_n-P)(A)| \to 0$ a.s., $n \to \infty$.

6.4 Central limit theorems for classes of unbounded functions.

Let $(X,\mathcal{A},P)$ be a probability space and $\mathcal{F}$ a class of measurable functions on X such that for some measurable real F on X, $|f(x)| \leq F(x)$ for all $f\in\mathcal{F}$ and $x\in X$. One may ask: if $F\in\mathcal{L}^p(X,\mathcal{A},P)$ and $N_{[]}^{(r)}(\varepsilon,\mathcal{F},P)$ satisfies suitable bounds as $\varepsilon \downarrow 0$, must $\mathcal{F}$ be a functional Donsker class? Here is one result:

6.4.1 <u>Theorem</u>. If $F\in\mathcal{L}^p(X,\mathcal{A},P)$ for some $p > 2$, and for some γ with $0 < \gamma < 1 - 2/p$, and some $M < \infty$, $N_{[]}^{(1)}(\varepsilon,\mathcal{F},P) \leq \exp(M\varepsilon^{-\gamma})$ for ε small enough, then $\mathcal{F}$ is a functional Donsker class for P.

This can be proved as in Dudley (1981a, Theorem 3.1 pp. 347-350), using Theorem 4.1.10 above. I do not reproduce the proof here partly because I do not know whether the result is close to optimal.

If $2 < p < 3$ and $\gamma > 1 - 1/(p-1)$ then $\mathcal{F}$ may not be a functional Donsker class by examples of Yukich (1982, Sec. 2.4), who also gives examples of classes which are and are not Donsker classes under conditions on $N_{[]}^{(r)}$ for other values of $r \geq 1$.

Note that as $p \downarrow 2$, $1 - 1/(p-1) \downarrow 0$, so that there are counter-examples for arbitrarily small γ. For $1 - 2/p \leq \gamma \leq 1 - 1/(p-1)$, $2 < p < 3$, it is not known whether $\mathcal{F}$ must be a functional Donsker class.

Central limit theorems with other hypotheses (Kolčinskii-Pollard entropy), where $p = 2$ becomes possible, will be given in Sec. 11.3 below.

Chapter 7 Approximation of functions and sets.

7.0 Introduction; the Hausdorff metric.

In this chapter upper and lower bounds will be shown for the metric entropies of various concrete classes of functions on Euclidean spaces and sets in such spaces. Some metric entropies with bracketing are treated, and some without. Metrics for functions are in $\mathcal{L}^p$, $1 \leq p \leq \infty$. For sets we use d_p metrics $d_p(B,C) := P(B\Delta C)$ or the Hausdorff metric, defined as follows.

For any metric space (S,d), $x \in S$, and a non-empty $B \subset S$, let

$$d(x,B) := \inf\{d(x,y)\colon y \in B\}.$$

For non-empty B, $C \subset S$ the Hausdorff pseudo-metric is defined by

$$h(B,C) := \max(\sup_{x\in B} d(x,C), \sup_{x\in C} d(x,B)).$$

Then h is a metric on the collection of bounded, closed, non-empty sets.

For any set $H \subset \mathbb{R}^{d-1}$ and function f from H into $[0,\infty[$ let

$$J_f := J(f) := \{x \in \mathbb{R}^d\colon 0 \leq x_d \leq f(x_{(d)}),\ x_{(d)} \in H\}$$

where $x_{(d)} := (x_1,\ldots,x_{d-1})$. For any other function $g \geq 0$ on H, clearly $h(J_f,J_g) \leq d_{\sup}(f,g)$. Thus for any collection $\mathcal{F}$ of real functions ≥ 0 on H, and any $\varepsilon > 0$,

(7.0.1) $$D(\varepsilon,\{J_f\colon f \in \mathcal{F}\},h) \leq D(\varepsilon,\mathcal{F},d_{\sup}).$$

If $d_{\sup}(f,g) \leq \varepsilon$ and $j := \max(f-\varepsilon,0)$, where $g \geq 0$, then $0 \leq j \leq g \leq f+\varepsilon$, so $J_j \subset J_g \subset J_{f+\varepsilon}$. If P is a law on $H \times [0,\infty[$ having a density p with respect to Lebesgue measure on $\mathbb{R}^d$ with $p(x) \leq M < \infty$ for all x, then

(7.0.2) $$N_I(2M\varepsilon,\{J_f\colon f \in \mathcal{F}\},P) \leq 2D(\varepsilon,\mathcal{F},d_{\sup}).$$

In tne converse direction there are corresponding estimates for Lipschitz functions. Let $||f||_L := \sup_{x\neq y}|f(x)-f(y)|/|x-y|$. Then we have:

7.0.3 Lemma. If $||f||_L \leq K$ and $||g||_L \leq K$ on H, then $h(J_f,J_g) \geq \min(1,1/K)d_{\sup}(f,g)/2$.

Proof. Let $t := d_{\sup}(f,g)$. Let $0 < s < t$. By symmetry, assume that for some $x \in H$, $f(x) \geq g(x) + s$. Then for any $y \in H$, either $|x-y| \geq s/(2K)$ or $g(y) \leq g(x) + s/2 \leq f(x) - s/2$. In either case $|\langle x,f(x)\rangle - \langle y,g(y)\rangle| \geq \min(1,1/K)s/2$. Let $s \uparrow t$. The closest point to $\langle x,f(x)\rangle$ in $J(g)$ must be of the form <y,g(y)>, and the result follows. □

7.1 Spaces of differentiable functions and sets with differentiable boundaries.

For any $\alpha > 0$, spaces of functions will be defined having "bounded derivatives through order α." If β is the greatest integer $< \alpha$, the functions will have partial derivatives through order β bounded, and the derivatives of order β will satisfy a uniform Hölder condition of order $\alpha - \beta$. Still more specifically: for $x := (x_1,\ldots,x_d) \in \mathbb{R}^d$ and $p = (p_1,\ldots,p_d) \in \mathbb{N}^d$ (where $\mathbb{N}$ is the set of nonnegative integers) let $[p] := p_1 + \ldots + p_d$,

$$x^p := x_1^{p_1} x_2^{p_2} \ldots x_d^{p_d}, \quad D^p := \partial^{[p]}/\partial x_1^{p_1} \ldots \partial x_d^{p_d}.$$

For a function f on an open set $U \subset \mathbb{R}^d$ having all partial derivatives $D^p f$ of orders $[p] \leq \beta$ defined everywhere on U, let

$$||f||_\alpha := ||f||_{\alpha,U} := \max_{[p]\leq\beta} \sup\{|D^p f(x)|: x \in \mathbb{R}^d\}$$
$$+ \max_{[p]=\beta} \sup_{x\neq y}\{|D^p f(x) - D^p f(y)|/|x-y|^{\alpha-\beta}\}$$

where $|u| := (u_1^2 + \ldots + u_d^2)^{1/2}$, $u \in \mathbb{R}^d$, and $x,y \in U$.

Let I^d denote the unit cube $\{x \in \mathbb{R}^d: 0 \leq x_j \leq 1,\ j = 1, \ldots, d\}$, and $x_{(d)} := (x_1,\ldots,x_{d-1})$, $x \in \mathbb{R}^d$.

Let $F \subset \mathbb{R}^d$ be a closed set which is the closure of its interior U. Let $\mathcal{F}_{\alpha,K}(F)$ denote the set of all continuous $f: F \to \mathbb{R}$ with $||f||_{\alpha,U} \leq K$. Let $\mathcal{G}_{\alpha,K,d} := \mathcal{F}_{\alpha,K}(I^d)$. Let $\mathcal{C}(\alpha,K,d)$ be the collection of all sets

$$J_f := \{x \in I^d: 0 \leq x_d \leq f(x_{(d)})\}, \quad f \in \mathcal{G}_{\alpha,K,d-1}, \quad f \geq 0.$$

If g and h are two functions defined for (small enough) $y > 0$, then $g \asymp h$ (as $y \downarrow 0$) means that

$$0 < \liminf_{y\downarrow 0}(g/h)(y) \leq \limsup_{y\downarrow 0}(g/h)(y) < +\infty.$$

Kolmogorov (1955) stated the following for the metric entropy in the supremum distance $d_{sup}(f,g) := \sup_x|(f-g)(x)|$:

7.1.1 Theorem. For $0 < K < \infty$ and $d \geq 1$, as $\varepsilon \downarrow 0$ $\log D(\varepsilon, \mathcal{G}_{\alpha,K,d}, d_{sup}) \asymp \varepsilon^{-d/\alpha}$.

For some $J := J(\alpha,K,d)$, any law P on I^d, $1 \leq r \leq \infty$ and $0 < \varepsilon < 1$,

$$\log N_{[]}^{(r)}(\varepsilon, \mathcal{G}_{\alpha,K,d}, P) \leqslant J\varepsilon^{-d/\alpha}, \quad \text{and}$$

$$\log D(\varepsilon, \mathcal{C}(\alpha,K,d+1), h) \leqslant J\varepsilon^{-d/\alpha}.$$

If Q is a law on I^{d+1} having a density with respect to Lebesgue measure bounded by M, then for some $M_1 = M_1(M,d,K,\alpha)$,

$$\log N_I(\varepsilon, \mathcal{C}(\alpha,K,d+1), Q) \leqslant M_1\varepsilon^{-d/\alpha}, \quad 0 < \varepsilon \leqslant 1.$$

Proof. The inequality on $N_{[]}^{(r)}$ follows from the first using (6.1.2) and (6.1.3). The inequalities for classes of sets (in h and for N_I) follow directly from that for functions in d_{sup} by (7.0.1) and (7.0.2). So only the bounds for d_{sup} need to be proved.

For each $f \in \mathcal{G}_{\alpha,K,d}$, $x \in I^d$ and $x + h \in I^d$ write the Taylor series with remainder

$$f(x+h) - \textstyle\sum_{k=0}^{\beta} Q_k(x,h) = R(x,h) \tag{7.1.2}$$

where for each x, $Q_k(x,\cdot)$ is a homogeneous polynomial of degree k in h and by the mean value theorem

$$|R(x,h)| \leqslant C|h|^{\alpha} \tag{7.1.3}$$

for some constant $C = C(d,K,\alpha)$, $|\cdot|$ = Euclidean norm. We take C large enough so that whenever $[p] \leqslant \beta$ we also have

$$\begin{aligned} |D^p f(x+h) - \textstyle\sum_{k=0}^{\beta-[p]} Q_{k,p}(x,h)| &:= |R_p(x,h)| \\ &\leqslant C|h|^{\alpha-[p]} \end{aligned} \tag{7.1.4}$$

where for each k, p and x, $Q_{k,p}(x,\cdot)$ is also a homogeneous polynomial of degree k.

To prove one half of 7.1.1 it needs to be shown that

$$\limsup_{\varepsilon\downarrow 0} \log D(\varepsilon, \mathcal{G}_{\alpha,K,d}, d_{sup})\varepsilon^{d/\alpha} < \infty. \tag{7.1.5}$$

Given $\varepsilon > 0$, let $\Delta := (\varepsilon/(4C))^{1/\alpha}$. Let $x_{(1)},\ldots,x_{(s)}$ be a $\Delta/2$-net in I^d, i.e., $\sup\{\inf_{j\leqslant s}|x-x_{(j)}|: x \in I^d\} \leqslant \Delta/2$. Here we can take $s \leqslant M_2\varepsilon^{-d/\alpha}$ for some constant $M_2 = M_2(d,C)$; specifically, let us choose the $x_{(j)}$ as centers of cubes of a decomposition of I^d into cubes of side $1/m$ where m is the least integer $\geqslant 2d/\Delta$. For each multi-index p with $[p] = k \leqslant \beta$ let $p! := \prod_{j=1}^{d} p_j!$.

Then for $f \in \mathcal{G}_{\alpha,K,d}$ let

$$Q^p(x,h) \quad := (D^p f)(x)h^p/p!.$$

Then in (7.1.2)

(7.1.6) $$Q_k(x,h) \;=\; \sum_{[p]=k} Q^p(x,h).$$

Let $\varepsilon_k := \varepsilon/(2\Delta^k e^d)$, $k = 0,1,\ldots,$

$$A_{i,p} := A_{i,p}(f) := \left[D^p f(x_{(i)})/\varepsilon_k\right],\ i=1,\ldots,s,\ [p] = k \leqslant \beta,$$

where $[x]$ denotes the greatest integer $\leqslant x$, $-\infty < x < \infty$.

Given some $A := \{A_{i,p} : i \leqslant s,\ [p] \leqslant \beta\}$ let

$$\mathcal{G}_{\alpha,K,d}(A) := \{f \in \mathcal{G}_{\alpha,K,d} : A_{i,p}(f) = A_{i,p} \text{ for all } i \leqslant s,\ [p] \leqslant \beta\}.$$

7.1.7 <u>Lemma</u>. If $f,g \in \mathcal{G}_{\alpha,K,d}(A)$ for some A then $\sup\{|(f-g)(x)| : x \in I^d\} \leqslant \varepsilon$.

<u>Proof</u>. Let $F := f-g$. Whenever $[p] \leqslant \beta$, $i \leqslant s$, we have $|(D^pF)(x_i)| \leqslant \varepsilon_{[p]}$. Also

$$|D^pF(x+h) - D^pF(x)| \;\leqslant\; 2K|h|^{\alpha-\beta} \quad \text{if } [p] = \beta.$$

For each $y \in I^d$ take an $x_{(i)}$ with $|y-x_{(i)}| \leqslant \Delta/2$. Then from (7.1.2), (7.1.3), and (7.1.6) with $h := y-x_{(i)}$,

$$\begin{aligned} |F(y)| &\leqslant \textstyle\sum_{k=0}^{\beta} \varepsilon_k \left|\sum_{[p]=k} h^p/p!\right| + 2C|h|^\alpha \\ &\leqslant \textstyle\sum_{k=0}^{\beta} \varepsilon_k \Delta^k \left(\sum_{[p]=k} 1/p!\right) + 2C\Delta^\alpha \\ &\leqslant e^d \max_{k\leqslant\beta} \varepsilon_k \Delta^k + \varepsilon/2 \;\leqslant\; \varepsilon. \end{aligned}$$

□

It follows that $D(\varepsilon, \mathcal{G}_{\alpha,K,d}, d_{\sup}) \leqslant N_{\alpha,K,d}$ where $N_{\alpha,K,d}$ is the number of distinct non-empty sets $\mathcal{G}_{\alpha,K,d}(A)$. Let the $x_{(j)}$ be ordered so that for $1 < j \leqslant s$, $|x_{(i)}-x_{(j)}| \leqslant \Delta$ for some $i < j$. Such an ordering clearly exists for $d = 1$. Then by induction on d, we enumerate sub-cubes beginning and ending with sub-cubes at vertices of I^d, where we first enumerate cubes on one face I^{d-1}, then on the adjoining level, etc.

Now suppose $f \in \mathcal{G}_{\alpha,K,d}(A)$ (so that this set is non-empty) and suppose given the values $A_{i,r}$ for $i < j$ for some $j \leqslant s$. Choose $i < j$ such that $|x_{(i)}-x_{(j)}| \leqslant \Delta$. Take the Taylor expansion as in (7.1.2) with $x = x_{(i)}$, $h = x_{(j)} - x_{(i)}$. By (7.1.4),

$$|D^pf(x_{(j)}) - \textstyle\sum_{k=0}^{\beta-[p]} \sum_{[q]=k} D^{p+q} f(x_{(i)})h^q/q!| \;\leqslant\; C|h|^{\alpha-[p]}.$$

Thus

$$|D^p f(x_{(j)})\varepsilon_{[p]}^{-1} - \sum_{k=0}^{\beta-[p]}(\varepsilon_{k+[p]}/\varepsilon_{[p]})\sum_{[q]=k} A_{i,p+q}\, h^q/q!|$$

$$\leq C\Delta^{\alpha-[p]}\varepsilon_{[p]}^{-1} + \sum_{k=0}^{\beta-[p]} (\varepsilon_{[p]+k}\Delta^k/\varepsilon_{[p]})\sum_{[q]=k} 1/q!$$

$$\leq 2Ce^d/4C + e^d \leq 3e^d/2.$$

So for $f \in \mathcal{G}_{\alpha,K,d}$ and given the $A_{i,r}(f)$ for $i < j$ there are at most $4e^d$ possible values of $A_{j,p}$ for a given p. The number of different $p \in \mathbb{N}^d$ with $[p] \leq \beta$ is at most $(\beta+1)^d$. Thus the number of possible sets of values $\{A_{j,p}\}_{[p]\leq\beta}$ for the given j is at most

$$\exp((d + \log 4)(\beta + 1)^d).$$

The number of possible values of the vectors

$$\{A_{1,p}\}_{[p]\leq\beta} \text{ is at most } ((2K+1)/\varepsilon_\beta)^{(\beta+1)^d}.$$

Thus

$$N_{\alpha,K,d} \leq ((2K+1)4e^{ds}/\varepsilon_\beta)^{(\beta+1)^d}$$

$$\leq \exp((\beta+1)^d\{dM_2\varepsilon^{-d/\alpha} + \log((8K+4)/\varepsilon_\beta)\})$$

$$\leq \exp(J\varepsilon^{-d/\alpha})$$

for some $J = J(d,\alpha,K)$ not depending on ε, proving (7.1.5).

In the other direction we need to prove

$$\liminf_{\varepsilon\downarrow 0} \log D(\varepsilon, \mathcal{G}_{\alpha,K,d}, d_{sup})\varepsilon^{d/\alpha} > 0. \tag{7.1.8}$$

Let f be a C^∞ function on $\mathbb{R}^d$, 0 outside I^d and positive on its interior, such as

(7.1.9)
$$f(x) := \Pi_{j=1}^d \; g(x_j), \text{ where}$$
$$g(t) := \begin{cases} \exp(-1/t)\exp(-1/(1-t)), & 0 < t < 1 \\ 0 \text{ elsewhere.} \end{cases}$$

For $m = 1,2,\ldots,$ decompose I^d into m^d sub-cubes A_{mi} of side $1/m$, $i = 1,\ldots,m^d$. Let $x_{(i)}$ be the vertex of A_{mi} closest to the origin. Given $\alpha > 0$ set

$$f_i(x) := m^{-\alpha}f(m(x - x_{(i)})).$$

Let $s := \sup_x f(x)$ ($= e^{-4d}$ for our f). For any $S \subset \{1,\ldots,m^d\}$ let $f_S := \sum_{i\in S} f_i$. Then $||f_S||_\alpha \leq ||f||_\alpha := B$ while for $S \neq T$, $\sup_x |(f_S - f_T)(x)| = m^{-\alpha}s$. Thus for any $\varepsilon_m := Km^{-\alpha}s/(3B)$ we have $D(\varepsilon_m, \mathcal{G}_{\alpha,K,d}, d_{\sup}) \geq 2^{m^d} \geq \exp(C\varepsilon_m^{-d/\alpha})$ for some $C = C(K,d,\alpha,s)$ not depending on m. Since $\varepsilon_{m+1}/\varepsilon_m \to 1$ as $m \to \infty$ this is enough to prove (7.1.8), and so finish the proof of Theorem 7.1.1. □

To fill out the picture, some lower bounds for metric entropies in the L^1 norm will be given. For a collection $\mathcal{F} \subset \mathcal{L}^1(A, \mathcal{A}, P)$, we have the $\mathcal{L}^1$ distance $d_{1,P}(f,g) := P(|f-g|)$.

7.1.10 <u>Theorem</u>. Let P be a law on I^d having a density with respect to Lebesgue measure bounded below by $\gamma > 0$. Then for some $C = C(\gamma,\alpha,K,d) > 0$, and $1 \leq r \leq \infty$, $N_{[]}^{(r)}(\varepsilon, \mathcal{G}_{\alpha,K,d}, P) \geq N_{[]}^{(1)}(\varepsilon, \mathcal{G}_{\alpha,K,d}, P) \geq D(\varepsilon, \mathcal{G}_{\alpha,K,d}, d_{1,P}) \geq \exp(C\varepsilon^{-d/\alpha})$ for ε small enough, and if $d > 2$

$$N_I(\varepsilon, \mathcal{C}(\alpha,K,d), P) \geq D(\varepsilon, \mathcal{C}(\alpha,K,d), d_P)$$

$$\geq \exp(C\varepsilon^{-(d-1)/\alpha}) \quad \text{for small enough } \varepsilon > 0.$$

<u>Proof</u>. The following combinatorial fact will be used:

7.1.11 <u>Lemma</u>. Let B be a set with n elements, $n = 0, 1, \ldots$. Then there exist subsets $E_i \subset B$, $i = 1, \ldots, k$, where $k \geq e^{n/6}$, such that for $i \neq j$, the symmetric difference $E_i \Delta E_j$ has at least $n/5$ elements.

<u>Proof</u>. For any set $E \subset B$, the number of sets $F \subset B$ such that $\text{card}(E\Delta F) \leq n/5$ is $2^n B(n/5, 1/2)$ as defined in sec. 2.2. If S_n is the sum of n independent Rademacher variables taking values ± 1 with probability $1/2$ each, then by one of Hoeffding's inequalities (2.2.4 above; or, for the final result, 2.2.9),

$$B(n/5,n,1/2) = \Pr(S_n \geq 3n/5) \leq \exp(-9n/50) < e^{-n/6}.$$

This implies the Lemma. □

Now to prove Theorem 7.1.10, the first inequality follows from (6.1.2) and the second is also straightforward. For the lower bound on D, let us use again the construction in the proof of (7.1.8). Let λ denote Lebesgue measure on I^d and $\delta := \gamma\int f\, d\lambda$ for the f in (7.1.9). Then for each i, $\int f_i dP \geq \delta m^{-\alpha-d}$. Applying

Lemma 7.1.9 and obtaining sets S with $\mathrm{card}(S) \geq m^d/5$ gives $\int f_S dP > \delta m^{-\alpha}/6$, and $\int |f_S - f_T| dP = \int f_{S\Delta T} dP$. Thus

$$D(\delta m^{-\alpha}/6, \mathcal{G}_{\alpha,K,d}, d_{1,P}) \geq \exp(m^d/6).$$

Thus if $0 < \varepsilon \leq \delta/6$,

$$D(\varepsilon, \mathcal{G}_{\alpha,K,d}, d_{1,P}) \geq \exp([(\delta/(6\varepsilon))^{1/\alpha}]^d/6)$$
$$\geq \exp(2^{-d}(\delta/(6\varepsilon))^{d/\alpha}/6),$$

proving the statement about $\mathcal{G}_{\alpha,K,d}$.

If $C = J(f)$ and $D = J(g)$ then

$$d_P(C,D) := P(C\Delta D) \geq \gamma\lambda(C\Delta D)$$
$$= \gamma\int|f-g|d\lambda, \text{ so for } \varepsilon > 0$$
$$N_I(\varepsilon, \mathcal{C}(\alpha,K,d), P) \geq D(\varepsilon, \mathcal{C}(\alpha,K,d), d_P)$$
$$\geq D(\varepsilon/\gamma, \mathcal{G}_{\alpha,K,d-1}, d_{1,\lambda})$$

which implies the result. □

To get lower bounds for the Hausdorff metric, the following will help:

7.1.12 <u>Lemma</u>. If $\alpha \geq 1$ and $f,g \in \mathcal{G}_{\alpha,K,d}$, then $h(J_f, J_g) \geq d_{sup}(f,g)/(2\max(1,Kd))$.

<u>Proof</u>. Let $t := d_{sup}(f,g)$ and say $f(x) \geq g(x) + t$ for some x. Then for any y, $g(y) \leq g(x) + Kd|y-x|$. So if $|y-x| \leq t/(2Kd)$, then $g(y) \leq f(x) - t/2$. Hence $(f(x) - g(y))^2 + |x-y|^2 \geq t^2/(4\max(1,K^2d^2))$, and $d(\langle x,f(x)\rangle, J_g) \geq t/(2\max(1,Kd))$. □

7.1.13 <u>Corollary</u>. If $\alpha \geq 1$ then as $\varepsilon \downarrow 0$, $\log D(\varepsilon, \mathcal{C}(\alpha,K,d+1), h) \asymp \varepsilon^{-d/\alpha}$.

<u>Proof</u>. This follows from 7.1.12 and the first and third statements in 7.1.1. □

<u>Remark</u>. Forming a set E of one point each from a grid of cubes of side $\varepsilon/(2d^{1/2})$ we get $D(\varepsilon/2, I^d, \sigma) \leq (1 + 2d^{1/2}/\varepsilon)^d$ where σ is the usual Euclidean metric. Thus for every $A \in 2^{I^d}$ (i.e. $A \subset I^d$), $h(A,B) \leq \varepsilon/2$ for some $B \subset E$. Thus

$$D(\varepsilon, 2^{I^d}, h) \leq 2^{(1+2d^{1/2}/\varepsilon)^d}.$$

Hence for $\alpha < d/(d+1)$, Corollary 7.1.13 cannot hold, nor can the upper bound for h in Theorem 7.1.1 be sharp - contrary to Dudley (1974, (3.2), even with the correction, 1979), which is only proved for $\alpha \geq 1$. I am very grateful to J. Fu for this remark.

Notes. The proof of Theorem 7.1.1 is essentially that of Kolmogorov and Tikhomirov (1959). Lorentz (1966, p.920) sketches another proof. Theorem 7.1.8 is essentially due to Clements (1963, Theorem 3); the proof here is adapted from Dudley (1974, Lemmas 3.5, 3.6).

7.2 Lower layers. A set $B \subset \mathbb{R}^d$ is called a lower layer if and only if for all $x = (x_1,\ldots,x_d) \in B$ and $y = (y_1,\ldots,y_d)$ with $y_j \leq x_j$, $j = 1,\ldots,d$, we have $y \in B$. Let $\mathcal{LL}_d$ denote the collection of all lower layers in $\mathbb{R}^d$. Let $\emptyset$ be the empty set and

$$\mathcal{LL}_{d,1} := \{L \cap I^d : L \in \mathcal{LL}_d,\ L \cap I^d \neq \emptyset\}.$$

Let λ denote Lebesgue measure on I^d. Then, where $f \sim g$ means $f/g \to 1$, we have:

7.2.1 Theorem. For each $d = 1,2,\ldots$, as $\varepsilon \downarrow 0$,

$$\log D(\varepsilon, \mathcal{LL}_{d,1}, h) \asymp \log D(\varepsilon, \mathcal{LL}_d, d_\lambda)$$
$$\asymp \log N_I(\varepsilon, \mathcal{LL}_{d,1}\lambda) \asymp \varepsilon^{1-d}, \quad d \geq 2;$$
$$D(\varepsilon, \mathcal{LL}_{1,1}, h) \sim D(\varepsilon, \mathcal{LL}_1, d_\lambda) \sim N_I(\varepsilon, \mathcal{LL}_{1,1}, \lambda) \sim \varepsilon^{-1}.$$

Proof. Let us induct on d. For $d = 1$, sets in $\mathcal{LL}_{1,1}$ are intervals $[0,t[$, $0 < t \leq 1$, or $[0,t]$, $0 \leq t \leq 1$, which can be approximated within $1/m$ for h or d_λ by $m+1$ intervals $[0,k/m]$, $k = 0,1,\ldots,m$, also with inclusion as needed. The result follows with $m = 1 + [1/\varepsilon]$, $\varepsilon \downarrow 0$.

Now let $d \geq 2$ and assume the result for dimension $d - 1$. Let $A_0 := \emptyset$, $h(\emptyset,\emptyset) := 0$, $h(\emptyset,A) := +\infty$, $A \neq \emptyset$. For each $j = 1,\ldots,d$ and each $L \in \mathcal{LL}_{d,1}$, let $L_j := \{x \in L : x_j = 0\}$ and $L_{d+j} := \{x \in L : x_j = 1\}$. Then with respect to the coordinates x_i, $i \neq j$, $L_j \in \mathcal{LL}_{d-1,1}$ and $L_{d+j} \in \mathcal{LL}_{d-1,1}$ or $L_{d+j} = \emptyset$. Let $\{A_1,\ldots,A_m\}$ be a maximal collection in $\mathcal{LL}_{d-1,1}$ with $h(A_i,A_j) > \varepsilon/4$, $i \neq j$. Then for each $L \in \mathcal{LL}_{d,1}$ there are $i(j,L)$, $j = 1,\ldots,2d$, with $h(L_j, A_{i(j,L)}) \leq \varepsilon/4$, $j = 1,\ldots,2d$. The total number of different 2d-tuples $\{i(j,L)\}_{j=1}^{2d}$ is at most $(m+1)^{2d}$. There is a constant $C = C_d < \infty$ such that for $0 < \varepsilon \leq 1$,

$$(7.2.2)\qquad D(\varepsilon/4, \mathcal{LL}_{d-1,1}, h) \leq \begin{cases} \exp(C\varepsilon^{2-d}), & d \geq 3; \\ C/\varepsilon, & d = 2; \end{cases}$$

$$(7.2.3)\qquad (m+1)^{2d} \leq \begin{cases} 4^d \exp(2dC\varepsilon^{2-d}), & d \geq 3; \\ 4^d (C/\varepsilon)^{2d}, & d = 2. \end{cases}$$

Now let U be a rotation of $\mathbb{R}^d$ which takes $v := (1,1,\ldots,1)$ into $(0,0,\ldots,d^{1/2})$. The angle between v and each hyperplane $x_j = 0$ is $\cos^{-1}(((d-1)/d)^{1/2}) = \sin^{-1} d^{-1/2} = \tan^{-1}(d-1)^{-1/2}$. Thus for any lower layer B and point p on its boundary, $U(B)$ includes the cone

$$\{x: |x_{(d)} - q_{(d)}| < (q_d - x_d)(d-1)^{-1/2}\},$$

where $x_{(d)} := (x_1,\ldots,x_{d-1})$, $q := U(p)$.

Hence the boundary of $U(B)$ is the graph of a function $f: \mathbb{R}^{d-1} \to \mathbb{R}$ where for any $s, t \in \mathbb{R}^{d-1}$,

$$f(s) \geq f(t) - K|s-t|, \quad K := (d-1)^{1/2}.$$

Hence, interchanging s and t,

$$|f(s) - f(t)| \leq K|s-t|.$$

Let $||f||_L := \sup_{s \neq t} |f(s) - f(t)|/|s-t|$. Thus U carries $\mathcal{LL}_d$ into the collection of all sets $\{x \in \mathbb{R}^d: x_d \leq f(x_{(d)})\} := J(f)$ with $||f||_L \leq K$.

Now $U(I^d)$ is included in a cube M, of side $d^{1/2}$, parallel to the axes, with $M = F \times [0,d^{1/2}]$ for a cube $F \subset \mathbb{R}^{d-1}$. If $g := \min(d^{1/2}, \max(f,-1))$, then $||g||_L \leq ||f||_L$ and $||g||_{\sup} \leq d^{1/2}$. Thus for each $A \in \mathcal{LL}_{d,1}$ we have

$$U(A) = U(I^d) \cap J(g)$$

where $g: \mathbb{R}^{d-1} \to \mathbb{R}$ satisfies $||g||_L \leq K$ and $||g||_{\sup} \leq d^{1/2}$.

Claim. $U^{-1}(J(g))$ is a lower layer.

Proof of Claim. For any constant C, $U^{-1}(J(C))$ is a lower layer $\{x: x_1 + \ldots + x_d \leq Cd^{1/2}\}$. The union or intersection of any two lower layers is a lower layer. Let $A = B \cap I^d$, $B \in \mathcal{LL}_d$, $U(B) = J(f)$. Then

$$U^{-1}(J(g)) = (B \cup U^{-1}(J(-1))) \cap U^{-1}(J(d^{1/2})),$$

a lower layer, proving the claim.

Let $D := 2d^{1/2}$. Then in the notation of Sec. 7.1, the set of all g arising above is included in $\mathcal{F}_{1,D}(F)$. Theorem 7.1.1 extends from I^d to

$M = F \times [0,d^{1/2}]$ by rescaling, so that for some constant $T = T_d < \infty$,

$$\log D(\varepsilon/4, \mathcal{C}(1,D,d), h) \leqslant T\varepsilon^{1-d}, \quad 0 < \varepsilon \leqslant 1.$$

Choose a maximal collection of sets $E_1, \ldots, E_r$ in $\mathcal{C}(1,D,d)$ with $h(E_i, E_j) > \varepsilon/4$, $i \neq j$, so $r \leqslant \exp(T\varepsilon^{1-d})$.

(Note that possibly $h(E_i \cap H, E_j \cap H) > h(E_i, E_j)$, so that there is more to do, using the induction; $H := U(I^d)$.)

Now for each $B \in \mathcal{LL}_{d,1}$ we have the 2d-tuple as in (7.2.3) and a $k \leqslant r$ such that $U(B) = E \cap U(I^d)$ with $h(E, E_k) \leqslant \varepsilon/4$, $E \in \mathcal{C}(1,D,d)$. The total number of (2d+1)-tuples $\{<\{i(j,L)\}_{j=1}^{2d}, k>\}$ is thus bounded above by $\exp(N\varepsilon^{1-d})$ for some $N < \infty$ and for $0 < \varepsilon \leqslant 1$. It suffices now to show that if B and C have the same (2d+1)-tuple then $h(B,C) \leqslant \varepsilon$.

Let $U(B) = J(g) \cap U(I^d)$ and $U(C) = J(f) \cap U(I^d)$ where $f, g \in \mathcal{F}_{1,D}(F)$. Suppose $h(B,C) > \varepsilon$. By symmetry, we can suppose $d(x,B) > \varepsilon$ for some $x \in C$. For some k, $h(J(f), E_k) \leqslant \varepsilon/4$ and $h(J(g), E_k) \leqslant \varepsilon/4$. Thus $|U(x) - w| \leqslant \varepsilon/2$ for some $w \in J(g)$. Let $y := U^{-1}(w)$. Then $|x-y| \leqslant \varepsilon/2$. Hence $y \notin B$, so $y \notin I^d$.

Let $y_i' := \min(y_i, 1)$, $i = 1, \ldots, d$. Then by the Claim, $y' \in U^{-1}(J(g))$ and $|x-y'| \leqslant |x-y|$ so we may assume $y' = y$. Then for some z on the closed line segment joining x to y, z is on the boundary of I^d, i.e. for some j, $z_j = 0$ or 1.

If $z_j = 0$, let $x_i' = x_i$ for $i \neq j$ and $x_j' = 0$. Then $x' \in C \in \mathcal{LL}_{d,1}$. By assumption, for some $y' \in B$ with $y_j' = 0$ (so $y' \in B_j$), $|x'-y'| \leqslant \varepsilon/2$ and $|x-x'| \leqslant |x-z| \leqslant |x-y| \leqslant \varepsilon/2$, so $|x-y'| \leqslant \varepsilon$ and $d(x,B) \leqslant \varepsilon$, a contradiction.

So $z_j = 1$. Then since $y_j \leqslant 1$ we have $x_j = y_j = 1$. Hence for some $q \in B_{d+j}$, $|x-q| \leqslant \varepsilon/4 < \varepsilon$, again a contradiction. So the proof of the upper bound in the Hausdorff metric is done.

To prove the upper bounds involving λ, let us again use the transformation U and the corresponding parts of Theorem 7.1.1. Let μ be Lebesgue measure on $\mathbb{R}^d$. Then U preserves μ. Let $C_i \in \mathcal{LL}_{d,1}$ with $U(C_i) = U(I^d) \cap J(g_i)$, $g_i \in \mathcal{F}_{1,D}(F)$, $i = 1,2,3$. Then

$$\mu(C_1 \Delta C_3) \leqslant \mu(J(g_1) \Delta J(g_3)),$$

where Δ denotes symmetric difference. If $g_1 \leqslant g_2 \leqslant g_3$ then $C_1 \subset C_2 \subset C_3$.

Thus the upper bounds for λ and d_λ follow.

The lower bounds for $d = 1$ are straightforward. For $d \geq 2$ let us use the same transformation U as above. The angle between v and each coordinate axis is

$$\cos^{-1}d^{-1/2} = \sin^{-1}(((d-1)/d)^{1/2}) = \tan^{-1}(d-1)^{1/2}.$$

Thus if $f: \mathbb{R}^{d-1} \to \mathbb{R}$ satisfies $||f||_L \leq (d-1)^{-1/2}$, then $U^{-1}(J(f))$ is a lower layer. Now let us apply (7.1.8) and its proof, for $\alpha = 1$, replacing d by $d-1$ and $\mathcal{G}_{\alpha,K,d}$ by $\mathcal{F}_{1,\delta}(H)$ where $\delta = (d-1)^{-1/2}d^{-1}$ and H is a small enough cube in $\mathbb{R}^{d-1}$, located so that for each $f \in \mathcal{F}_{1,\delta}(H)$, $\frac{1}{2}v + U^{-1}(J(f)) \subset I^d$. For such f, $J(f) \longleftrightarrow \frac{1}{2}v + U^{-1}(J(f))$ is an isometry for h and for d_λ and preserves inclusion. Each f can be extended to $\mathbb{R}^{d-1}$, preserving $||f||_L \leq \delta$ (Kirszbraun, 1934; McShane, 1934), by the non-linear part of the proof of the Hahn-Banach theorem. So the lower bound with $d_{\sup}$ in 7.1.1 gives, via (7.0.3), the lower bound with h in 7.2.1. Theorem 7.1.10, likewise adapted from I^{d-1} to H, gives the lower bounds with λ, proving 7.2.1. □

To make Theorem 7.2.1 more precise one might try to evaluate constants K_d and M_d such that as $\varepsilon \downarrow 0$, $\log D(\varepsilon, \mathcal{LL}_{d,1}, h) \sim K_d \varepsilon^{1-d}$ and $\log D(\varepsilon, \mathcal{LL}_d, d_\lambda) \sim M_d \varepsilon^{1-d}$, $d \geq 2$. Here is a step in that direction for $d = 2$:

7.2.4 __Theorem.__ For $m = 1,2,\ldots$, and $0 < t < 1/m$, $\max(N_I(2/m, \mathcal{LL}_2, \lambda)$, $D(2^{1/2}/m, \mathcal{LL}_{2,1}, h)) \leq \binom{2m-2}{m-1} \leq D(t, \mathcal{LL}_{2,1}, h)$. For $0 < \varepsilon \leq 1$, $N_I(\varepsilon, \mathcal{LL}_{2,1}, \lambda) \leq 4^{1+2/\varepsilon}$ and $D(\varepsilon, \mathcal{LL}_{2,1}, h) \leq 4 \exp((2^{1/2}\log 4)/\varepsilon)$.

__Proof.__ Let $J := [0,1[$. Decompose $J \times J$ into a union of m^2 squares $[(i-1)/m, i/m[\times [(j-1)/m, j/m[$, $i,j = 1,\ldots,m$. For any $L \in \mathcal{LL}_{2,1}$, let ${}_mL$ be the union of the squares in the grid included in L and L_m the union of the squares which intersect L. Then ${}_mL \subset L \subset L_m$ and both ${}_mL$ and L_m are in $\mathcal{LL}_{2,1} \cup \{\emptyset\}$.

For each m and each function f from $\{2,3,\ldots, 2m-1\}$ into $\{0,1\}$ taking the value 1 exactly $m-1$ times, define a sequence $S(f)(j)$, $j = 1,\ldots,2m$, of squares in the grid as follows. Let $S(f)(1)$ be the upper left square $[0,1/m[\times [(m-1)/m, 1[$. Given $S(f)(j-1)$, let $S(f)(j)$ be the square just to its right if $f(j) = 1$, otherwise the square just below it, $j = 2,\ldots,2m-1$. Then

$S(f)(2m-1)$ is always the lower right square $[(m-1)/m, 1[\times [0, 1/m[$. Let $B_m(f) := \bigcup_{j=1}^{2m} S(f)(j)$. Let $A_m(f)$ be the union of the squares below and to the left of $B_m(f)$, and $C_m(f) := A_m(f) \cup B_m(f)$. Here $A_m(f)$ and $C_m(f)$ belong to $\mathcal{LL}_{2,1} \cup \{\emptyset\}$. Then one can show that for any $L \in \mathcal{LL}_{2,1} \cup \{\emptyset\}$ there is an f as above such that

$$A_m(f) \subset {}_mL \subset L \subset L_m \subset C_m(f),$$

so that $L_m \setminus {}_mL \subset B_m(f)$, and $\lambda(B_m(f)) = (2m-1)/m^2 < 2/m$, while $1/m \leqslant h(A_m(f), C_m(f)) \leqslant 2^{1/2}/m$. Since the number of such functions f is $\binom{2m-2}{m-1}$, the first sentence of the Theorem is proved. Then since $\binom{2m-2}{m-1} \leqslant 4^{m-1}$ the second sentence follows. □

To get a lower bound for $N_I(\varepsilon, \mathcal{LL}_2, \lambda)$ let us consider squares in the same grid along the diagonal $x + y = 1$. If we include all squares below the diagonal, and an arbitrary set of squares on the diagonal, we always get a lower layer. Thus Lemma 7.1.11 gives, for any $\alpha < 0.2$,

$$N_I(\alpha/m, \mathcal{LL}_{2,1}, \lambda) \geqslant D(\alpha/m, \mathcal{LL}_{2,1}, d_\lambda) \geqslant e^{m/6}.$$

This and the proof of Theorem 7.2.4, together with $\binom{2k}{k} \geqslant 4^k k^{-1/2}/3$ from Stirling's formula, give

$$\begin{aligned} \log 4 &\leqslant \underline{\lim}_{\varepsilon \downarrow 0} \varepsilon \log D(\varepsilon, \mathcal{LL}_{2,1}, h) \\ &\leqslant \overline{\lim}_{\varepsilon \downarrow 0} \varepsilon \log D(\varepsilon, \mathcal{LL}_{2,1}, h) \leqslant 2^{1/2} \log 4, \\ 1/30 &\leqslant \underline{\lim}_{\varepsilon \downarrow 0} \varepsilon \log N_I(\varepsilon, \mathcal{LL}_{2,1}, \lambda) \\ &\leqslant \overline{\lim}_{\varepsilon \downarrow 0} \varepsilon \log N_I(\varepsilon, \mathcal{LL}_{2,1}, \lambda) \leqslant \log 16. \end{aligned}$$

Thus if the numbers K_d and M_d exist (as I conjecture they do) then $1 \leqslant K_2/\log 4 \leqslant 2^{1/2}$ and $1/30 \leqslant M_2 \leqslant \log 16$. For $d \geqslant 3$, some explicit but even less sharp bounds could be found from the proofs of Theorems 7.1.1 and 7.1.10. Improvement of these bounds may involve results on packing and covering (e.g. Rogers, 1964).

__Notes.__ Theorem 7.2.1 and its proof for $d \geqslant 3$ are due to Lucien Birgé (1982, unpublished), who kindly agreed to their inclusion here. For Theorem 7.2.4 I am indebted to earlier conversations with Mike Steele. Any errors, however, are mine.

A lower bound for empirical processes on lower layers in the plane will be given in Sec. 8.4 below, where P is uniform on the unit square. For other, previous results, e.g. laws of large numbers uniformly over $\mathcal{LL}_d$ for more general P, and on the statistical interest of lower layers (monotone regression), see Wright (1981) and references given there.

7.3 Metric entropy of classes of convex sets.

Let $\mathcal{C}_d$ denote the collection of all non-empty closed convex subsets of the unit cube $I^d := \{x: 0 \leq x_j \leq 1,\ j = 1,\dots,d\} \subset \mathbb{R}^d$. Let λ be the uniform Lebesgue measure on $\mathbb{R}^d$. Upper and lower bounds will be given for the metric entropy of $\mathcal{C}_d$ for the metric d_λ and for the Hausdorff metric h. Here I^d could be replaced by any bounded set in $\mathbb{R}^d$ with non-empty interior, such as the unit ball $B_1 := \{x: |x| \leq 1\}$.

7.3.1. **Theorem.** For each $d \geq 2$ we have

$$\log D(\varepsilon, \mathcal{C}_d, d_\lambda) \asymp \log D(\varepsilon, \mathcal{C}_d, h) \asymp \varepsilon^{(1-d)/2}, \quad \varepsilon \downarrow 0.$$

Remark. For $d = 1$ we instead have, rather trivially,

$$D(\varepsilon, \mathcal{C}_1, d_\lambda) \asymp D(\varepsilon, \mathcal{C}_1, h) \asymp \varepsilon^{-2}.$$

Before discussing the proof of 7.3.1, let us prove from it:

7.3.2. **Corollary.** If P is a law with a density f with respect to Lebesgue measure on I^d, $d \geq 2$, and $0 < \inf f \leq \sup f < +\infty$ then

$$\log N_I(\varepsilon, \mathcal{C}_d, P) \asymp \varepsilon^{(1-d)/2}, \quad \varepsilon \downarrow 0.$$

Proof. Let $0 < v < f(x) \leq V < +\infty$ for all x. For $B \subset R^d$ and $\delta > 0$ let $_\delta B := \{x: y \in B$ whenever $|x-y| < \delta\}$, $B^\delta := \{x: d(x,B) < \delta\}$. Then $_\delta B \subset B \subset B^\delta$. If B, C are convex and $h(B,C) < \delta$, then $C \subset B^\delta$. Also, $_\delta B \subset C$: if not, let $x \in {_\delta B} \backslash C$ and take a half-space $J \supset C$, $x \notin J$. On the line through x perpendicular to the hyperplane bounding J, opposite J there are points $y \in B \backslash C^\delta$, a contradiction.

So $_\delta B \subset C \subset B^\delta$. Next, here is a

Claim. There is a $K = K(d) < \infty$ such that for any $B \in \mathcal{C}_d$,

$$\lambda(B^\delta \backslash {_\delta B}) \leq K\delta, \quad 0 < \delta \leq 1.$$

Proof of Claim. If B is convex and has null interior then it is included in some hyperplane. Then $\lambda(B^\delta \setminus_\delta B) = \lambda(B^\delta) \leq K_1\delta$ where K_1 is twice the $(d-1)$-dimensional volume of a ball of radius $d+2$ in $\mathbb{R}^{d-1}$.

Now let B have an interior. Let ∂B denote the boundary of B. Then $B^\delta \setminus_\delta B = (\partial B)^\delta$. Let

$$S := \lim_{\delta \downarrow 0} \lambda(B^\delta \setminus B)/\delta.$$

Then S is finite and equals the surface area, one of the mixed volumes of B (Bonnesen and Fenchel, 1934, pp. 38, 46-47, or Eggleston, 1958, pp. 82-89). Then

$$\lambda(B^{\varepsilon/2} \setminus B) \geq \frac{1}{2} v_d (\varepsilon/2)^d D(\varepsilon, \partial B, \rho)$$

where ρ is the Euclidean distance and $v_d = \lambda(B_1)$ is the volume of the unit ball in $\mathbb{R}^d$. Thus

$$\limsup_{\varepsilon \downarrow 0} D(\varepsilon, \partial B, \rho)\varepsilon^{d-1} \leq 2^d S/v_d.$$

Now

$$\lambda((\partial B)^\delta) \leq D(\delta, \partial B, \rho) v_d \delta^d < 2^{d+1} S\delta$$

for δ small enough, and the Claim follows.

Now assuming Theorem 7.3.1, given $0 < \delta \leq 1$, let $\mathcal{N}_d(\delta)$ be a δ-net in $\mathcal{C}_d$ for h with cardinality

$$\operatorname{card} \mathcal{N}_d(\delta) \leq \exp(A_d \delta^{(1-d)/2})$$

where A_d is a large enough constant.

The sets $_\delta B$ and B^δ, $B \in \mathcal{N}_d(\delta)$, form a collection of at most double this cardinality. For any $C \in \mathcal{C}_d$ there is such a B with $_\delta B \subset C \subset B^\delta$ (as seen just before the Claim) and $\lambda(B^\delta \setminus_\delta B) \leq K\delta$, so $P(B^\delta \setminus_\delta B) \leq KV\delta$, and

$$N_I(\varepsilon, \mathcal{C}_d, P) \leq 2D(\varepsilon/(KV), \mathcal{C}_d, h) \leq 2\exp(A_d(\varepsilon/KV)^{(1-d)/2})$$

for ε small enough, giving an upper bound as desired.

A lower bound is easier: we have

$$N_I(\varepsilon, \mathcal{C}_d, P) \geq D(\varepsilon, \mathcal{C}_d, d_P) \geq D(\varepsilon/v, \mathcal{C}_d, d_\lambda) \geq \exp(c_d(\varepsilon/v)^{(1-d)/2})$$

for some $c_d > 0$ and all ε small enough, which finishes the proof of 7.3.2 from 7.3.1. □

Now the proof of Theorem 7.3.1, due to Bronšhtein (1976), will be sketched. Here, take $\mathcal{C}_d$ to be the collection of all closed convex subsets of the unit ball

B_1. Let $0 < \varepsilon < 1$.

For any $C \in \mathcal{C}_d$ let $\mathcal{N}_\varepsilon(C) := \{D \in \mathcal{C}_d: h(C,D) \leq \varepsilon\}$. For $r \geq 0$ let $C^{r]} := \{x \in \mathbb{R}^d: d(x,C) \leq r\}$, $B_r := \{0\}^{r]} := \{x: |x| \leq r\}$, $\mathcal{C}_d(B_r) := \{C: C$ convex, closed, and $C \subset B_r\}$. Note that for all $C, D \in \mathcal{C}_d$ and $r \geq 0$, $h(C^{r]},D^{r]}) = h(C,D)$. Thus

(7.3.3) $\quad \varphi_r: D \to D^{r]}$ is an isometry for h from $\mathcal{C}_d(B_1)$ into $\mathcal{C}_d(B_{1+r})$.

For $r > 0$ φ_r is a useful smoothing, as it takes a convex set D, which may have a sharply curved boundary (vertices, edges, ...) to a set $D^{r]}$ whose boundary is no more curved than a sphere of radius r, and so is easier to approximate. The main remaining step is:

7.3.4 <u>Lemma</u>. For each $d \geq 2$ there is a constant $M_d < \infty$ such that for any $C \in \mathcal{C}_d$ and $0 < \varepsilon \leq 1/4$, $D(\varepsilon, \mathcal{N}_{2\varepsilon}(C), h) \leq \exp(M_d \varepsilon^{(1-d)/2})$.

Let us next prove Theorem 7.3.1 from Lemma 7.3.4. First, $K_d := D(1/4, \mathcal{C}_d, h)$ is some finite number. Let $\varepsilon(n) := 2^{-n-2}$, $n = 0,1,\ldots$; we use induction on n. Let $K(0,d) := K_d$. Let $\{C_{ni}\}_{1 \leq i \leq K(n,d)}$ be a maximal $\varepsilon(n)$-distinguishable set in $\mathcal{C}_d$, $K(n,d) := D(\varepsilon(n), \mathcal{C}_d, h)$. Then $\mathcal{C}_d = \cup_i \mathcal{N}_{\varepsilon(n)}(C_{ni})$ and

$$\begin{aligned} D(\varepsilon(n)/2, \mathcal{C}_d, h) &\leq \textstyle\sum_{i=1}^{K(n,d)} D(\varepsilon(n+1), \mathcal{N}_{\varepsilon(n)}(C_{ni}), h) \\ &\leq K(n,d) \exp(M_d \varepsilon(n+1)^{(1-d)/2}) \end{aligned}$$

by Lemma 7.3.4 for $\varepsilon = \varepsilon(n+1)$. Thus

$$\begin{aligned} K(n+1,d) &\leq K_d \exp(\textstyle\sum_{j=0}^{n} (2^{-j-3})^{(1-d)/2} M_d) \\ &\leq K_d \exp(M_d 2^{(n+3)(d-1)/2}/(1-2^{(1-d)/2})). \end{aligned}$$

For $0 < \varepsilon \leq 1/4$ we have $2^{-n-3} < \varepsilon \leq 2^{-n-2}$ for some n, and

$$D(\varepsilon, \mathcal{C}_d, h) \leq K_d \exp(M_d 2^{(d+3)/2} \varepsilon^{(1-d)/2})$$

giving the desired upper bound for h, hence also for d_λ by the proof of 7.3.2.

For a lower bound Lemma 7.1.11 can be used as in the last two sections. On the unit sphere $S^{d-1} := \{x \in \mathbb{R}^d: |x| = 1\}$, $d \geq 2$, take the usual Euclidean metric ρ. For some $a_d > 0$ we have

$$D(\varepsilon, S^{d-1}, \rho) \geq a_d \varepsilon^{1-d}, \quad 0 < \varepsilon \leq 1.$$

Given ε take a 2ε-distinguishable set $\{x_i\}_{1 \leq i \leq D(2\varepsilon)}$ in S^{d-1} of maximal cardinality $D(2\varepsilon) := D(2\varepsilon, S^{d-1}, \rho)$. Let $\angle abc$ denote the angle at b in the triangle abc. Then for $i \neq j$, $\angle x_i 0 x_j \geqslant 2\varepsilon$. Let K_i be the half-line from 0 through x_i. Let C_i be the spherical cap cut from the unit ball B_1 by a hyperplane perpendicular to K_i at distance $\cos \varepsilon$ from 0. Then the caps C_i are disjoint.

For any set I, let $D_I := B_1 \setminus \cup_{i \in I} C_i$. Then each D_I is convex. The d-dimensional volume of each C_i is larger than $b_d \varepsilon^{d+1}$ for some constant $b_d > 0$. By Lemma 7.1.11 there are at least $e^{D(2\varepsilon)/6}$ sets $I = I(j)$ with $\mathrm{card}(I(i) \Delta I(j)) \geqslant D(2\varepsilon)/5$, $i \neq j$, and then

$$\lambda^d(D_{I(i)} \Delta D_{I(j)}) \geqslant \alpha_d \varepsilon^{1-d+1+d} = \alpha_d \varepsilon^2$$

for some $\alpha_d > 0$ where λ^d is d-dimensional Lebesgue measure. Thus for some $\beta_d > 0$,

$$D(\delta, \mathcal{C}_d, d_\lambda) \geqslant \exp(\beta_d \delta^{(1-d)/2}), \quad 0 < \delta \leqslant 1,$$

completing the proof of the lower bound for d_λ, hence also for h, by the proof of 7.3.2.

It remains to prove Lemma 7.3.4. For this one can first use φ_1 in (7.3.3) to replace $\mathcal{N}_{2\varepsilon}(C)$ by $\{D^1 : D \in \mathcal{N}_{2\varepsilon}(C)\}$. Such sets D^1 are all included in the ball $B(0,3)$ with center at 0 and radius 3 ($\varepsilon \leqslant 1/4$). Fixing a point $p \in C \subset B(0,1)$ and shifting to assume $p = 0$, we then have $D^1 \subset B(0,4)$.

For $\delta > 0$ we take a δ-net $\{p_i\}_{i=1}^{m(\delta)} \subset S^{d-1}$ where the p_i are vertices of a polyhedron all whose faces are simplices of diameter $< \delta$. Specifically, take a cube centered at 0, decompose its $(d-1)$-dimensional faces into small enough subcubes, and let a p_i be on each half-line from 0 through a sub-cube vertex. Then one can decompose each $(d-1)$-cube into simplices without adding new vertices: to see this it suffices by induction to do it for $S \times [0,1]$ where S is a simplex with vertices $(v_0, \ldots, v_p)$. Let $a_i := \langle v_i, 0\rangle$, $b_i := \langle v_i, 1\rangle$. Then for each $i = 0, \ldots, p$, $(a_0, \ldots, a_i, b_i, \ldots, b_p)$ are vertices of a $(p+1)$-dimensional simplex S_i, and these S_i give the desired decomposition of $S \times [0,1]$. I thank J. Munkres for telling me this construction.

Remark. The resulting polyhedra are not necessarily convex for $d \geqslant 3$.

Let H_i be the half-line with an end at 0 passing through p_i. Let v_i be the point at which H_i intersects the boundary ∂D^1 (note that $D^1 \supset B(0,1-2\varepsilon) \supset B(0,1/2)$, so that the open line segment from 0 to v_i is in the interior of D^1 and v_i is unique). Let T be the polyhedron with vertices v_i as constructed above for the p_i. To bound the lengths of edges of T we can use:

7.3.5 <u>Lemma</u>. Let $u, v \in \partial D^1$ where $d(0,D) \leqslant 1/2$, $D \subset B(0,3)$, $\angle u0v := \Theta$, $0 \leqslant \Theta < \pi/2$. Then $|u-v| \leqslant 32\Theta$.

<u>Proof</u>. Let L be the line through u and v (if u = v we are done). Let p be the point on L closest to 0. If p is not on the closed line segment with endpoints u and v, we can suppose the points are in the order p, u, v (Fig.1).

<u>Fig.1</u>

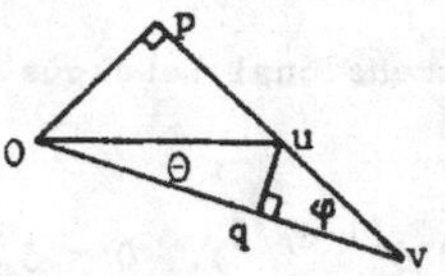

Then $|p| \geqslant 1/2$ since $v \in D^{1]}$ and $u \in \partial D^1$, while $|v| \leqslant 4$. Let q be the point on the line from 0 to v closest to u. Then for $\varphi := \angle 0vp$,

$$|u-v| \leqslant |u-q|/\sin\varphi \leqslant 8|u-q| = 8|u|\sin\Theta \leqslant 32\Theta$$

as desired.

If p is between u and v then for $\Theta = \Theta_1+\Theta_2$, $\Theta_1 \geqslant 0$,

$$|u-v| = |u-p| + |p-v| \leqslant 4(\sin\Theta_1 + \sin\Theta_2) \leqslant 4\Theta \leqslant 32\Theta. \qquad \square$$

The rest of the proof of Lemma 7.3.4 will just be briefly sketched. First let d = 2; see Fig.2.

Let $q \in D^1\backslash T$. Then the closest point of T to q is on an edge $\overline{v_1v_2}$ where v_1 and v_2 are consecutive vertices of the polygon T with $\angle v_1 0 v_2 < \delta$,

<u>Fig.2</u>

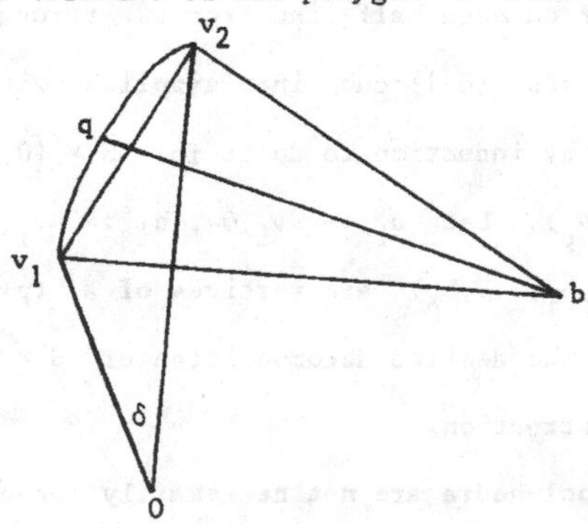

Then by Lemma 7.3.5, $|v_1-v_2| < 32\delta$. We have $|q-b| \leqslant 1$ for some $b \in D$ with $|b-v_i| \geqslant 1$, $i = 1,2$. Hence if $\delta \leqslant 1/16$, then

$$d(q,T) \leqslant 1 - \cos(\sin^{-1}(16\delta)) = 1 - (1-256\delta^2)^{1/2} \leqslant 256\delta^2.$$

Hence to approximate D^1 within $\varepsilon/2$ for h by T we can take $\delta = \varepsilon^{1/2}/32$.

In higher dimensions one would replace the segment $\overline{v_1v_2}$ by a simplex whose edges are all of length $\leqslant 32\delta$ by Lemma 7.3.5, and obtain $d(q,T) \leqslant A_d\delta^2$ for some $A_d < \infty$.

Thus for a given small ε the number $m(\delta)$ of rays H_i is $\leqslant B_d\varepsilon^{(1-d)/2}$ for some $B_d < \infty$. Let $D \in \mathcal{N}_{2\varepsilon}(C)$ and for a ray H_i let $\{v\} = \partial D^1 \cap H_i$, $\{w\} = \partial C^1 \cap H^i$. Since $0 \in C$, $C^1 \supset B(0,1)$. Thus the support hyperplane L to C^1 at w forms an angle at least $\sin^{-1}(1/4)$ with H_i, and $|v-u| \leqslant 2\varepsilon$ for some $u \in L$, if $w \in D^1$. Then $|v-w| \leqslant 8\varepsilon$. Or, if $v \in C^1$, then likewise since $C^1 \subset B(0,3)$ and $D^1 \supset B(0,1/2)$, we have $|v-w| \leqslant 12\varepsilon$. Hence the possible vertices of the polyhedron T for D^1 are in an interval of length 20ε with center at w. To approximate such polyhedra within $\varepsilon/2$ for h it suffices to approximate their vertices, using 41 vertices on each ray. The number of polyhedra formed (by the same triangulation as above) by all possible such vertices is thus $\leqslant \exp(B_d(\log 41)\varepsilon^{(1-d)/2})$, which implies Lemma 7.3.4.

__Notes.__ Theorem 7.3.1 is due to Bronštein (1976). Specifically, the smoothing $D \to D^r$ and the use of Lemma 7.3.4 and induction are his ideas, allowing improvement on the result of Dudley (1974, Theorem 4.1) where in the upper bound $\varepsilon^{(1-d)/2}$ was multiplied by $|\log \varepsilon|$. The lower bound in Dudley (1974) has been reproduced here.

Bronštein's proof of Lemma 7.3.4 involves several lemmas and pages which have been omitted here. Further simplification of his proof would be desirable. See also Gruber (1981).

Chapter 8 Classes of sets or functions too large for central limit theorems.

8.1 Universal lower bounds. This chapter is primarily about asymptotic lower bounds for $||P_n-P||_{\mathcal{F}}$ on certain classes of functions $\mathcal{F}$ as treated in the last chapter, mainly classes of indicators of sets. Section 8.2 will give some upper bounds which indicate the sharpness of some of the lower bounds. Section 8.4 gives some relatively difficult lower bounds on classes such as the convex sets in $\mathbb{R}^3$ and lower layers in $\mathbb{R}^2$. In preparation for this, Section 8.3 treats Poissonization and random "stopping sets" analogous to stopping times. The present section gives lower bounds in some cases which hold not only with probability converging to 1, but for all possible P_n. Definitions are as in Secs. 2.1 and 7.1, with $P = \lambda^d$ = Lebesgue measure on I^d.

8.1.1. Theorem. For any $d = 1,2,\ldots$ and $\alpha > 0$, there is a $\gamma = \gamma(d,\alpha) > 0$ such that for all possible values of P_n,

$$||P_n-P||_{\mathcal{G}(\alpha,1,d)} \geq \gamma n^{-\alpha/d}.$$

Remark. When $\alpha < d/2$, this shows that $\mathcal{G}(\alpha,K,d)$, $K > 0$, is not a functional Donsker class. For $\alpha > d/2$ the lower bound in 8.1.1 becomes smaller than the average size of $||P_n-P||_{\mathcal{G}(\alpha,1,d)}$ which is at least of order $n^{-1/2}$ (even for one function f not constant a.e. P, $E|(P_n-P)(f)| \geq cn^{-1/2}$ for some $c > 0$).

Theorem 8.1.1 gives information about accuracy of possible methods of numerical integration in several dimensions, or "cubature", using the values of a function $f \in \mathcal{G}(\alpha,K,d)$ at just n points chosen in advance (from the proof, it will be seen that one has the same lower bound even if one can use any partial derivatives of f at the n points). It was in this connection that Bakhvalov (1959) proved the theorem.

Proof. Given n let $m := m(n) := [(2n)^{1/d}] + 1$. Decompose the unit cube I^d into m^d cubes C_i of side $1/m$. Then $m^d > 2n$. For any P_n let $S := \{i: P_n(C_i) = 0\}$. Then $\mathrm{card}(S) \geq n$. For f_S as defined after (7.1.9), we then have $|(P_n-P)(f_S)| = P(f_S) \geq m^{-\alpha-d}nP(f) \geq cn^{-\alpha/d}$ for some constant $c > 0$, while $||f_S||_\alpha \leq ||f||_\alpha \leq 1$ (dividing the original f by some constant depending on d and α if necessary). □

8.1.2 **Theorem.** For any $K > 0$ and $0 < \alpha < d-1$ there is a $\delta = \delta(\alpha,K,d) > 0$ such that for all possible values of P_n, $||P_n - P||_{\mathcal{C}(\alpha,K,d)} > \delta n^{-\alpha/(d-1+\alpha)}$.

Remark. Since $\alpha/(d-1+\alpha) < 1/2$ for $\alpha < d-1$, the classes $\mathcal{C}(\alpha,K,d)$ are then not functional Donsker classes. For $\alpha > d-1$, $\mathcal{C}(\alpha,K,d)$ is a functional Donsker class by Theorems 7.1.1 and 6.2.1. For $\alpha = d-1$, it is not a functional Donsker class (Sec. 8.4 below).

Proof. Again, we use the construction and notation around (7.1.9), now on I^{d-1}. Let $c := K\lambda^{d-1}(f)/(2||f||_\alpha)$, $m := m(n) := \lceil (2nc)^{1/(\alpha+d-1)} \rceil$. Then $1/(2n) \leq cm^{1-d-\alpha}$. Take $n \geq r$ (the result holds for $n < r$) for an r with

$$\sup_{n\geq r} 2nc\, m(n)^{1-d-\alpha} := M < \infty.$$

Let $\Theta_n := m(n)^{\alpha+d-1}/(2cn)$. Then $1/M \leq \Theta_n \leq 1$. Let $g_i := K\Theta_n f_i/(2||f||_\alpha)$. Then $\lambda^{d-1}(g_i) = 1/(2n)$ and $||g_i||_\alpha \leq K/2$. Let $B_i := \{x \in I^d: 0 < x_d < g_i(x_{(d)})\}$ where $x_{(d)} := (x_1,\ldots,x_{d-1})$. Then for each i, $\lambda^d(B_i) := P(B_i) = 1/(2n)$ and either $P_n(B_i) = 0$ or $P_n(B_i) \geq 1/n$. Either at least half the B_i have $P_n(B_i) = 0$, or at least half have $P_n(B_i) \geq 1/n$. In either case,

$$||P_n - P||_{\mathcal{C}(\alpha,K,d)} \geq m^{d-1}/(4n) \geq cm^{-\alpha}/(2M) \geq \delta n^{-\alpha/(\alpha+d-1)} \quad \text{for some}$$

$$\delta(\alpha,K,d) > 0. \qquad \square$$

The method of the last proof is due to W. Schmidt (1975), who applied it to convex sets and lower layers. On convex sets, for example, take disjoint spherical caps as in the proof of the lower bound in Theorem 7.3.1. Let each cap have volume exactly $1/(2n)$. Then as $n \to \infty$, the angular radius ε_n of such caps is asymptotic to $c_d n^{-1/(d+1)}$ for some constant c_d. Thus the number of such disjoint caps is of the order of $n^{(d-1)/(d+1)}$ $(d \geq 2)$ and we have:

8.1.3 **Theorem.** For the collection $\mathcal{C}_d$ of closed convex subsets of a bounded open set U in $\mathbb{R}^d$ there is a constant $b_d > 0$ such that for P = Lebesgue measure normalized on U, and all P_n,

$$\sup\{|(P_n - P)(C)|: C \in \mathcal{C}_d\} \geq b_d n^{-2/(d+1)}.$$

Thus for $d \geq 4$, $\mathcal{C}_d$ is not a functional Donsker class for P. If $d = 3$ it is not either, but the proof is harder (Sec. 8.4 below). $\mathcal{C}_2$ is a functional Donsker class for λ^2 by Theorems 7.3.1 and 6.2.1.

<u>Notes</u>. Bakhvalov (1959) proved Theorem 8.1.1. W. Schmidt (1975) proved Theorem 8.1.3. Theorem 8.1.2, proved by the same method as theirs, also is in Dudley (1983, Theorem 1).

8.2 <u>An upper bound</u>. Here, using metric entropy with bracketing N_I as in (6.1.4), is an upper bound for $||\nu_n||_{\mathcal{C}} := \sup_{B\in\mathcal{C}}|\nu_n(B)|$ which applies in many cases where the hypotheses of Theorem 6.2.1 fail. Let $(X,\mathcal{A},Q)$ be a probability space, $\nu_n := n^{1/2}(Q_n-Q)$, and recall N_I as defined above (6.1.4).

8.2.1 <u>Theorem</u>. Let $1 \leq \zeta < \infty$, $\eta > 2/(\zeta+1)$ and $\Theta := (\zeta-1)/(2\zeta+2)$. If for some $K < \infty$, $N_I(\varepsilon,\mathcal{C},Q) \leq \exp(K\varepsilon^{-\zeta})$, $0 < \varepsilon \leq 1$, then

$$\lim_{n\to\infty} \Pr^*\{||\nu_n||_{\mathcal{C}} > n^{\Theta}(\log n)^{\eta}\} = 0.$$

<u>Remarks</u>. The classes $\mathcal{C} = \mathcal{C}(\alpha,K,d)$ satisfy the hypothesis of 8.2.1 for $\zeta = (d-1)/\alpha \geq 1$, i.e. $\alpha \leq d-1$, by the last inequality in Theorem 7.1.1. Then $\Theta = \frac{1}{2} - \frac{\alpha}{d-1+\alpha}$. Thus Theorem 8.1.2 shows that the exponent Θ is sharp for $\zeta > 1$. Conversely, 8.2.1 shows that the exponent on n in Theorem 8.1.2 cannot be improved. In 8.2.1 we cannot take $\zeta < 1$, for then $\Theta < 0$, which is impossible even for a single set, $\mathcal{C} = \{C\}$, with $0 < P(C) < 1$.

<u>Proof</u>. The chaining method will be used as in sec. 6.2. Let $\log_2$ be logarithm to the base 2. For each $n \geq 3$ let $k(n) := [(\frac{1}{2} - \Theta)\log_2 n - \eta \log_2 \log n]$.

Let $N(k) := N_I(2^{-k},\mathcal{C},Q)$, $k = 1,2,\ldots$. Then for some $A_{ki} \in \mathcal{A}$, $i = 1,\ldots N(k)$, and any $A \in \mathcal{C}$, there are $i, j \leq N(k)$ with $A_{ki} \subset A \subset A_{kj}$ and $Q(A_{kj}\setminus A_{ki}) \leq 2^{-k}$. Let $A_{01} := \emptyset$ (the empty set) and $A_{02} := X$. Choose such $i = i(k,A)$ and $j = j(k,A)$, $k = 0,1,\ldots$ Then for each $k \geq 0$,

$$Q(A_{ki(k,A)} \,\Delta\, A_{k-1,i(k-1,A)}) \leq 2^{2-k}$$

where $\Delta :=$ symmetric difference.

For $k \geq 1$ let $\mathcal{B}(k)$ be the collection of sets B, with $Q(B) \leq 2^{2-k}$, of the form $A_{ki}\setminus A_{k-1,j}$ or $A_{k-1,j}\setminus A_{ki}$ or $A_{kj}\setminus A_{ki}$. Then

$$\operatorname{card}(\mathcal{B}(k)) \leq 2N(k-1)N(k) + N(k)^2 \leq 3\exp(2k2^{k\zeta}).$$

For each $B \in \mathcal{B}(k)$, Bernstein's inequality (2.2.3) implies, for any $t > 0$,

$$\Pr\{|\nu_n(B)| > t\} \leq \exp(-t^2/(2^{3-k} + tn^{-1/2})). \tag{8.2.2}$$

Choose $\delta > 0$ such that $\delta < 1$ and $(2+2\delta)/(1+\zeta) < \eta$. Let $c := \delta/(1+\delta)$ and $t := t_{n,k} := cn^{\Theta}(\log n)^{\eta}k^{-1-\delta}$.

Then for each $k = 1, \ldots, k(n)$, $2^{3-k} \geq 8n^{\Theta-1/2}(\log n)^{\eta} \geq t_{n,k}n^{-1/2}$.

Hence by (8.2.2),

$$\Pr\{|\nu_n(B)| > t_{n,k}\} \leq \exp(-t^2_{n,k}/2^{4-k}),$$

$$\begin{aligned} P_{nk} &:= \Pr\{\sup_{B \in \mathcal{B}(k)} |\nu_n(B)| > t_{n,k}\} \\ &\leq 3\exp(2K2^{k\zeta} - 2^{k-4}t^2_{n,k}) \\ &= 3\exp(2K2^{k\zeta} - 2^{k-4}c^2n^{2\Theta}(\log n)^{2\eta}k^{-2-2\delta}). \end{aligned}$$

For $k \leq k(n)$ we have $2^k(\log n)^{\eta} \leq n^{1/2-\Theta}$. Since $\Theta < 1/2$ and $2\Theta/(\frac{1}{2} - \Theta) = \zeta-1$, we have $n^{2\Theta} \geq 2^{k(\zeta-1)}(\log n)^{\eta(\zeta-1)}$, and

$$P_{nk} \leq 3\exp(2K2^{k\zeta} - 2^{k\zeta-4}c^2(\log n)^{\eta(\zeta+1)}k^{-2-2\delta}).$$

Let $\gamma := \eta(\zeta+1) - 2 - 2\delta > 0$ by choice of δ. Since $\frac{1}{2} < \log 2$, $k(n) \leq \log n$ and $P_{nk} \leq 3\exp(2^{k\zeta}(2K - 2^{-4}c^2(\log n)^{\gamma}))$.

For n large, $(\log n)^{\gamma} > 64K/c^2$. Then

$$\sum_{k=1}^{k(n)} P_{nk} \leq 3(\log n)\exp(-2^{-5}c^2(\log n)^{\gamma}) \longrightarrow 0 \quad \text{as} \quad n \to \infty.$$

Let $\mathcal{E}_n := \{\omega: \sup_{B \in \mathcal{B}(k)} |\nu_n(B)| \leq t_{n,k}, k = 1, \ldots, k(n)\}$. Then $\lim_{n\to\infty}\Pr(\mathcal{E}_n) = 1$. For any $A \in \mathcal{C}$ and n, let $k := k(n)$, $i := i(k,A)$ and $j := j(k,A)$. Then for each $\omega \in \mathcal{E}_n$,

$$|\nu_n(A_{ki})| \leq 2\sum_{r=1}^{k} t_{n,r} = 2n^{\Theta}(\log n)^{\eta}\sum_{r\geq 1} cr^{-1-\delta} \leq 2n^{\Theta}(\log n)^{\eta},$$

$$|\nu_n(A_{kj}\setminus A_{ki})| \leq n^{\Theta}(\log n)^{\eta}, \text{ and}$$

$$n^{1/2}Q(A_{kj}\setminus A_{ki}) \leq n^{1/2}/2^k < 2n^{\Theta}(\log n)^{\eta}.$$

Hence $n^{1/2}Q_n(A_{kj}\setminus A_{ki}) \leq 3n^{\Theta}(\log n)^{\eta}$, and $|\nu_n(A\setminus A_{ki})| \leq 3n^{\Theta}(\log n)^{\eta}$. So on $\mathcal{E}_n$, $|\nu_n(A)| \leq 5n^{\Theta}(\log n)^{\eta}$. As $\eta \downarrow 2/(\zeta+1)$ the factor of 5 can be dropped. Since $A \in \mathcal{C}$ is arbitrary, the proof is done. □

Note. Walter Philipp (unpublished) proved Theorem 8.2.1 for $\zeta = 1$, with a refinement ($\eta = 1$ and a power of $\log\log n$ factor). The proof for $\zeta > 1$ follows the same scheme. Theorem 8.2.1 is Theorem 2 of Dudley (1983).

8.3 <u>Poissonization and random sets.</u> Section 8.4 gives some lower bounds $||\nu_n||_{\mathcal{C}} \geq f(n)$ with probability converging to 1 as $n \to \infty$ where f is a product of powers of logarithms or iterated logarithms. Such an f has the following property. A real-valued function f defined for large enough $x > 0$ is called <u>slowly varying</u> (in the sense of Karamata) iff for every $c > 0$, $f(cx)/f(x) \to 1$ as $x \to +\infty$.

8.3.1 <u>Lemma</u>. If f is continuous and slowly varying then for every $\varepsilon > 0$ there is a $\delta = \delta(\varepsilon) > 0$ such that whenever $x > 1/\delta$ and $|1 - \frac{y}{x}| < \delta$ we have $|1 - \frac{f(y)}{f(x)}| < \varepsilon$.

<u>Proof</u>. For each $c > 0$ and $\varepsilon > 0$ there is an $x(c,\varepsilon)$ such that for $x \geq x(c,\varepsilon)$, $|\frac{f(cx)}{f(x)} - 1| \leq \frac{\varepsilon}{4}$. By the Baire category theorem, for fixed $\varepsilon > 0$ there is an $n < \infty$ such that $x(c,\varepsilon) \leq n$ for all c in a set dense in some interval $[a,b]$, where $0 < a < b$, and thus by continuity for all c in $[a,b]$. Then for $c,d \in [a,b]$ and $x \geq n$, $|(f(cx) - f(dx))/f(x)| \leq \varepsilon/2$.

Let $u := cx$. Then

$$|\frac{f(ud/c)}{f(u)} - 1||\frac{f(u)}{f(u/c)}| \leq \frac{\varepsilon}{2}, \quad u \geq nc.$$

As $u \to +\infty$, $f(u)/f(u/c) \to 1$. Let $c := (a+b)/2$. Then there is a $\delta > 0$ such that for $u \geq 1/\delta$ and all r with $|r-1| < \delta$ we have $|\frac{f(ru)}{f(u)} - 1| < \varepsilon$ $(\delta < (b-a)/(b+a))$. □

Let $P_c(k) := e^{-c}c^k/k!$, $k = 0,1,\ldots$ (Poisson law on $\mathbb{N}$ with parameter $c \geq 0$). Given a probability space $(X,\mathcal{A},P)$, let U_c be a Poisson point process on $(X,\mathcal{A})$ with intensity measure cP. That is, for any disjoint $A_1,\ldots,A_m$ in $\mathcal{A}$, $U_c(A_j)$ are independent random variables, $j = 1, \ldots, m$, and for any $A \in \mathcal{A}$, $U_c(A)(\cdot)$ has law $P_{cP(A)}$.

Let $Y_c(A) := (U_c - cP)(A)$, $A \in \mathcal{A}$. Then Y_c has mean 0 on all A and still has independent values on disjoint sets.

Let $x(1), x(2), \ldots$ be coordinates for the product space $(X^\infty,\mathcal{A}^\infty,P^\infty)$. For $c > 0$ let $n(c)$ be a random variable with law P_c, independent of the $x(i)$. Then for $P_n := n^{-1}(\delta_{x(1)} + \ldots + \delta_{x(n)})$, $n \geq 1$, $P_0 := 0$ we have:

8.3.2 Lemma. The process $Z_c := n(c)P_{n(c)}$ is a Poisson process with intensity measure cP.

Proof. We have laws $\mathcal{L}(U_c(X)) = \mathcal{L}(Z_c(X)) = P_c$. If X_i are independent Poisson variables with $\mathcal{L}(X_i) = P_{c(i)}$ and $\sum_{i=1}^m c(i) = c$, then given $n := \sum_{i=1}^m X_i$, the conditional distribution of $\{X_i\}_{i=1}^m$ is multinomial with total n and probabilities $p_i = c(i)/c$. Thus U_c and Z_c have the same conditional distributions on disjoint sets given their values on X. This implies the Lemma. □

Below, the version $U_c \equiv Z_c$ will be used. Thus for each ω, $U_c(\cdot)(\omega)$ is a countably additive integer-valued measure of total mass $U_c(X)(\omega) = n(c)(\omega)$. Then

$$Y_c = n(c)P_{n(c)} - cP = n(c)(P_{n(c)}-P) + (n(c)-c)P,$$

$$(8.3.3) \qquad Y_c/c^{1/2} = (n(c)/c)^{1/2}\nu_{n(c)} + (n(c)-c)c^{-1/2}P.$$

The following shows that the empirical process ν_n is asymptotically "as large" as a corresponding Poisson process.

8.3.4 Lemma. Let $(X,\mathcal{A},P)$ be a probability space and $\mathcal{C} \subset \mathcal{A}$. Assume that for each n and constant t, $\sup_{A\in\mathcal{C}}(P_n - tP)(A)$ is measurable. Let f be a continuous, slowly varying function such that as $x \to +\infty$, $f(x) \to +\infty$ and $f(x)/x \to 0$. For $b > 0$ let

$$g(b) := \liminf_{x\to+\infty} \Pr\{\sup_{A\in\mathcal{C}} Y_x(A) \geq bf(x)x^{1/2}\}.$$

Then for any $a < b$,

$$\liminf_{n\to\infty} \Pr\{\sup_{A\in\mathcal{C}} \nu_n(A) \geq af(n)\} \geq g(b).$$

Proof. From (8.3.3), $\sup_{A\in\mathcal{C}} Y_x(A)$ is measurable. As $x \to +\infty$, $\Pr(n(x) > 0) \to 1$ and $(n(x)/x)^{1/2} \to 1$ in probability.

If the Lemma is false there is a $\Theta < g(b)$ and a sequence $m = m_k \to +\infty$ with

$$\Pr\{\sup_{A\in\mathcal{C}} \nu_m(A) \geq af(m)\} \leq \Theta.$$

Choose $0 < \varepsilon < 1/3$ such that $a(1 + 7\varepsilon) < b$. Then let $0 < \delta < 1/2$ be such that $\delta \leq \delta(\varepsilon)$ in Lemma 8.3.1 and $(1+\delta)(1+5\varepsilon) < 1 + 6\varepsilon$. We may assume that for all $k = 1,2,\ldots$, $m = m_k \geq 2/\delta$ and $1 + 2\varepsilon < a\varepsilon f(m)^{1/2}$.

Set $\delta_m := (f(m)/m)^{1/2}$. Then since $f(x)/x \to 0$ we may assume $\delta_m < \delta/2$ for all $m = m_k$. Then for any $m = m_k$, if $(1 - \delta_m)m \leq n \leq m$, then for all $A \in \mathcal{A}$,

$mP_m(A) \geqslant nP_n(A)$, $m(P_m-P)(A) \geqslant n(P_n-P)(A) - m\delta_m$, and if $\nu_n(A) \geqslant 0$, then

$$\nu_m(A) \geqslant (n/m)^{1/2}\nu_n(A) - f(m)^{1/2} \geqslant (1+\delta)^{-1}\nu_n(A) - f(m)^{1/2},$$

so $\nu_n(A) \leqslant (1+\delta)(\nu_m(A) + f(m)^{1/2})$. Next, $|1 - \frac{m}{n}| = \frac{m}{n} - 1 < 2\delta_m < \delta$ implies $|\frac{f(m)}{f(n)} - 1| < \varepsilon$, so $1/f(n) < (1+\varepsilon)/f(m)$. Hence if $\nu_n(A) \geqslant 0$ then

$$\nu_n(A)/f(n) < (1+2\varepsilon)(\nu_m(A)f(m)^{-1} + f(m)^{-1/2}).$$

Thus since $(1 + 2\varepsilon)f(m)^{-1/2} < a\varepsilon$, $\Pr\{\sup_{A\in\mathcal{C}}\nu_n(A) \geqslant af(n)(1+3\varepsilon)\} \leqslant \Theta$.

For each $m = m_k$, set $c = c_m = (1 - \frac{1}{2}\delta_m)m$. Then as $k \to \infty$, since $f(m_k) \to \infty$,

$$\Pr\{(1-\delta_m)m \leqslant n(c) \leqslant m\} \to 1.$$

Then for any y with $\Theta < y < g := g(b)$ and k large enough, since the $X(i)$ are independent of $n(c)$,

$$\Pr\{\sup_{A\in\mathcal{C}}\nu_{n(c)}(A) \geqslant af(n(c))(1+3\varepsilon)\} < y.$$

Since $\delta_m \leqslant \delta(\varepsilon)/2$, for $(1-\delta_m)m \leqslant n \leqslant m$ we have $|1 - \frac{f(n)}{f(c)}| < \varepsilon$. Thus for k large enough we may assume

$$\Pr\{\sup_{A\in\mathcal{C}}\nu_{n(c)}(A) \geqslant af(c)(1+5\varepsilon)\} < y.$$

By (8.3.3), for k large enough

$$\Pr\{\sup_{A\in\mathcal{C}}Y_c(A) \geqslant ac^{1/2}f(c)(1+7\varepsilon)\} < (y+g)/2 < g,$$

a contradiction, proving Lemma 8.3.4. □

Next, the Poisson process's independence property on disjoint sets will be extended to suitable random sets. Let $(X,\mathcal{A})$ be a measurable space, and $(\Omega,\mathcal{B},\Pr)$ a probability space. A stochastic process $Y: \langle A,\omega\rangle \to Y(A)(\omega)$, $A \in \mathcal{A}$, $\omega \in \Omega$, will be said to have <u>independent pieces</u> iff for any disjoint $A_1,\ldots,A_m \in \mathcal{A}$, $Y(A_j)$ are independent, $j = 1,\ldots,m$, and $Y(A_1 \cup A_2) = Y(A_1) + Y(A_2)$ almost surely.

Clearly each Y_c has independent pieces.

For any $C \in \mathcal{A}$ let $\mathcal{B}_C$ be the smallest σ-algebra for which every $Y(A)(\cdot)$ is measurable for $A \subset C$, $A \in \mathcal{A}$.

A function G from Ω into $\mathcal{A}$ will be called a <u>stopping set</u> iff for all $C \in \mathcal{A}$, $\{\omega: G(\omega) \subset C\} \in \mathcal{B}_C$. Given a stopping set $G(\cdot)$, let $\mathcal{B}_G$ be the σ-algebra of all sets $B \in \mathcal{B}$ such that for every $C \in \mathcal{A}$, $B \cap \{G \subset C\} \in \mathcal{B}_C$.

(Note that if G is not a stopping set, then $\Omega \notin \mathcal{B}_G$, so $\mathcal{B}_G$ would not be a σ-algebra.)

8.3.5 Lemma. Suppose Y has independent pieces and for all $\omega \in \Omega$ and $j = 0,\ldots,m$, $G_j(\omega) := G(j)(\omega) \in \mathcal{A}$ and $A(\omega) \in \mathcal{A}$.

Assume that:

i) $G_0(\omega)$ is a fixed set $G_0 \in \mathcal{A}$;

ii) For all ω, $G_0 \subset G_1(\omega) \subset \ldots \subset G_m(\omega)$, and $G_m(\omega)$ is disjoint from $A(\omega)$;

iii) Each $G_j(\omega)$, and $A(\omega)$, has just countably many possible values $G(j,i) := G_{ji} \in \mathcal{A}$ and $C(i) := C_i \in \mathcal{A}$ respectively;

iv) For all i and for $j = 1,\ldots,m$, $\{G_j(\cdot) = G_{ji}\} \in \mathcal{B}_{G(j-i)}$ and $\{A(\cdot) = C_i\} \in \mathcal{B}_{G(m)}$.

Then each $G_j(\cdot)$ is a stopping set, $j = 0,1,\ldots,m$, and the conditional probability

$$\Pr\{Y(A(\cdot)) \leq t \mid \mathcal{B}_{G(m)}\} = \sum_i 1_{\{A(\cdot) = C(i)\}} \Pr(Y(C_i) \leq t)$$

almost surely, for each real t.

Proof. First let us show by induction on j that each G_j is a stopping set. For $j = 0$ this is clear. Let $j \geq 1$ and assume G_{j-1} is a stopping set. Let $C \in \mathcal{A}$. Then

$$\{G_j \subset C\} = \cup_i \{\{G_j(\cdot) = G_{ji}\} : G_{ji} \subset C\}.$$

For each i, $\{G_j = G_{ji}\} \in \mathcal{B}_{G(j-1)}$ by (iv).

If $G_{ji} \subset C$, then by (ii),

$$\{G_j = G_{ji}\} = \{G_j = G_{ji}\} \cap \{G_{j-1} \subset C\} \in \mathcal{B}_C$$

by definition of $\mathcal{B}_{G(j-1)}$. Thus $\{G_j \subset C\}$ is a countable union of sets in $\mathcal{B}_C$, so is in $\mathcal{B}_C$, and G_j is a stopping set.

If for some ω, $A(\omega) = C_i$ and $G_m(\omega) = G_{mj}$, then by (ii), $C_i \cap G_{mj} = \emptyset$ (empty set). Thus $Y(C_i)$ is independent of $\mathcal{B}_{G(m,j)}$. Let $B_i := \{A(\cdot) = C_i\} \in \mathcal{B}_{G(m)}$ by (iv). For each j, also by (iv), $\{G_m = G_{mj}\} \in \mathcal{B}_{G(m-1)}$, so by (ii)

$$\{G_m = G_{mj}\} = \{G_m = G_{mj}\} \cap \{G_{m-1} \subset G_{mj}\} \in \mathcal{B}_{G(m,j)}.$$

For any $B \in \mathcal{B}_{G(m)}$,

$$B \cap \{G_m = G_{mj}\} = B \cap \{G_m \subset G_{mj}\} \cap \{G_m = G_{mj}\} \in \mathcal{B}_{G(m,j)}.$$

Thus almost surely

$$\begin{aligned} \Pr(Y(A) \leqslant t|\mathcal{B}_{G(m)}) &= \textstyle\sum_{i,j} \Pr(Y(A) \leqslant t|\mathcal{B}_{G(m)}) 1_{B(i)} 1_{\{G(m) = G(m,j)\}} \\ &= \textstyle\sum_{i,j} \Pr(Y(C_i) \leqslant t|\mathcal{B}_{G(m,j)}) 1_{B(i)} 1_{\{G(m) = G(m,j)\}} \\ &= \textstyle\sum_i \Pr(Y(C_i) \leqslant t) 1_{B(i)}. \qquad \square \end{aligned}$$

<u>Notes</u>. Lemma 8.3.1 is classical, see e.g. Feller (vol.II, VIII.8, Lemma 2). Lemma 8.3.2, and with it the main idea of Poissonization are due to Kac (1949). Pyke (1968) gives relations between Poissonized and non-Poissonized cases along the line of Lemma 8.3.4. Evstigneev (1977, Theorem 1) proves a Markov property for random fields indexed by closed subsets of a Euclidean space, somewhat along the line of Lemma 8.3.5. This entire section is essentially Sec.3 of Dudley (1983) with proofs of Lemmas 8.3.1 and 8.3.2 supplied here.

8.4 <u>Lower bounds in borderline cases.</u>

Recalling the classes $\mathcal{C}(\alpha,K,d)$ defined in Sec. 7.1, we had lower bounds for $P_n - P$ on them in Theorem 8.1.2 which imply that for $\alpha < d-1$, $||\nu_n||_{\alpha,K,d} \longrightarrow \infty$ surely as $n \longrightarrow \infty$. For $\alpha > d-1$, $\mathcal{C}(\alpha,K,d)$ is a functional Donsker class by Theorems 6.2.1 and 7.1.1. So $\alpha = d-1$ is a borderline case. A collection of sets of the same size as these is given by the class $\mathcal{LL}_2$ of lower layers in $\mathbb{R}^2$ (Sec. 7.2) or the class $\mathcal{C}_3$ of convex sets in $\mathbb{R}^3$ (Sec. 7.3), for λ^d = Lebesgue measure on the unit cube I^d.

Letting $P = \lambda^d$ and recalling the Poisson process Y_c from Sec. 8.3, the following lower bound holds for all the above borderline classes;

8.4.1 <u>Theorem</u>. For any $K > 0$ and $\delta > 0$ there is a $\gamma = \gamma(d,K,\delta) > 0$ such that

$$\lim_{x \to +\infty} \Pr\{||Y_x||_{\mathcal{C}} > \gamma(x \log x)^{1/2}(\log\log x)^{-\delta-1/2}\} = 1$$

and

$$\lim_{n \to \infty} \Pr\{||\nu_n||_{\mathcal{C}} > \gamma(\log n)^{1/2}(\log\log n)^{-\delta-1/2}\} = 1$$

where $\mathcal{C} = \mathcal{C}(d-1,K,d)$, $d \geqslant 2$, or $\mathcal{C} = \mathcal{LL}_2$, or $\mathcal{C} = \mathcal{C}_3$.

For a proof of this in all cases I refer to Dudley (1983, Theorems 3,4, Secs. 4,5); other results needed from Sec. 3 of that paper were in Sec. 8.3 above.

Here I will just give a proof in the lower layers case ($\mathcal{C} = \mathcal{LL}_2$, d = 2) which also covers $\mathcal{C}(1,K,2)$ as seen in Sec. 7.2. This proof, although it does not seem to extend to $d \geq 3$, is perhaps more intuitive.

To see that the supremum of Y_c or an empirical measure ν_n over $\mathcal{LL}_2$ is measurable, note that for each $F \subset \{1,\ldots,n\}$ and each ω, there is a smallest lower layer $L_F(\omega)$ containing the x_j for $j \in F$. For any c, $\omega \to (P_n - cP)(L_F(\omega))(\omega)$ is measurable. The supremum of $P_n - cP$ over $\mathcal{LL}_2$, as the maximum of these 2^n measurable functions, is measurable. Letting $n = n(c)$ as in Lemma 8.3.2 and (8.3.3) then shows $\sup\{Y_c(A)\colon A \in \mathcal{LL}_2\}$ is measurable.

For each ω a lower layer L_ω will be constructed recursively as follows. Let $A_{01} :=]0,1[\times]0,1[$. For each $k \geq 0$ there will be a fixed number $r_k > 0$ and random sequences of open rectangles A_{ki} and B_{ki}, $i = 1,\ldots,2^k$, such that each B_{ki} is similar, parallel and concentric to the A_{ki} with same indices. The ratio of each side of B_{ki} to a parallel side of A_{ki} will be $r_k < 1$.

In the statement of the Theorem we can assume $\delta < 1/2$. Let $r_j := 1/4$, $j = 0,1,2$, and $r_j := j^{-1/2}/(2(\log j)^{\delta+1/2})$, $j \geq 3$. Then $\sum_j r_j^2$ converges, but slowly, while $\sum_j r_j$ diverges. Already B_{01} is determined.

Now recursively in k, suppose given the A_{ki}, B_{ki} and $Y_x(B_{ki})$. For given k and i let $A_{ki} =]a,b[\times]c,d[$ and $B_{ki} =]e,f[\times]g,h[$ (see Fig.3). Then $a < e < f < b$ and $c < g < h < d$, $e-a = b-f$, $d-h = g-c$, and $(f-e)/(b-a) = (h-g)/(d-c) = r_k$.

Fig.3

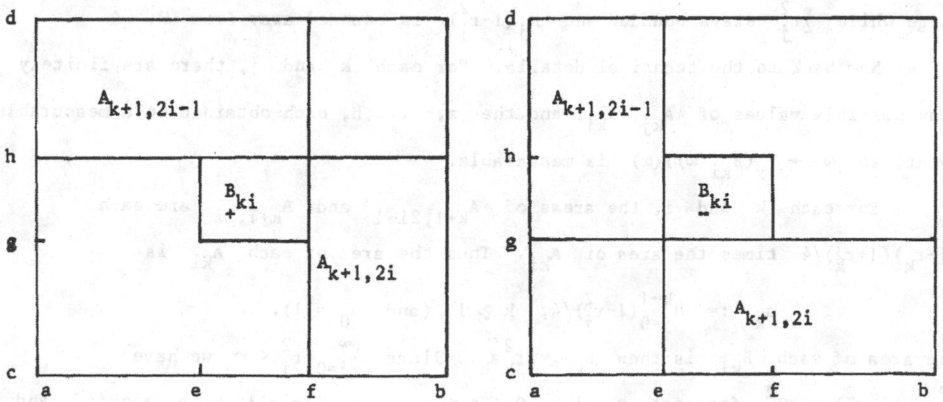

$Y_x(B_{ki}) \geq 0$ $\qquad\qquad$ $Y_x(B_{ki}) < 0$

If $Y_x(B_{ki}) \geq 0$, B_{ki} will be included in L_ω. Then let

$$A_{k+1,2i-1} :=]a,f[\times]h,d[$$

and

$$A_{k+1,2i} :=]f,b[\times]c,h[.$$

Or, if $Y_x(B_{ki}) < 0$, B_{ki} will be excluded from L_ω. Then let

$$A_{k+1,2i-1} :=]a,e[\times]g,d[,$$

$$A_{k+1,2i} :=]e,b[\times]c,g[.$$

In all cases, $\langle e,g\rangle \in L_\omega$ and $\langle f,h\rangle \notin L_\omega$. Each $B_{k+1,j}$ is a concentric, parallel subrectangle of $A_{k+1,j}$ with similarity ratio r_{k+1} as defined above. Now all the $A_{kj}(\omega)$ and $B_{kj}(\omega)$ are defined, $k = 0,1,\ldots$, $j = 1,\ldots,2^k$. For each ω, all the rectangles $B_{kj}(\omega)$ are disjoint.

The idea of the construction is as follows. We seek a lower layer L_ω with $Y_x(L_\omega)$ large. Let the boundary of L_ω pass through $\langle 0,1\rangle$ and $\langle 1,0\rangle$. If the boundary crosses the main diagonal at $\langle t,t\rangle$, the rest of the boundary in I^2 will be in two rectangles $[0,t] \times [t,1]$ and $[t,1] \times [0,t]$, each of area $A_1 = t(1-t) \leq 1/4$. If we choose t as one of $(1\pm r)/2$ then $A_1 = (1-r^2)/4$. A central square B of area r^2 will be included in L_ω, choosing $t = (1+r)/2$, if $Y_x(B) > 0$. Otherwise B is excluded. Thus we obtain a contribution $\max(0,Y_x(B))$ which (if $xr^2 \geq 1$) has mean proportional to r but at a cost of only $r^2/4$ in area. Since the two rectangles are disjoint from B we can proceed recursively, keeping the independence properties of Y_x. Then $\sum r_j$ can be made large while $\sum r_j^2$ stays bounded and $\Pi_j(1-r_j^2)$ is bounded away from 0.

Now back to the technical details. For each k and j, there are finitely many possible values of A_{kj}, B_{kj}, and the $a,b,\ldots,h$, each obtained on a measurable event, so $\omega \longrightarrow Y_x(B_{ki}(\omega))(\omega)$ is measurable.

For each k and i, the areas of $A_{k+1,2i-1}$ and $A_{k+1,2i}$ are each $(1-r_k)(1+r_k)/4$ times the area of A_{ki}. Thus the area of each A_{ki} is

$$a_k := \Pi_{j=0}^{k-1}(1-r_j^2)/4, \quad k \geq 1 \quad (\text{and } a_0 = 1).$$

The area of each B_{ki} is then $b_k := r_k^2 a_k$. Since $\sum_{j=0}^{\infty} r_j^2 < \infty$ we have $\Pi_{j=0}^{\infty}(1-r_j^2) = \rho^2$ for some ρ with $0 < \rho < 1$. Then for all k, $a_k > \rho^2/4^k$ and

$$xb_k \geqslant x\rho^2/(4^{k+1}k(\log k)^{1+2\delta}).$$

For $m = 1,2,\ldots$, let $L_{m,\omega}$ be the smallest lower layer including all those $B_{ki}(\omega)$ for which $Y_x(B_{ki}) \geqslant 0$, $k \leqslant m$, and containing the lower left vertex $\langle e,g\rangle$ of every B_{ki}, $k \leqslant m$. Let $G_{m,\omega} := \bigcup_{k\leqslant m} \bigcup_i B_{ki}$, $C_{m,\omega} := L_{m,\omega} \setminus G_{m,\omega}$. For any Y let $Y^+ := \max(Y,0)$. Let

$$S_{k,\omega} := \sum_{i=1}^{2^k} Y_x(B_{ki})^+.$$

For fixed k, each $Y_x(B_{ki}) = z_{ki} - xb_k$ where the z_{ki} are independent Poisson variables with parameter xb_k, $i = 1,\ldots,2^k$. If z is any Poisson variable with parameter m, then $E(((z-m)^+)^2) \leqslant E((z-m)^2) = m$.

For each integer k with $xb_k \geqslant 1$, Lemma 2.2.10 gives

$$ES_{k,\omega} = 2^k EY_x(B_{k1})^+ \geqslant 2^{k-3}(xb_k)^{1/2} \geqslant x^{1/2}\rho r_k/8.$$

Also, the variance $\sigma^2(S_{k,\omega}) \leqslant 2^k xb_k$. So by Chebyshev's inequality

$$\Pr\{S_{k,\omega} \leqslant x^{1/2}\rho r_k/16\} \leqslant \Pr\{S_{k,\omega} \leqslant 2^{k-4}(xb_k)^{1/2}\}$$

$$\text{(8.4.2)} \qquad \leqslant 2^k xb_k/(4^{k-4}xb_k) = 2^{8-k}.$$

Let $[y]$ be the greatest integer $\leqslant y$.

Then for

$$3 \leqslant k \leqslant [(\log x + 2\log(\rho/2))/\log 12] := [\alpha\log x - \beta] := k(x)$$

for some $\alpha, \beta > 0$, we have $(\log k)^{1+2\delta} \leqslant (\log k)^2$ $(\delta < 1/2)$, so

$$k(\log k)^2 4^k \leqslant k^3 4^k \leqslant 12^k \leqslant x\rho^2/4$$

and hence, $xb_k \geqslant 1$, $3 \leqslant k \leqslant k(x)$. Let $J = 3,4,\ldots,k(x)$. Except on an event with probability at most $\sum_{k\geqslant J} 2^{8-k} \leqslant 2^{9-J}$, we have by (8.4.2)

$$S_\omega^{(x)} := \sum_{k=0}^{k(x)} S_{k,\omega} \geqslant \rho x^{1/2} 2^{-4} \sum_{k=J}^{k(x)} r_k$$

$$\text{(8.4.3)} \qquad = 2^{-5}\rho x^{1/2} \sum_{k=J}^{k(x)} k^{-1/2}(\log k)^{-\delta-1/2}$$

$$\geqslant 2^{-5}\rho x^{1/2}(\log k(x))^{-\delta-1/2} \sum_{k=J}^{k(x)} k^{-1/2}$$

which is asymptotic as $x \to +\infty$ to

$$\text{(8.4.4)} \qquad 2^{-4}\rho x^{1/2}(\log\log x)^{-\delta-1/2}(\alpha\log x)^{1/2}.$$

Given x, let $m := k(x)$. Then $Y_x(L_{m,\omega}) = S_\omega^{(x)} + Y_x(C_{m,\omega})$.

If $C_{m,\omega}$ were a lower layer, the proof could be finished directly, but in general it is not. Instead, Lemma 8.3.5 will be used. Now $C_{m,\omega}$ is a finite union of the following rectangles: for each pair of rectangles

$$B_{ki} =]e,f[\times]g,h[\subset A_{ki} =]a,b[\times]c,d[,$$

as above, $k < m$, $1 \leqslant i \leqslant 2^k$, the lower left rectangle $[a,e] \times [c,g] \subset C_{m,\omega}$, and if $B_{ki} \subset L_{m,\omega}$, i.e. if $Y_x(B_{ki}) \geqslant 0$, then the rectangles $[e,f[\times [c,g] \subset C_{m,\omega}$ and $[a,e] \times [g,h[\subset C_{m,\omega}$. Let $G(m) := G_{m,\omega}$. Then for $m \geqslant 1$, $C_{m,\omega}$ has finitely many possible values, each on a $\mathcal{B}_{G(m-1)}$ measurable set. The same holds for $G_{m,\omega}$ by construction, and $C_{m,\omega} \cap G_{m,\omega} = \emptyset$. Thus for each m, Lemma 8.3.5 applies with $A(\omega) = C_{m,\omega}$. Hence conditional on $\mathcal{B}_{G(m)}$, $Y_x(C_{m,\omega})$ has the law of $Y_x(A)$ with $P(A) = P(C_{m,\omega})$. For any $A \in \mathcal{A}$ and number $K > 0$,

$$\text{(8.4.5)} \qquad \Pr(Y_x(A) \leqslant -Kx^{1/2}) \leqslant K^{-2}.$$

For any $t > 0$,

$$\Pr(Y_x(L_{m,\omega}) \leqslant t/4) \leqslant \Pr(S_\omega^{(x)} \leqslant t/2) + \Pr(Y_x(C_{m,\omega}) \leqslant -t/4),$$

so for x large enough, by (8.4.3), (8.4.4) and (8.4.5)

$$\Pr\{Y_x(L_{m,\omega}) \leqslant 2^{-7}\rho x^{1/2}(\alpha \log x)^{1/2}/(\log\log x)^{\delta+1/2}\}$$

$$\leqslant 2^{9-J} + 2^{14}\rho^{-2}(\log\log x)^{1+2\delta}/(\alpha \log x) \longrightarrow 0$$

as $x \longrightarrow \infty$ and then $J \longrightarrow \infty$. Thus the Theorem holds with $\gamma = 2^{-7}\rho\alpha^{1/2}$, for lower layers $(d = 2)$. □

<u>Notes.</u> This proof of Theorem 8.4.1 for $d = 2$ has not been published elsewhere. For $d \geqslant 3$ see Dudley (1983).

Chapter 9 Vapnik-Červonenkis classes.

9.1 Introduction. This chapter will deal with some combinatorial questions for collections of subsets in preparation for limit theorems in Chapter 11.

Let X be any set and $\mathcal{C}$ a collection of subsets of X. For $A \subset X$ let $A \sqcap \mathcal{C} := \{A \cap B: B \in \mathcal{C}\} := \mathcal{C} \sqcap A$. Let card(A) denote the cardinality (number of elements) of A. Let $2^A := \{B: B \subset A\}$ and $\Delta^{\mathcal{C}}(A) := \mathrm{card}(A \sqcap \mathcal{C})$. If $A \sqcap \mathcal{C} = 2^A$, or equivalently if $\Delta^{\mathcal{C}}(A) = 2^{\mathrm{card}(A)}$, then $\mathcal{C}$ is said to shatter A.

Let $m^{\mathcal{C}}(n) := \max\{\Delta^{\mathcal{C}}(F): \mathrm{card}\ F = n\}$, $n = 0,1,\ldots$. Then $m^{\mathcal{C}}(n) \leq 2^n$. Let

$$V(\mathcal{C}) := \begin{cases} \inf\{n: m^{\mathcal{C}}(n) < 2^n\} \\ +\infty \ \text{ if } m^{\mathcal{C}}(n) = 2^n \text{ for all } n < \infty, \end{cases}$$

$$S(\mathcal{C}) := \begin{cases} \sup\{n: m^{\mathcal{C}}(n) = 2^n\} \\ -1 \ \text{ if } \mathcal{C} \text{ is empty} \end{cases}$$

$$\equiv V(\mathcal{C}) - 1.$$

So $S(\mathcal{C})$ is the largest cardinality of a set shattered by $\mathcal{C}$, or $+\infty$ if arbitrarily large finite sets are shattered. If $V(\mathcal{C}) < +\infty$, or equivalently if $S(\mathcal{C}) < +\infty$, then I call $\mathcal{C}$ a Vapnik-Červonenkis class.

Let ${}_N C_{\leq k} := \sum_{j=0}^{k} \binom{N}{j}$, where

$$\binom{N}{j} := \begin{cases} N!/(j!(N-j)!), & j = 0,1,\ldots, N; \\ 0, & j > N. \end{cases}$$

Then ${}_N C_{\leq k}$ is the number of combinations of N things, at most k at a time. (In an older notation ${}_N C_k := \binom{N}{k}$.)
Pascal's "triangle" of identities for binomial coefficients extends to the ${}_N C_{\leq k}$:

9.1.1 Proposition. ${}_N C_{\leq k} = {}_{N-1} C_{\leq k} + {}_{N-1} C_{\leq k-1}$ for $k = 1,2,\ldots,$ and $N = 1,2,\ldots$.

Proof. For each $j = 1,2,\ldots,N$, we have

$$\binom{N}{j} = \binom{N-1}{j} + \binom{N-1}{j-1}, \text{ and } \binom{N}{0} = \binom{N-1}{0} = 1. \qquad \square$$

9.1.2 Theorem. If $m^{\mathcal{C}}(n) > {}_n C_{\leq k-1}$, where $k \geq 1$, then $m^{\mathcal{C}}(k) = 2^k$. Hence if $S(\mathcal{C}) < +\infty$, then $m^{\mathcal{C}}(n) \leq {}_n C_{S(\mathcal{C})}$ for all n.

<u>Proof</u>. We use induction on k and n. For $k = 1$, ${}_nC_{\leq 0} = 1 < m^{\mathcal{C}}(n)$ implies that $\mathcal{C}$ contains at least two elements, so for some singleton $G = \{x\}$, $\Delta^{\mathcal{C}}(G) = 2$ as desired. If $k > n$, then ${}_nC_{\leq k-1} = 2^n \geq m^{\mathcal{C}}(n)$, so our assumption implies $k \leq n$.

Now assume the theorem holds whenever $k \leq K$ and $n \geq k$, for all $\mathcal{C}$. Fix $k := K + 1$. We will use induction on n. For $n < k$, as noted, the theorem holds vacuously. Suppose it holds for $n \leq N$ and let us prove it for $n = N + 1$. Let $H_n := \{x_1,\ldots,x_n\}$ be a set with n elements. Suppose $\Delta^{\mathcal{C}}(H_n) > {}_nC_{\leq K}$. Let $H_N := \{x_1,\ldots,x_N\}$. If $\Delta^{\mathcal{C}}(H_N) > {}_NC_{\leq K}$, then by induction assumption, $m^{\mathcal{C}}(k) = 2^k$ as desired. So assume

$$\Delta^{\mathcal{C}}(H_N) \leq {}_NC_{\leq K}. \tag{9.1.3}$$

Let $\mathcal{C}_n := H_n \sqcap \mathcal{C} := \{A \cap H_n : A \in \mathcal{C}\}$. Call a set $E \subset H_N$ <u>full</u> iff both E and $E \cup \{x_n\}$ belong to $\mathcal{C}_n$. Let f be the number of full sets. Then

$$\Delta^{\mathcal{C}}(H_n) = \Delta^{\mathcal{C}}(H_N) + f. \tag{9.1.4}$$

Let $\mathcal{F}$ be the collection of all full sets. Suppose $f = \Delta^{\mathcal{F}}(H_N) > {}_NC_{\leq K-1}$. Then by induction assumption there is a $G \subset H_N$ with $\operatorname{card} G = K$ and $\Delta^{\mathcal{F}}(G) = 2^K$. For $J := G \cup \{x_n\}$ we then have $\operatorname{card} J = k$ and $\Delta^{\mathcal{C}}(J) = 2^k$ as desired.

In the remaining case, $f \leq {}_NC_{\leq K-1}$. Then by (9.1.3) and (9.1.4),

$$\Delta^{\mathcal{C}}(H_n) \leq {}_NC_{\leq K} + {}_NC_{\leq K-1} = {}_nC_{\leq K}$$

by Prop. 9.1.1, a contradiction. Thus the first sentence of the Theorem is proved. Then the second follows from the definition of $S(\mathcal{C})$. □

9.1.5 <u>Proposition</u>. For $n \geq k + 2 \geq 2$, ${}_nC_{\leq k} \leq 1.5n^k/k!$.

<u>Proof</u>. First, by the binomial theorem, $n^{k-1}(k+n) \leq (n+1)^k$, so $n^{k-1}/(k-1)! + n^k/k! \leq (n+1)^k/k!$. Thus by Prop. 9.1.1 and induction, it is enough to treat the case $k = 1$ and the case $n = k+2$. For $k = 1$, ${}_nC_{\leq k} = n + 1 \leq 1.5n$ since $n \geq 2$. For $n = k+2$, the desired inequality is

$$2^n - n - 1 \leq 1.5n^{n-2}/(n-2)! = 1.5(n-1)n^{n-1}/n!.$$

This can be checked directly for $n = 3, 4, 5, 6$. By Stirling's formula with error bound (e.g. Feller, vol. I, p.54, (9.15)) $n! \leq (n/e)^n(2\pi n)^{1/2}e^{1/(12n)}$.

Thus it is enough to prove

$$2^n(n/e)^n(2\pi n)^{1/2}e^{1/(12n)} \leq 1.5(n-1)n^{n-1}, \quad n \geq 7,$$

which follows from $(e/2)^n \geq 2n^{1/2}$, $n \geq 7$; the latter is easily checked using derivatives. □

Combining Theorem 9.1.2 and Prop. 9.1.5 gives $m^{\mathcal{C}}(n) \leq 1.5n^k/k!$ for $n \geq k+2$ where $k = S(\mathcal{C})$.

To see that Theorem 9.1.2 is sharp, let X be an infinite set and $\mathcal{C}$ the collection of all subsets of X with cardinality k. Then $S(\mathcal{C}) = k$ and the inequality in the second sentence of the theorem becomes an equality for all n.

Let

$$\text{dens}(\mathcal{C}) := \inf\{r > 0: \text{ for some } K < \infty, \\ m^{\mathcal{C}}(n) \leq Kn^r \text{ for all } n \geq 1\}.$$

Then we have

9.1.6 **Corollary.** For any set X and $\mathcal{C} \subset 2^X$, $\text{dens}(\mathcal{C}) \leq S(\mathcal{C})$. Conversely if $\text{dens}(\mathcal{C}) < \infty$ then $S(\mathcal{C}) < \infty$.

Proof. The first sentence follows from 9.1.2 and 9.1.5. If $\text{dens}(\mathcal{C}) < \infty$ then since $m^{\mathcal{C}}(n) \leq Kn^r < 2^n$ for n large we have $S(\mathcal{C}) < \infty$. □

Remark. If $\mathcal{C} = 2^X$ for $\text{card}\, X = n < \infty$ then $S(\mathcal{C}) = n$ while $\text{dens}(\mathcal{C}) = 0$.

It is elementary that if $\mathcal{C} \subset \mathcal{D} \subset 2^X$ then $S(\mathcal{C}) \leq S(\mathcal{D})$ and $\text{dens}(\mathcal{C}) \leq \text{dens}(\mathcal{D})$. If $\mathcal{A} \subset 2^X$ and $\mathcal{C} \subset 2^X$ then clearly $\text{dens}(\mathcal{A} \cup \mathcal{C}) = \max(\text{dens}(\mathcal{A}), \text{dens}(\mathcal{C}))$.

9.1.7 **Proposition.** For any $\mathcal{A} \subset 2^X$ and $\mathcal{C} \subset 2^X$, $S(\mathcal{A} \cup \mathcal{C}) \leq S(\mathcal{A}) + S(\mathcal{C}) + 1$.

Proof. By Theorem 9.1.2, if $n \geq S(\mathcal{A}) + S(\mathcal{C}) + 2$,

$$m^{\mathcal{A} \cup \mathcal{C}}(n) \leq m^{\mathcal{A}}(n) + m^{\mathcal{C}}(n) \leq {}_nC_{\leq S(\mathcal{A})} + {}_nC_{\leq S(\mathcal{C})}$$

$$= \sum_{k=0}^{S(\mathcal{A})} \binom{n}{k} + \sum_{k=n-S(\mathcal{C})}^{n} \binom{n}{k} < 2^n,$$

so $S(\mathcal{A} \cup \mathcal{C}) < n$. □

Remark. If $\text{card}\, X = m$, $k = 0,1,\ldots,m$,

$$\mathcal{A} = \{Y \subset X: \text{card}\, Y \leq k\}, \text{ and}$$
$$\mathcal{C} = \{Y \subset X: \text{card}\, Y > k\},$$

then $S(\mathcal{A}) = k$, $S(\mathcal{C}) = m-k-1$, and $S(\mathcal{A} \cup \mathcal{C}) = m$, so Prop. 9.1.7 is sharp for all values of $S(\mathcal{A})$ and $S(\mathcal{C})$.

The following is immediate:

9.1.8 Fact. If $\mathcal{C} \subset 2^X$ and $\mathcal{D} := \{X \setminus A: A \in \mathcal{C}\}$ then $S(\mathcal{D}) = S(\mathcal{C})$, indeed $m^{\mathcal{C}}(n) = m^{\mathcal{D}}(n)$ for all n, and for all $B \subset X$, $\Delta^{\mathcal{C}}(B) = \Delta^{\mathcal{D}}(B)$.

Notes. The definitions of $\Delta^{\mathcal{C}}$, $m^{\mathcal{C}}$ and (in effect) $V(\mathcal{C})$ appeared in Vapnik and Červonenkis (1968). In their 1971 paper they had a weaker form of Theorem 9.1.2 with "$\geq {}_nC_{\leq k}$" instead of "$> {}_nC_{\leq k-1}$". The theorem as stated appears in Sauer (1972, Theorem 1). Extended results for possibly infinite cardinals are said to have been found independently by Shelah (1972). Vapnik and Červonenkis in their 1974 book also gave Theorem 9.1.2, then proved Prop. 9.1.5. Prop. 9.1.7 is due to Assouad (1981, 3.2; 1982, Prop. 2.4).

9.2 Generation of classes and evaluation of $S(\mathcal{C})$.

Classes with $S(\mathcal{C})$ finite can be formed in a variety of ways. Here is one. Let G be a collection of real-valued functions on a set X. Let

$$\text{pos}(g) := \{x: g(x) > 0\}, \quad g \in G,$$

$$\text{pos}(G) := \{\text{pos}(g): g \in G\}.$$

9.2.1 Theorem. Let H be an m-dimensional real vector space of functions on a set X, f any real function on X, and $H_1 := \{f+h: h \in H\}$. Then $S(\text{pos}(H_1)) = m$.

Proof. Clearly card $X \geq m$. If card $X = m$, then $H = H_1 = \mathbb{R}^X :=$ all real functions on X, so the result holds.

Otherwise, let $A \subset X$ with card $A = m+1$. Let G be the vector space $\{af+h: a \in \mathbb{R}, h \in H\}$. Let $r_A: G \to \mathbb{R}^A$ be the restriction of functions in G to A. If r_A is not onto, take $0 \neq v \in \mathbb{R}^A$ where v is orthogonal to $r_A(G)$ for the usual inner product $(\cdot,\cdot)_A$.
Let $A_+ := \{x \in A: v(x) > 0\}$. We may assume A_+ is nonempty, replacing v by $-v$. If $A_+ = A \cap \text{pos}(g)$, $g \in G$, then $(r_A(g), v)_A > 0$, a contradiction. So A is not shattered by $\text{pos}(G)$.

Suppose instead that $r_A(G) = \mathbb{R}^A$. Then r_A is 1-1 on G, $f \notin H$, and $r_A(H_1)$ is an affine hyperplane in $\mathbb{R}^A$ not containing 0. Thus for some $v \in \mathbb{R}^A$,

$(j,v)_A = -1$ for all $j \in r_A(H_1)$. Again let $A_+ := \{x \in A: v(x) > 0\}$. If $A_+ = A \cap \mathrm{pos}(f+h)$, $h \in H$, then $(f+h,v)_A \geq 0$, a contradiction (A_+ may be empty). Thus $\mathrm{pos}(H_1)$ never shatters A, so $S(\mathrm{pos}(H_1)) \leq m$.

For some $B \subset X$ with card $B = m$, we have $r_B(H) = \mathbb{R}^B$. Hence $r_B(H_1) = \mathbb{R}^B$ and $\mathrm{pos}(r_B(H_1)) = 2^B$, so $S(\mathrm{pos}(H_1)) = m$. □

Let X be a real vector vector space of dimension m. Let H be the space of all real <u>affine</u> functions on X of the form $h+c$ where h is real linear and c is any real constant. Then H has dimension $m+1$ and $\mathrm{pos}(H)$ is the collection of all open half-spaces of X. Then letting $f = 0$ in Theorem 9.2.1 gives a special case known as <u>Radon's Theorem</u> (Radon, 1921, p.114; Danzer, Grünbaum and Klee, 1963, p.103). On the other hand, Theorem 9.2.1 for $f = 0$ can be deduced from Radon's theorem via the following stability result.

9.2.2 <u>Theorem.</u> If X and Y are sets, F is a function from X into Y, $\mathcal{C} \subset 2^Y$, and $F^{-1}(\mathcal{C}) := \{F^{-1}(A): A \in \mathcal{C}\}$, then $S(F^{-1}(\mathcal{C})) \leq S(\mathcal{C})$. If F is onto Y then $S(F^{-1}(\mathcal{C})) = S(\mathcal{C})$.

<u>Proof</u>. Let $F^{-1}(\mathcal{C})$ shatter $\{x_1,\ldots,x_m\}$ where $x_i \neq x_j$, $i \neq j$. Then $F(x_i) \neq F(x_j)$ for $i \neq j$ and $\mathcal{C}$ shatters $\{F(x_1),\ldots, F(x_m)\}$. If F is onto $Y \supset H$ with card $H = m$, choose $G \subset X$ where F takes G 1-1 onto H. Then if $\mathcal{C}$ shatters H, $F^{-1}(\mathcal{C})$ shatters G, so $S(F^{-1}(\mathcal{C})) = S(\mathcal{C})$. □

Now given any set X and a finite-dimensional real vector space G of real functions on X, there is a natural map $F: x \to \delta_x$ from X into the space of linear functions on G. Then using Theorem 9.2.2 one can deduce Theorem 9.2.1 from its special case where X is an m- or $(m+1)$-dimensional real vector space and f and all functions in H are linear, so that sets in $\mathrm{pos}(H_1)$ are open half-spaces.

Next we look at the effect of finitely many Boolean operations.

<u>9.2.3</u>. Let X be a set, $\mathcal{C} \subset 2^X$, $k = 1,2,\ldots$, and let $\mathcal{C}^{(k)}$ be the union of all (Boolean) algebras generated by k or fewer elements of $\mathcal{C}$. Then $\mathrm{dens}(\mathcal{C}^{(k)}) \leq k\cdot\mathrm{dens}(\mathcal{C})$, so if $S(\mathcal{C}) < \infty$ then $S(\mathcal{C}^{(k)}) < \infty$.

Proof. Let $\text{dens}(\mathcal{C}) = r$ so that for any $\varepsilon > 0$ there is some $M < \infty$ with $m^{\mathcal{C}}(n) \leq Mn^{r+\varepsilon}$ for all n.

For any $A \subset X$ we have $A \sqcap \mathcal{C}^{(k)} = (A \sqcap \mathcal{C})^{(k)}$. An algebra with k generators has at most 2^k atoms and 2^{2^k} elements. Thus $\text{card}(A \sqcap \mathcal{C}^{(k)}) \leq \text{card}(A \sqcap \mathcal{C})^{k} 2^{2^k} \leq 2^{2^k} M(\text{card}A)^{k(r+\varepsilon)}$. Letting $\varepsilon \downarrow 0$ gives $\text{dens}(\mathcal{C}^{(k)}) \leq k\cdot\text{dens}(\mathcal{C})$. If $S(\mathcal{C}) < \infty$ then using Cor. 9.1.6, $S(\mathcal{C}^{(k)}) < \infty$. □

Remarks. Let X be an infinite set, $r = 1,2,\ldots$, and $\mathcal{C}_r$ the collection of all subsets of X with at most r elements. Then clearly $\text{dens}(\mathcal{C}_r) = S(\mathcal{C}_r) = r$ while for any $A \subset X$ and $B \in A \sqcap \mathcal{C}_r^{(k)}$, either B or $A\backslash B$ has at most kr elements. Thus for $\mathcal{C} = \mathcal{C}_r^{(k)}$, $m^{\mathcal{C}}(n) \leq 2\,{}_nC_{\leq kr}$. So by 9.1.5, $\text{dens}(\mathcal{C}_r^{(k)}) \leq kr$ and the inequality $\text{dens}(\mathcal{C}^{(k)}) \leq k\cdot\text{dens}(\mathcal{C})$ is sharp. On the other hand it does not always hold for $S(\cdot)$ in place of $\text{dens}(\cdot)$: if $\mathcal{C}$ is the collection of open half-spaces in $\mathbb{R}^d$, $d \geq 1$, then $S(\mathcal{C}) = d+1$ by Radon's theorem while $S(\mathcal{C}^{(d)}) \geq 2^d > d(d+1)$ for $d \geq 5$.

Next let us consider to what extent classes with $S(\mathcal{C}) < \infty$ can be characterized and their general structure determined.

We have $V(\mathcal{C}) = 0$ if and only if $\mathcal{C} = \emptyset$ ($\mathcal{C}$ is empty) in which case we have set $S(\mathcal{C}) = -1$. Next, $V(\mathcal{C}) = 1 = S(\mathcal{C}) + 1$ if and only if $\text{card}\,\mathcal{C} = 1$. Collections with $S(\mathcal{C}) = 1$ are less trivial:

9.2.4 Theorem. If $\mathcal{C}$ is linearly ordered by inclusion and $\text{card}(\mathcal{C}) \geq 2$ then $S(\mathcal{C}) = 1$. Conversely if $S(\mathcal{C}) = 1$ and for every $A \subset X$ with $\text{card}\,A = 2$, $A \sqcap \mathcal{C} \supset \{\emptyset, A\}$, then $\mathcal{C}$ must be linearly ordered by inclusion.

Remark. If $\mathcal{C} \supset \{\emptyset, X\}$ then $A \sqcap \mathcal{C} \supset \{\emptyset, A\}$ for all A.

Proof. As just noted, $\text{card}\,\mathcal{C} \geq 2$ implies $S(\mathcal{C}) \geq 1$. For $\mathcal{C}$ linearly ordered by inclusion and $x \neq y$, $\mathcal{C} \sqcap \{x,y\}$ contains at most one of the singletons $\{x\}$, $\{y\}$, so $S(\mathcal{C}) = 1$.

Conversely if $S(\mathcal{C}) = 1$, suppose $\mathcal{C}$ is not linearly ordered by inclusion. Then for some $x, y \in X$ and $B, C \in \mathcal{C}$, we have $x \in B\backslash C$ and $y \in C\backslash B$. Let $A := \{x,y\}$. Then $\{x\} \in A \sqcap \mathcal{C}$ and $\{y\} \in A \sqcap \mathcal{C}$. By assumption $\{\emptyset, A\} \subset A \sqcap \mathcal{C}$ so $\mathcal{C}$ shatters A, a contradiction. □

Let us say that four sets A, B, C, D form a <u>diamond</u> if $A \subset B \subset C$, $A \subset D \subset C$, $B \not\subset D$ and $D \not\subset B$.

9.2.5 <u>Proposition</u>. For any set X and $\mathcal{C} \subset 2^X$, $S(\mathcal{C}) \geq 2$ if and only if for some $Y \subset X$, $Y \sqcap \mathcal{C}$ includes a diamond.

<u>Proof</u>. If $S(\mathcal{C}) \geq 2$ then for some $Y \subset X$ with card $Y = 2$, $\mathcal{C} \sqcap Y = 2^Y$, a diamond.

Conversely if $\mathcal{C} \sqcap Y$ includes a diamond A 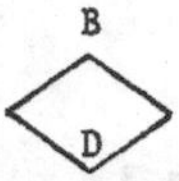

C, take $b \in B \backslash D$ and $d \in D \backslash B$. Then $\mathcal{C}$ shatters $\{b,d\}$. □

Clearly $S(\mathcal{C}) = 1$ for any collection of disjoint sets with card $\mathcal{C} \geq 2$.

Next we turn to unions, intersections and Cartesian products of sets. Given $\mathcal{C} \subset 2^X$ and $\mathcal{D} \subset 2^X$ let

$$\mathcal{C} \sqcap \mathcal{D} := \{C \cap D: C \in \mathcal{C}, D \in \mathcal{D}\},$$

$$\mathcal{C} \sqcup \mathcal{D} := \{C \cup D: C \in \mathcal{C}, D \in \mathcal{D}\}.$$

If $\mathcal{A} \subset 2^Y$ let $\mathcal{C} \boxtimes \mathcal{A} := \{C \times A: C \in \mathcal{C}, A \in \mathcal{A}\}$.

9.2.6 <u>Theorem</u>. If $S(\mathcal{C}) < \infty$, $S(\mathcal{D}) < \infty$ and $S(\mathcal{A}) < \infty$ then $S(\mathcal{C} \sqcap \mathcal{D}) < \infty$, $S(\mathcal{C} \sqcup \mathcal{D}) < \infty$ and $S(\mathcal{C} \boxtimes \mathcal{A}) < \infty$.

<u>Proof</u>. We have by 9.1.2 and 9.1.5 $m^{\mathcal{C} \sqcap \mathcal{D}}(n) \leq m^{\mathcal{C}}(n) m^{\mathcal{D}}(n) < 2^n$ for n large and likewise for the other two classes. □

For k, m = 0,1,2,..., let

$$S(k,m) := \max\{S(\mathcal{C} \boxtimes \mathcal{A}): S(\mathcal{C}) = k, S(\mathcal{A}) = m\}.$$

9.2.7 <u>Theorem.</u> For all k, m = 0,1,..., $S(k,m) = T(k,m) := \max\{S(\mathcal{C} \sqcap \mathcal{D}): S(\mathcal{C}) = k, S(\mathcal{D}) = m\} = \max\{S(\mathcal{C} \sqcup \mathcal{D}): S(\mathcal{C}) = k, S(\mathcal{D}) = m\}$.

<u>Proof</u>. The latter equality follows by taking complements (9.1.8). We have $T(k,m) \leq S(k,m)$ by considering $X \times X$ and restricting to the diagonal $\{\langle x,x\rangle: x \in X\}$.

Conversely let Π_X and Π_Y be the projections of $X \times Y$ onto X and Y respectively. Let

$$\mathcal{C}' := \{\Pi_X^{-1}(C): C \in \mathcal{C}\}, \quad \mathcal{A}' := \{\Pi_Y^{-1}(A): A \in \mathcal{A}\}.$$

By 9.2.2, $S(\mathcal{C}') = S(\mathcal{C})$ and $S(\mathcal{A}') = S(\mathcal{A})$.

Since $\Pi_X^{-1}(C) \cap \Pi_Y^{-1}(A) \equiv C \times A$ we have $S(\mathcal{C}' \sqcap \mathcal{D}') \geq S(\mathcal{C} \underline{\times} \mathcal{D})$, so $T(k,m) \geq S(k,m)$. □

Next let us investigate the values of $S(k,m)$. Clearly $S(m,k) \equiv S(k,m)$. Since $S(\mathcal{C}) = 0$ just for card $\mathcal{C} = 1$, it is easy to see that $S(k,0) = k$ for all k. For the density, since $m^{\mathcal{C} \underline{\cup} \mathcal{D}}(n) \leq m^{\mathcal{C}}(n) m^{\mathcal{D}}(n)$, clearly $\operatorname{dens}(\mathcal{C} \underline{\cup} \mathcal{D}) \leq \operatorname{dens} \mathcal{C} + \operatorname{dens} \mathcal{D}$. But this subadditivity fails for $S(\cdot)$, as the following shows.

9.2.8 Proposition. $S(1,1) = 3$.

Proof. Let $X = \{1,2,3\}$, $\mathcal{C} = \{A \subset X: \text{card } A \leq 1\}$, $\mathcal{D} = \{\emptyset, \{1\}, \{2\}, \{2,3\}\}$. Then clearly $S(\mathcal{C}) = 1$, $S(\mathcal{D}) = 1$, and $S(\mathcal{C} \underline{\cup} \mathcal{D}) = 3$. So $S(1,1) \geq 3$.

Suppose $S(1,1) \geq 4$. Then for $X = \{1,2,3,4\}$ take $\mathcal{C} \subset 2^X$ and $\mathcal{D} \subset 2^X$ with $S(\mathcal{C}) = S(\mathcal{D}) = 1$ and $S(\mathcal{C} \underline{\cup} \mathcal{D}) = 4$. Then $\emptyset \in \mathcal{C} \cap \mathcal{D}$. We may assume $\mathcal{C}$ contains at least as many singletons $\{k\}$ as $\mathcal{D}$ does. Then we have three cases:

Case 1. $\mathcal{C}$ contains all four singletons. Then $\mathcal{C} = \{\emptyset, \{1\}, \{2\}, \{3\}, \{4\}\}$. To obtain $\{1,2,3,4\} \in \mathcal{C} \underline{\cup} \mathcal{D}$ we may assume either $\{1,2,3\} \in \mathcal{D}$ or $\{1,2,3,4\} \in \mathcal{D}$. Then since $\{1,2\} \in \mathcal{C} \underline{\cup} \mathcal{D}$, either:

Case 1a. $\mathcal{D}$ contains $\{1\}$. Then $\mathcal{D}$ contains either $\{2\}$, $\{2,3\}$, or $\{3\}$, and shatters either $\{1,2\}$ or $\{1,3\}$, a contradiction;

Case 1b. $\mathcal{D}$ contains $\{2\}$: likewise; or

Case 1c. $\mathcal{D}$ contains $\{1,2\}$. Then if $\mathcal{D}$ contains $\{3\}$ it shatters $\{2,3\}$ or if $\{2,3\} \in \mathcal{D}$, $\mathcal{D}$ shatters $\{1,3\}$ so $\{2\} \in \mathcal{D}$. Then $\{1,3\} \in \mathcal{D}$ so $\mathcal{D}$ shatters $\{2,3\}$, again a contradiction, finishing Case 1.

Case 2 $\mathcal{C} \supset \{\emptyset, \{1\}, \{2\}, \{3\}\}$, while $\{4\} \in \mathcal{D} \setminus \mathcal{C}$.

Case 2a. $\mathcal{C} = \{\emptyset, \{1\}, \{2\}, \{3\}\}$. Then $\mathcal{D}$ contains either $\{1,2,3,4\}$ or some 3-element set containing 4, say $\{1,2,4\}$. Then to obtain $\{1,2\}$, $\mathcal{D}$ must contain $\{1\}$ or $\{1,2\}$ - and shatter $\{1,4\}$ - or $\{2\}$, and shatter $\{2,4\}$, a contradiction.

Case 2b. $\mathcal{C}$ contains some other set, which must be a two-element set containing 4, say $\{1,4\}$. Not to shatter any 2-element set, $\mathcal{C}$ must contain no other sets, so $\mathcal{C} = \{\emptyset, \{1\}, \{2\}, \{3\}, \{1,4\}\}$. To obtain $\{2,3,4\}$, $\mathcal{D}$ must contain either it or one of {2,4} or {3,4}, say {2,4}. In either case, for $\mathcal{D}$ not to shatter {2,4} while $\{1,2\} \in \mathcal{C} \cup \mathcal{D}$, we have $\{1\} \in \mathcal{D}$. Likewise $\{2,3\} \in \mathcal{C} \cup \mathcal{D}$ implies $\{3\} \in \mathcal{D}$. Then $\mathcal{D} = \{\emptyset, \{1\}, \{3\}, \{4\}, A\}$ where $A = \{2,3,4\}$ - but then $\mathcal{D}$ shatters {3,4} - or $A = \{2,4\}$ - but then $\{1,2,3,4\} \notin \mathcal{C} \cup \mathcal{D}$, a contradiction, finishing Case 2.

Case 3. $\mathcal{C} \supset \{\emptyset, \{1\}, \{2\}\}$, $\mathcal{D} \supset \{\emptyset, \{3\}, \{4\}\}$, and neither $\mathcal{C}$ nor $\mathcal{D}$ contains other singletons. Then $\{1,2\} \in \mathcal{D} \setminus \mathcal{C}$ and $\{3,4\} \in \mathcal{C} \setminus \mathcal{D}$. Not to shatter any 2-element set, neither $\mathcal{C}$ nor $\mathcal{D}$ can contain further elements, but then $\{1,2,3\} \notin \mathcal{C} \cup \mathcal{D}$, completing the proof. □

To strengthen 9.2.6 one can bound $S(j,k)$ above as follows. Let

$$\Theta(j,k) := \sup\{r \in \mathbb{N}: {}_rC_{\leq j}\, {}_rC_{\leq k} \geq 2^r\},$$

$j,k \in \mathbb{N}$. Then $\Theta(j,k) < +\infty$ for each j,k and we have:

9.2.9 Proposition. $S(j,k) \leq \Theta(j,k)$ for all $j, k \in \mathbb{N}$.

Proof. Let $S(\mathcal{C}) = j$ and $S(\mathcal{D}) = k$. Then for any $n > \Theta(j,k)$, by Sauer's theorem (9.1.2),

$$m^{\mathcal{C} \cup \mathcal{D}}(n) \leq m^{\mathcal{C}}(n) m^{\mathcal{D}}(n) \leq {}_nC_{\leq j}\, {}_nC_{\leq k} < 2^n. \quad \square$$

We have the following superadditivity of $S(\cdot,\cdot)$:

9.2.10 Proposition. For any nonnegative integers j, k, m and n,

$$S(j+k, m+n) \geq S(j,m) + S(k,n).$$

Proof. Take X with $\mathcal{C}, \mathcal{D} \subset 2^X$, $S(\mathcal{C}) = j$, $S(\mathcal{D}) = m$, and $S(\mathcal{C} \cup \mathcal{D}) = S(j,m) = \text{card } X$. Take a set Y, disjoint from X, with $\mathcal{E}, \mathcal{F} \subset 2^Y$, $S(\mathcal{E}) = k$, $S(\mathcal{F}) = n$, and $S(\mathcal{E} \cup \mathcal{F}) = S(k,n) = \text{card } Y$. Then $\emptyset \in \mathcal{C} \cap \mathcal{D} \cap \mathcal{E} \cap \mathcal{F}$, unless $j = m = 0$ or $k = n = 0$, but then the assertion holds. Then in $Z = X \cup Y$ we have $S(\mathcal{C} \cup \mathcal{E}) = j+k$ and $S(\mathcal{D} \cup \mathcal{F}) = m+n$, while

$$S((\mathcal{C} \cup \mathcal{E}) \cup (\mathcal{D} \cup \mathcal{F})) = S((\mathcal{C} \cup \mathcal{D}) \cup (\mathcal{E} \cup \mathcal{F})) = S(j,m) + S(k,n). \quad \square$$

At this writing I do not know the value of $S(1,2)$. From (9.2.9), $S(1,2) \leq 8$. The following example, due to L. Birgé (unpublished) shows that $S(1,2) \geq 5$: let $\mathcal{C} = \{\emptyset, \{1\}, \{2\}, \{3\}, \{4\}, \{4,5\}\}$, $\mathcal{D} = \{\emptyset, \{1\}, \{2\}, \{3\}, \{5\}, \{1,2\}, \{1,3\}, \{1,5\}, \{2,3\}, \{2,5\}, \{2,3,4\}, \{2,3,5\}, \{2,3,4,5\}\}$.

Calculation of $S(j,m)$, even for the next few cases, would apparently require some theoretical method, and could not be crudely turned over to a computer. For example, the number of pairs $\mathcal{C}, \mathcal{D} \subset 2^X$ for card $X = 6$ is

$$(2^{2^6})^2/2 = 2^{2^7-1} = 2^{127} > 10^{38}.$$

9.2.11 <u>Theorem</u>. Let $\mathcal{C} \subset 2^X$ and $\mathcal{D} \subset 2^Y$ where X and Y are any sets and $\mathcal{C}$ is linearly ordered by inclusion. Then $S(\mathcal{C} \underline{\times} \mathcal{D}) \leq S(\mathcal{D}) + 1$.

<u>Proof</u>. Let $F \subset X \times Y$ with card $F = m+2$ where $m = S(\mathcal{D})$. Suppose $\mathcal{C} \underline{\times} \mathcal{D}$ shatters F. The subsets of $\Pi_X F \subset X$ induced by $\mathcal{C}$ are linearly ordered by inclusion. Let G be the next largest after $\Pi_X F$ itself and take $p \in \Pi_X F \backslash G$, $\langle p,y \rangle \in F$. Then all subsets of F containing $\langle p,y \rangle$ are cut by sets of the form $\Pi_X F \times D$, $D \in \mathcal{D}$. Thus $H := F \backslash \{\langle p,y \rangle\}$ is shattered by such sets. Now card $H = m+1$, and Π_Y must be 1-1 on H or H could not be shattered by sets of the given form. But then $\mathcal{D}$ shatters $\Pi_Y H$, a contradiction. □

9.2.12 <u>Corollary</u>. For any $\mathcal{C} \subset 2^X$ and $\mathcal{D} \subset 2^X$, if $\mathcal{C}$ is linearly ordered by inclusion, $S(\mathcal{C} \underline{\cap} \mathcal{D}) \leq S(\mathcal{D})+1$ and $S(\mathcal{C} \underline{\cup} \mathcal{D}) \leq S(\mathcal{D})+1$.

<u>Proof</u>. See the first paragraph of the proof of 9.2.7. □

Then by induction we obtain:

9.2.13 <u>Corollary</u>. Let $\mathcal{C} = \{\bigcap_{i=1}^n C_i : C_i \in \mathcal{C}_i, \ i = 1,\ldots,n\}$, where each $\mathcal{C}_i$ is linearly ordered by inclusion. Then $S(\mathcal{C}) \leq n$.

<u>Definition</u>. Call a Vapnik-Červonenkis class $\mathcal{C} \subset 2^X$ <u>bordered</u> if for some $F \subset X$, with card $F = S(\mathcal{C})$, and $x \in X \backslash F$, F is shattered by sets in $\mathcal{C}$ all containing x.

9.2.14 <u>Theorem</u>. Let $\mathcal{C}_i \subset 2^{X(i)}$ be bordered Vapnik-Červonenkis classes, $i = 1,2$. Then $S(\mathcal{C}_1 \underline{\times} \mathcal{C}_2) \geq S(\mathcal{C}_1) + S(\mathcal{C}_2)$.

<u>Proof</u>. Choose $F_i \subset X(i)$ and x_i as in the definition of "bordered". Put $H = (\{x_1\} \times F_2) \cup (F_1 \times \{x_2\})$. Then card $H = S(\mathcal{C}_1) + S(\mathcal{C}_2)$. For any $V_i \subset F_i$, $i = 1,2$, take $C_i \in \mathcal{C}_i$ with $C_i \cap F_i = V_i$, $x_i \in C_i$. Then $(C_1 \times C_2) \cap H = (\{x_1\} \times V_2) \cup (V_1 \times \{x_2\})$. Thus $\mathcal{C}_1 \underline{\times} \mathcal{C}_2$ shatters H. □

Theorem 9.2.14 extends by induction to any number of factors. In particular it implies:

9.2.15 Corollary. In $\mathbb{R}^m$ let $\mathcal{C}$ be the collection of rectangles $\mathcal{C} := \{X_{i=1}^m]a_i, b_i[: -\infty \leq a_i \leq b_i \leq +\infty,\ i = 1,\ldots,m\}$. Then $S(\mathcal{C}) = 2m$. If we take $a_i = -\infty$ for all i then $S(\mathcal{C}) = m$.

Proof. The class $\mathcal{D}$ of intervals $]-\infty, b[\subset \mathbb{R}$ is clearly a bordered class linearly ordered by inclusion. So is the class of intervals $]a, +\infty[$. The class $\mathcal{I}$ of all intervals $]a,b[\subset \mathbb{R}$ is bordered with $S(\mathcal{I}) = 2$. The results now follow from Corollary 9.2.13 with $n = m$ and $2m$, and Theorem 9.2.14 and induction. □

Here is an extension:

9.2.16 Proposition. Let $\mathcal{C}_1 = \{]a,b[: -\infty \leq a \leq b \leq +\infty\}$. Then for any set Y and $\mathcal{C}_2 \subset 2^Y$, with $Y \in \mathcal{C}_2$, $S(\mathcal{C}_1 \underline{\times} \mathcal{C}_2) \leq 2 + S(\mathcal{C}_2)$.

Proof. Let $F \subset \mathbb{R} \times Y$ with card $F = 3 + S(\mathcal{C}_2)$ and suppose $\mathcal{C}_1 \underline{\times} \mathcal{C}_2$ shatters F (if $S(\mathcal{C}_2) = +\infty$ there is no problem). Let $\langle x_i, y_i \rangle$ be the points of F. Let $u = \min_i x_i$ and $v = \max_j x_j$. Let $p = \langle u, y_i \rangle \in F$ and $q = \langle v, y_j \rangle \in F$. Now all subsets of F which include $\{p,q\}$ must be induced by sets of the form $\mathbb{R} \times C$, $C \in \mathcal{C}_2$. Thus Π_Y must be 1-1 on $F \backslash \{p,q\}$, so $\mathcal{C}_2$ shatters $\Pi_Y(F \backslash \{p,q\})$ of cardinality $S(\mathcal{C}_2) + 1$, a contradiction. □

Notes. Theorems 9.2.1 and 9.2.4 and Corollaries 9.2.12 and 9.2.15 are from Wenocur and Dudley (1981). Theorems 9.2.2 and 9.2.3 are from Assouad (1981, 1983). I do not have references for the other results except 9.2.9. Prop. 9.2.8 is a counterexample to Assouad (1981, (3.3), "Dens" case); in a letter, Assouad kindly sent me the correction which is Prop. 9.2.9.

9.3 Vapnik-Červonenkis classes, independence and probability laws.

Let $(X, \mathcal{A}, P)$ be a probability space. Recall the pseudo-metric $d_P(A,B) := P(A \Delta B)$ on $\mathcal{A}$ and the function $D(\varepsilon, S, d)$ (Sec. 6.0).

Let

$$s(\mathcal{C}) := \inf\{w: \text{there is a } K = K(w,\mathcal{C}) < \infty \text{ such that for every law } P \text{ on } \mathcal{A} \text{ and } 0 < \varepsilon \leq 1,\ D(\varepsilon, \mathcal{C}, d_P) \leq K\varepsilon^{-w}\}.$$

9.3.1 Theorem. For any measurable space $(X, \mathcal{A})$ and $\mathcal{C} \subset \mathcal{A}$, $\mathrm{dens}(\mathcal{C}) = s(\mathcal{C})$.

Proof. Suppose $A_1, \ldots, A_m \in \mathcal{C}$ and $d_P(A_i, A_j) > \varepsilon > 0$ for $i \neq j$. Let $X_1, X_2, \ldots$ be i.i.d. P. For $n = 1,2,\ldots$, we have

$$\Pr\{\text{for some } i \neq j,\ X_k \notin A_i \Delta A_j \text{ for all } k \leq n\} \leq m(m-1)(1-\varepsilon)^n/2 < 1$$

for n large enough, $n > -\log(m(m-1)/2)/\log(1-\varepsilon)$. For such n, we have with positive probability that $P_n(A_i \Delta A_j) > 0$ for all $i \neq j$ and so $m^{\mathcal{C}}(n) \geq m$. For any $r > \text{dens}(\mathcal{C})$ there is an $M = M(r,\mathcal{C}) < \infty$, $M \geq 2$, such that $m^{\mathcal{C}}(n) \leq Mn^r$ for all n. Note that $-\log(1-\varepsilon) \geq \varepsilon$. Thus for $m \geq 2$

$$m \leq M(2\cdot\log(m^2))^r \varepsilon^{-r}, \text{ or}$$

$$m(\log m)^{-r} \leq M_1 \varepsilon^{-r} \quad \text{for some } M_1 = M_1(r,\mathcal{C}) = 4^r M.$$

Now for any $\delta > 0$ and C large enough $(\log m)^r \leq Cm^\delta$ for all $m \geq 1$, so for all $m \geq 0$ $m^{1-\delta} \leq M_2 \varepsilon^{-r}$ for some $M_2(r,\mathcal{C},\delta)$.

Thus $m \leq (M_2 \varepsilon^{-r})^{1/(1-\delta)}$. Letting $r \downarrow \text{dens}(\mathcal{C})$ and $\delta \downarrow 0$ gives $\text{dens}(\mathcal{C}) \geq s(\mathcal{C})$.

In the converse direction let $|A| := \text{card}\, A$. For $r < t < \text{dens}(\mathcal{C})$ and $k = 1,2,\ldots$ let $A_k \subset X$ with $|A_k \sqcap \mathcal{C}| > k|A_k|^t$. Let $B_0 := A_1$. Given disjoint $B_0,\ldots,B_{n-1} \subset X$ with $|B_j \sqcap \mathcal{C}| \geq 2^j|B_j|^t$ and $|B_j| \geq 2^j$ for $j = 0,\ldots,n-1$, let $B_n := A_{k(n)} \setminus \bigcup_{j=0}^{n-1} B_j$ where $k(n) := k_n \geq 2^n|(\bigcup_{j=0}^{n-1} B_j) \sqcap \mathcal{C}|$ and $|B_n| \geq 2^n$. Then

$$|B_n \sqcap \mathcal{C}| \geq 2^n|A_{k(n)}|^t \geq 2^n|B_n|^t$$

and

$$B_n \cap B_j = \emptyset,\ j < n.$$

Let

$$\alpha_n := |B_n|^{-t/r}, \quad S := \sum_{n=0}^\infty |B_n|^{1-t/r} < \infty.$$

Let P be the law on $\bigcup_{n=1}^\infty B_n \subset X$ giving mass α_n/S to each point of B_n. The distinct sets in $\mathcal{C} \sqcap B_n$ are at d_P-distance at least α_n/S apart, hence so are a corresponding set of elements of $\mathcal{C}$. Thus for all n

$$D(\alpha_n/(2S),\mathcal{C},d_P) \geq |\mathcal{C} \sqcap B_n| \geq 2^n|B_n|^t$$

$$= 2^n\alpha_n^{-r} = 2^n S^{-r}(\alpha_n/S)^{-r}.$$

Since $2^n \to \infty$ as $\alpha_n \downarrow 0$ this implies $r \leq s(\mathcal{C})$. Then let $r \uparrow \text{dens}\,\mathcal{C}$. □

Next, in order to define independence of sets, let $A^1 := A$ and $A^{-1} := X\setminus A$ for any set $A \subset X$. Then sets $A_1,\ldots,A_m$ are called <u>independent</u> if and only if for all functions $s(\cdot)$ from $\{1,\ldots,m\}$ into $\{-1,+1\}$, $\bigcap_{j=1}^m A_j^{s(j)} \neq \emptyset$ (every possible intersection of some of the A_j, and the complements of the other A_j, is non-empty).

It is easy to see that if $A_1,\ldots,A_m$ are independent, then there is a probability law on the (σ-) algebra they generate for which the A_i are independent in the probabilistic sense and for which $P(A_i) = 1/2$, $i = 1,\ldots,m$ (or, if desired, $P(A_i) = p_i$ for any p_i, $0 \leq p_i \leq 1$). Then $P(A_i \Delta A_j) = 1/2$ for any $i \neq j$.

For a collection $\mathcal{C} \subset 2^X$ let $\mathcal{I}(C) := \sup\{m$: some $A_j \in \mathcal{C}$ are independent, $j = 1,\ldots,m\}$. Then we have:

9.3.2 Theorem. For any set X and $\mathcal{C} \subset 2^X$, $n = 1,2,\ldots$, if $S(\mathcal{C}) \geq 2^n$ then $\mathcal{I}(\mathcal{C}) \geq n$. On the other hand if $\mathcal{I}(\mathcal{C}) \geq 2^n$ then $S(\mathcal{C}) \geq n$. In both cases, 2^n cannot be replaced by 2^n-1. Thus $S(\mathcal{C}) < +\infty$ if and only if $\mathcal{I}(\mathcal{C}) < +\infty$.

Proof. Clearly, if a set Y has n or more independent subsets, then $\text{card}(Y) \geq 2^n$. Conversely if $\text{card}(Y) = 2^n$, let Y correspond to the set of all sequences of 0's and 1's (binary digits) of length n. Let A_j be the set of elements for which the jth digit is 1. Then the A_j are independent. It follows that if $S(\mathcal{C}) \geq 2^n$, then $\mathcal{I}(\mathcal{C}) \geq n$, while if $\text{card}(Y) = 2^n-1$ and $\mathcal{C} = 2^Y$, $S(\mathcal{C}) = 2^n-1$ but $\mathcal{I}(\mathcal{C}) < n$ as stated.

Conversely if B_j are independent, $j = 1,\ldots,2^n$, $B_j \in \mathcal{C}$, let $A_i := A(i)$ be independent subsets of $\{1,\ldots,2^n\}$, $i = 1,\ldots,n$, and choose

$x_i \in (\bigcap_{j\in A(i)} B_j) \cap \bigcap_{j\notin A(i)} (X\backslash B_j)$.

For each set $S \subset \{1,\ldots,n\}$,

$$\bigcap_{i\in S} A_i \cap \bigcap_{i\notin S} (\{1,\ldots,2^n\}\backslash A_i) = \{j\}$$

for some $j := j_S$, $1 \leq j \leq 2^n$. Then

$$B_j \cap \{x_1,\ldots,x_n\} = \{x_i : i \in S\}.$$

Thus $\mathcal{C}$ shatters $\{x_1,\ldots,x_n\}$ and $S(\mathcal{C}) \geq n$, as stated. If $\mathcal{C}$ consists of 2^n-1 (independent) sets, then clearly $S(\mathcal{C}) < n$. So again 2^n is best possible. □

Notes. Theorem 9.3.1 is partly in Dudley (1978, sec. 7), and was stated in the current form by Assouad (1981, 1983). I have no reference for 9.3.2.

Chapter 10. Measurability considerations.

10.1 Sufficiency. Let $(X,\mathcal{A})$ be a measurable space and $(X^n,\mathcal{A}^n)$ its n-fold product space. For any P in the family $\mathcal{P}$ of all probability laws on $\mathcal{A}$, we have the n-fold product P^n on $\mathcal{A}^n$. A sub-σ-algebra $\mathcal{B} \subset \mathcal{A}^n$ is called sufficient (for $\mathcal{P}$) if and only if for every $C \in \mathcal{A}^n$ there is some $\mathcal{B}$-measurable function g_C such that for every $P \in \mathcal{P}$, the conditional probability

(10.1.1) $$P^n(C|\mathcal{B}) = g_C \text{ a.s. } (P^n).$$

The essential point is that g_C does not depend on P.

The intuitive meaning of sufficiency is as follows. Suppose a statistician has observed $x_1,\ldots,x_n$ i.i.d. P and has no further information about P. She would like to make statistical inferences about the unknown P from the data $x := \langle x_1,\ldots,x_n\rangle$. Events $x \in A$, $A \in \mathcal{A}^n$, will be more probable for some laws, say P, than for others, say Q $(P^n(A) > Q^n(A))$. Suppose $\mathcal{B} \subset \mathcal{A}^n$ is a sufficient sub-σ-algebra. For simplicity, suppose $\mathcal{B}$ is the smallest σ-algebra for which a real function S(x) is measurable. Then, once given S(x), further information about x is of no use to the statistician in deciding which P's are more likely, since by (10.1.1) given S(x), all $P \in \mathcal{P}$ become indistinguishable: they all have the same conditional probabilities given S(x).

If, for example, we want to estimate the probability of success in some independent trials from the results of n of the trials (all with the same, unknown probability p of success), S may be the number of successes observed. This S - that is, the smallest σ-algebra for which it is measurable - is sufficient. The order in which the successes came is not informative since the trials are independent and had the same probability of success. In this case, X may be taken to have just two points, corresponding to success or failure. The 2^n possible sequences of outcomes are more efficiently represented by the function S with just n+1 possible values $0,1,\ldots,n$.

Now let $\mathcal{S}_n$ be the sub-σ-algebra of $\mathcal{A}^n$ consisting of sets invariant under all permutations of the coordinates. For each such permutation

$$f_\Pi: \langle x_1,\ldots,x_n\rangle \longrightarrow \langle x_{\Pi(1)},\ldots,x_{\Pi(n)}\rangle,$$

where $\Pi \in S_n$, the symmetric group of all permutations of $\{1,\ldots,n\}$, f_Π is a measurable isomorphism of $(X^n, \mathcal{A}^n)$ and preserves each P^n.

Thus for any $C \in \mathcal{A}^n$, we have

(10.1.2) $$P^n(C|\mathcal{S}_n) = (n!)^{-1} \sum_{\Pi \in S_n} 1_{f_\Pi(C)}(\cdot)$$

a.s. P^n since for any $B \in \mathcal{S}_n$,

$$P^n(B \cap C) = (n!)^{-1} \sum_{\Pi \in S_n} P^n(C \cap f_\Pi^{-1}(B))$$
$$= (n!)^{-1} \sum_{\Pi \in S_n} P^n(f_\Pi(C) \cap B).$$

Thus $\mathcal{S}_n$ is sufficient.

When X has just two points, say $X = \{0,1\}$, and $S = \sum_{i=1}^n x_i$, then $\mathcal{S}_n$ is the smallest σ-algebra for which S is measurable. In this case no σ-algebra strictly smaller than $\mathcal{S}_n$ is sufficient. ($\mathcal{S}_n$ is "minimal sufficient").

For each $B \in \mathcal{A}$ and $x = \langle x_1, \ldots, x_n \rangle \in X^n$ we have the empirical measure

$$P_n(B)(x) := n^{-1} \sum_{j=1}^n 1_B(x_j).$$

Note that in contrast to other uses of this notation, $x \to P_n(B)(x)$ is just a measurable function, or "statistic", not a random variable until laws for x are specified; $P_n(B)(x)$ is, here, just a function of B and x.

For a collection $\mathcal{F}$ of measurable functions on $(X^n, \mathcal{A}^n)$ let $\mathcal{S}_\mathcal{F}$ be the smallest σ-algebra making all functions in $\mathcal{F}$ measurable. Then $\mathcal{F}$ is called <u>sufficient</u> if and only if $\mathcal{S}_\mathcal{F}$ is sufficient.

10.1.3 <u>Theorem</u>. For each $n = 1,2,\ldots$, the empirical measure P_n is sufficient for $\mathcal{P}$, i.e. the set $\mathcal{F}$ of functions $x \to P_n(B)(x)$, for all $B \in \mathcal{A}$, is sufficient. In fact the σ-algebra $\mathcal{S}_\mathcal{F}$ is exactly $\mathcal{S}_n$.

<u>Proof</u>. This result follows from facts given by Neveu (1977, pp.267-268). For completeness a proof, shown to me by Sam Gutmann, will be given here.

Clearly $\mathcal{S}_\mathcal{F} \subset \mathcal{S}_n$. Let us prove the converse inclusion. For each set $B \in \mathcal{A}^n$ let $S(B) := \bigcup_{\Pi \in S_n} f_\Pi(B) \in \mathcal{S}_n$.

Then if $B \in \mathcal{S}_n$, $S(B) = B$.

Let $\mathcal{E} := \{C \in \mathcal{A}^n: S(C) \in \mathcal{S}_\mathcal{F}\}$.

We want to prove $\mathcal{E} = \mathcal{A}^n$. Now $\mathcal{E}$ is a monotone class: if $C_n \in \mathcal{E}$ and $C_n \uparrow C$ or $C_n \downarrow C$, then $C \in \mathcal{E}$. Also, any finite union of sets in $\mathcal{E}$ is in $\mathcal{E}$.

Thus it will suffice to prove

(10.1.4) $A_1 \times \ldots \times A_n \in \mathcal{E}$ for any $A_j \in \mathcal{A}$, $j = 1,\ldots,n$

(since the collection $\mathcal{C}$ of finite unions of such sets is an algebra; the smallest monotone class including $\mathcal{C}$ is $\mathcal{A}^n$).

Given $A_1,\ldots,A_n$ in $\mathcal{A}$, let $B_1,\ldots,B_m$ be the atoms of the algebra in X generated by $A_1,\ldots,A_n$, so that $m \leq 2^n$. Now it is enough to prove that for all $j(1),\ldots,j(n)$ with $1 \leq j(i) \leq m$, $S(B_{j(1)} \times\ldots\times B_{j(n)}) \in \mathcal{S}_{\mathcal{F}}$. Here $x \in S(B_{j(1)} \times\ldots\times B_{j(n)})$ if and only if for each $i = 1,\ldots,m$, $P_n(B_i) = k_i/n$ where k_i is the number of values of r such that $j(r) = i$. □

Next, one can ask: for what collections $\mathcal{C} \subset \mathcal{A}$ is the set of all $P_n(C)(\cdot)$, $C \in \mathcal{C}$, sufficient - for a given n? For some $\mathcal{C} \subset \mathcal{A}$, the values of $P_n(A)$ for each $A \in \mathcal{A}$ are determined by those of $P_n(C)$, $C \in \mathcal{C}$. For example, in $\mathbb{R}^1$, let $\mathcal{C} = \{]-\infty,t]: t \in \mathbb{R}\}$. Then P_n restricted to $\mathcal{C}$ is effectively the empirical distribution function, which of course completely determines $P_n(A)$ for all sets. If A is a Borel set, then $P_n(A)$ is a measurable function of the set of all $P_n(]-\infty,t])$ for t rational.

For sufficiency of $P_n(C)(\cdot)$, $C \in \mathcal{C}$, it is not enough that X be finite and $\mathcal{C}$ generate $2^X = \mathcal{A}$: let $X = \{1,2,3,4,5\}$, and $\mathcal{C} = \{\{1,2,3\}, \{2,3,4\}, \{3,4,5\}\}$. Then $\mathcal{C}$ generates $\mathcal{A}$ but the $P_n(C)$, $C \in \mathcal{C}$, do not determine $P_n(\{1\})$, for $n \geq 2$ ($P_2 = (\delta_1 + \delta_4)/2$ or $(\delta_2 + \delta_5)/2$ give the same values on $\mathcal{C}$ but not on 1).

In another example of dependence of sufficiency on n, in $\mathbb{R}^2$ let

$$\mathcal{C} = \{\{\langle x,y\rangle: x < t\}: t \in \mathbb{R}\} \cup \{\{\langle x,y\rangle: y < u\}: u \in \mathbb{R}\}.$$

Then although $\mathcal{C}$ generates the Borel σ-algebra, again $\{P_n(C)(\cdot): C \in \mathcal{C}\}$ are not sufficient for $n \geq 2$. Or if we let $\mathcal{D} := \mathcal{C} \cup \{\{\langle x,y\rangle: x+y < t\}: t \in \mathbb{R}\}$ then $\{P_n(D)(\cdot): D \in \mathcal{D}\}$ is not sufficient when n is large at least for $n \geq 14$ (let $P_{14} = (\sum_{i,j=1}^{4} \delta_{\langle i,j\rangle}) - \delta_{\langle 1,1\rangle} - \delta_{\langle 4,4\rangle})/14$; $Q = P_{14}$ on $\mathcal{D}$ gives 13 equations in 14 unknowns and must have a non-unique solution).

10.2 <u>Admissibility</u>. Let $(X,\mathcal{B})$ be a measurable space. The $(X,\mathcal{B})$ will be called <u>separable</u> if $\mathcal{B}$ is countably generated and contains all singletons $\{x\}$, $x \in X$. In this section $(X,\mathcal{B})$ is assumed to be such a space. Let $\mathcal{F}$ be a collection of real-valued functions on X. (The next definition has nothing to do with statistical admissibility.)

<u>Definition</u>. $\mathcal{F}$ is called <u>admissible</u> if and only if there is a σ-algebra $\mathcal{J}$ of subsets of $\mathcal{F}$ such that the evaluation map $\langle f,x\rangle \to f(x)$ is measurable from $(\mathcal{F},\mathcal{J}) \times (X,\mathcal{B})$ (with product σ-algebra) to $\mathbb{R}$ with Borel sets. Then $\mathcal{J}$ is called an <u>admissible structure</u> for $\mathcal{F}$.

$\mathcal{F}$ will be called <u>image-admissible via</u> $(Y,\mathcal{S},T)$ if $(Y,\mathcal{S})$ is a measurable space and T is a map from Y onto $\mathcal{F}$ such that the map $\langle y,x\rangle \to T(y)(x)$ is measurable from $(Y,\mathcal{S}) \times (X,\mathcal{B})$ (with product σ-algebra) to $\mathbb{R}$ with Borel sets.

To apply these definitions to a family $\mathcal{C}$ of sets let $\mathcal{F} = \{1_C: C \in \mathcal{C}\}$.

<u>Remarks</u>. If a family $\mathcal{F}$ is admissible, then it is image admissible, with T the identity. In regard to the converse direction here is an example. Let $X = [0,1]$ with usual Borel σ-algebra $\mathcal{B}$. Let $(Y,\mathcal{S})$ be a countable product of copies of $(X,\mathcal{B})$. For $y = \{y_n\}_{n=1}^{\infty} \in Y$ let $T(y)(x) = 1_J(\langle x,y\rangle)$ where $\langle x,y\rangle \in J$ if and only if $x = y_n$ for some n. Let $\mathcal{F} := \{T(y)(\cdot): y \in Y\}$, i.e. the set of indicator functions of countable subsets of X. Then it is easy to see that $\mathcal{F}$ is image admissible via $(Y,\mathcal{S},T)$. If we transport the σ-algebra $\mathcal{S}$ to $\mathcal{F}$ by letting $\mathcal{J} := \{F \subset \mathcal{F}: T^{-1}(F) \in \mathcal{S}\}$ then $\mathcal{J}$ is not countably generated (Freedman, 1966, Lemma (5)) although $\mathcal{S}$ is. This example shows how in some situations image admissibility may work better than admissibility.

Let $\mathcal{A} := \{A_n\}_{n\geq 1}$ be a countable set of generators of $\mathcal{B}$. We may assume that finite unions, finite intersections and complements of sets in $\mathcal{A}$ are in $\mathcal{A}$. The <u>Marczewski function</u>

$$M(x) := \sum_{n=1}^{\infty} 2 1_{A_n}(x)/3^n$$

is 1-1 from X into the interval $I := [0,1]$ and is a measurable isomorphism. Thus there is no loss of generality to assume that X is a subset of I with (relative) Borel measurability.

Let $\mathcal{F}_0$ be the collection of all finite sums $\sum_{i=1}^n a_i 1_{A_i}$, $a_i \in \mathbb{R}$, $n = 1,2,\ldots$. Then "Borel classes" or "Banach classes" are defined by recursion as follows. For each countable ordinal α, if $\mathcal{F}_\alpha$ is defined let $\mathcal{F}_{\alpha+1}$ be the set of all limits of everywhere convergent sequences of functions in $\mathcal{F}_\alpha$. If α is a limit ordinal let $\mathcal{F}_\alpha := \bigcup_{\beta<\alpha} \mathcal{F}_\alpha$. Then all measurable functions on $(X,\mathcal{B})$ belong to $\mathcal{F}_\Omega$, where Ω is the least uncountable ordinal.

On admissibility there is the following main theorem:

10.2.1 Theorem. Given $(X,\mathcal{B})$ separable and a family $\mathcal{F}$ of measurable functions on X, the following are equivalent:

(i) $\mathcal{F} \subset \mathcal{F}_\alpha$ for some $\alpha < \Omega$;

(ii) there is a jointly measurable function $G: I \times X \to \mathbb{R}$ such that for each $f \in \mathcal{F}$, $f = G(t,\cdot)$ for some $t \in I$;

(iii) there is a separable admissible structure for $\mathcal{F}$;

(iv) $\mathcal{F}$ is admissible;

(v) $2^{\mathcal{F}}$ is an admissible structure for $\mathcal{F}$;

(vi) $\mathcal{F}$ is image admissible via some $(Y,\mathcal{S},T)$.

Remark. While the specific classes $\mathcal{F}_\alpha$ depend on the choice of the countable family $\mathcal{A}$ of generators, condition (i) does not: if $\mathcal{A}'$ is another countable set of generators of $\mathcal{B}$ with corresponding classes $\mathcal{F}'_\alpha$ then for any $\alpha < \Omega$ there are $\beta < \Omega$ and $\gamma < \Omega$ with $\mathcal{F}_\alpha \subset \mathcal{F}'_\beta$ and $\mathcal{F}'_\alpha \subset \mathcal{F}_\gamma$.

Proof of 10.2.1. Clearly (ii) $\Rightarrow$ (iii) (choosing, for each $f \in \mathcal{F}$, a unique $t \in I$ and restricting the Borel σ-algebra to the set of t's chosen). Evidently (iii) $\Rightarrow$ (iv) $\Leftrightarrow$ (v). Now (iv) $\Rightarrow$ (iii) since a real-valued measurable function on $\mathcal{F} \times X$ is measurable with respect to some countably generated sub-σ-algebra in $\mathcal{F} \times X$, which in turn is included in a σ-algebra generated by countably many rectangles $A \times B$, $A \subset \mathcal{F}$, $B \subset X$.

For (iii) $\Rightarrow$ (ii), as above we may identify $(\mathcal{F},\mathcal{S})$ with a subset of I carrying Borel measurability. Then $\langle f,x\rangle \to f(x)$ extends to a Borel function G on $I \times X$, raising its Borel class α (the smallest α such that $G \in \mathcal{F}_\alpha$) by at most 1, as shown by Alexits, Sierpiński and Kuratowski, see Kuratowski (1966, pp. 434-435).

Next, (ii) $\Rightarrow$ (i) since G belongs to some $\mathcal{F}_\alpha$ on $I \times X$, $\alpha < \Omega$, hence so do all its sections $G(t,\cdot)$ on X (Kuratowski, 1966, p.377). We have (i) $\Rightarrow$ (ii) using Lebesgue's "universal class α function" (e.g. Natanson, 1957, p.137) and the Lebesgue-Hausdorff relation between Baire and Borel classes (Kuratowski, 1966, p. 393).

Clearly (iv) $\Rightarrow$ (vi). If (vi) holds, let Z be a subset of Y on which the map $z \to T(z)(\cdot)$ is one-to-one. Let $\mathcal{S}_Z$ and T_Z be the restrictions to Z of $\mathcal{S}$ and T respectively. Then $\mathcal{F}$ remains image admissible via $(Z,\mathcal{S}_Z,T_Z)$, and $\{T_Z(A): A \in \mathcal{S}_Z\}$ is an admissible structure for $\mathcal{F}$, giving (iv). □

For $0 \leq p \leq \infty$ and a probability law P on $(X,\mathcal{B})$ we have the space $\mathcal{L}^p(X,\mathcal{B},P)$ of measurable real-valued functions f on X such that $\int|f|^p dP < \infty$, with the pseudo-metric

$$d_{p,P}(f,g) := \begin{cases} (\int|f-g|^p dP)^{1/p}, & 1 \leq p < \infty; \\ \int|f-g|^p dP, & 0 < p < 1; \\ \inf\{\varepsilon > 0: P(|f-g|>\varepsilon) < \varepsilon, & p = 0. \end{cases}$$

Here is a useful property of admissible classes:

10.2.2 <u>Theorem</u>. Let $(X,\mathcal{B})$ be a separable measurable space, $0 \leq p < \infty$, and $\mathcal{F} \subset \mathcal{L}^p(X,\mathcal{B},P)$ where $\mathcal{F}$ is admissible. Then whenever $\mathcal{F}$ is image admissible via $(Y,\mathcal{S},T)$, $(Y,\mathcal{S})$ is separable, $U \subset \mathcal{F}$ and U is $d_{p,P}$-open, we have $T^{-1}(U) \in \mathcal{S}$.

<u>Proof</u>. Since $(X,\mathcal{B})$ is separable, the pseudo-metric spaces $\mathcal{L}^p(X,\mathcal{B},P)$ are separable, $0 \leq p < \infty$. Thus by Lindelöf's theorem, U is a countable union of balls

$$\{f: d_{p,P}(f,g) < r\}, \quad g \in \mathcal{F}, \; 0 \leq r < \infty.$$

So for $p > 0$ it is enough to show that each function $y \to \int|T(y)(\cdot) - g|^p dP$ is measurable. For g fixed, $\langle y,x\rangle \to |T(y)(x) - g(x)|^p$ is jointly measurable. Thus for $0 < p < \infty$ we can reduce to the case $p = 1$ and $g = 0$ with $T(y)(x) \geq 0$ for all x, y. Now, identifying Y with a subset of I as above (Theorem 10.2.1, proof of (iii) $\Rightarrow$ (ii)), Fubini's theorem implies the desired measurability.

For $p = 0$ we consider the indicator of the set where $|T(y)(x)-g(x)| > \varepsilon$ and apply the same argument. □

10.2.3 Corollary. If $\mathcal{F} \subset \mathcal{L}^1(X,\mathcal{B},P)$ and $\mathcal{F}$ is image admissible via $(Y,\mathcal{S},T)$ with $\mathcal{S}$ separable then $y \to \int T(y)\,dP$ is $\mathcal{S}$-measurable.

Proof. For any real u, $\{f: \int f\,dP > u\}$ is open for $d_{1,P}$. □

Remark. For any image admissible $\mathcal{F}$ there always is some separable admissible structure, even on $\mathcal{F}$ itself, by Theorem 10.2.1.

Notes. Theorem 10.2.1 is due to Aumann (1961), and the proof here, to B.V. Rao (1971), except that the "image admissible" part has been added here.

10.3 Suslin properties, selection, and a counter-example.

A Polish space is a topological space metrizable as a complete separable metric space. A separable measurable space $(Y,\mathcal{S})$ will be called a Suslin space if and only if there is a Polish space X with Borel σ-algebra $\mathcal{B}$ and a measurable map from X onto Y. (Such a space has also been called an analytic Borel space, see e.g. Christensen, 1974.) A subset of a measurable space Z is called a Suslin set if and only if it is a Suslin space with the relative σ-algebra (this terminology will only be used when Z is itself a Suslin space).

Given a measurable space $(X,\mathcal{B})$ and $M \subset X$, M is called universally measurable if and only if for every probability law P on $\mathcal{A}$, M is measurable for the completion of P, i.e. for some $A, B \in \mathcal{A}$, $A \subset M \subset B$ and $P(A) = P(B)$.

It is known that in a Polish space (or, with the above definitions, in any separable measurable space) all Suslin sets are universally measurable (e.g. Cohn, 1980, Theorem 8.4.1, in the light of Props. 8.1.6 and 8.2.3).

A real function f on X is called universally measurable if and only if for each Borel set $B \subset \mathbb{R}$, $f^{-1}(B)$ is universally measurable.

The following will be useful:

10.3.1 Selection Theorem. Let $(X,\mathcal{A})$ and $(Y,\mathcal{B})$ be Suslin measurable spaces and let H be a Suslin subset of $X \times Y$ with product σ-algebra. Let

$$\Pi_X H := \{x: \langle x,y\rangle \in H \text{ for some } y\}.$$

Let $\mathcal{S}(X)$ be the σ-algebra generated by all Suslin subsets of X. Then $\Pi_X H$ is a Suslin set and there is a function f from $\Pi_X H$ into Y whose graph $\{\langle x, f(x)\rangle : x \in \Pi_X H\} \subset H$ and such that for all $B \in \mathcal{B}$, $f^{-1}(B) \in \mathcal{S}(X)$.

For a proof of Theorem 10.3.1 see Cohn (1980, Theorem 8.5.3, p. 286).

Given a measurable space $(X, \mathcal{A})$ a collection $\mathcal{F}$ of measurable real functions on X will be called <u>admissible Suslin</u> if and only if $(X, \mathcal{A})$ is Suslin and there is an admissible structure $\mathcal{S}$ for $\mathcal{F}$ such that $(\mathcal{F}, \mathcal{S})$ is Suslin. $\mathcal{F}$ will be called <u>image admisssible Suslin</u> via $(Y, \mathcal{S}, T)$ if and only if we have:

a) $(X, \mathcal{A})$ and $(Y, \mathcal{S})$ are Suslin measurable spaces;
b) T maps Y onto $\mathcal{F}$, and
c) $\langle x,y\rangle \to T(y)(x)$ is jointly measurable.

Here is one way the Suslin property helps to make empirical measures measurable:

10.3.2 <u>Theorem.</u> Let $(X, \mathcal{A}, P)$ be a probability space and $\mathcal{F} \subset \mathcal{L}(X, \mathcal{A}, P)$ where $\mathcal{F}$ is image admissible Suslin via $(Y, \mathcal{S}, T)$. Let P_n and P'_m be independent copies of empirical measures for P (say $P'_m = ((n+m)P_{n+m} - nP_n)/m$). Let $\|G\|_{\mathcal{F}} := \sup_{f \in \mathcal{F}} |G(f)|$. Then for any real numbers a, b, $\|P_n - P\|_{\mathcal{F}}$ and $\|aP_n - bP'_m\|_{\mathcal{F}}$ are universally measurable random variables.

<u>Proof.</u> Take any $t \geqslant 0$ and let

$$H := \{\langle x,y\rangle : |(P_n - P)(T(y))| > t\}, \quad x := \langle x(1), \ldots, x(n)\rangle, \quad P_n := n^{-1}\textstyle\sum_{j=1}^{n} \delta_{x(j)}.$$

Then by image admissibility and a finite sum, $\langle x,y\rangle \to P_n(T(y))$ is jointly measurable. By Corollary 10.2.3, $y \to P(T(y))$ is measurable. Thus H is product measurable, hence Suslin (Cohn, 1980, Prop. 8.2.3). Its projection $\{x : \|P_n - P\|_{\mathcal{F}} > t\}$ is thus also Suslin, hence universally measurable as desired. The proof for $\|aP_n - bP'_m\|$ is similar. □

10.3.3 Next, here is an example showing that the Suslin property cannot simply be removed from 10.3.2. Let $\aleph_1$ be the least uncountable ordinal, a set well-ordered by a relation <. Let $\mathcal{C}$ be the collection of all the proper initial segments $\{\gamma : \gamma < \beta\}$, $\beta \in \aleph_1$. Then $\mathcal{C}$ consists of countable sets, and we can take $\aleph_1$ to be (in 1-1 correspondence with) a subset of $[0,1]$, by the axiom of choice (or, by the continuum hypothesis, all of $[0,1]$), where the ordering $<$ has no relation to any usual structure on $[0,1]$. Thus $\mathcal{C}$ is admissible by Theorem 10.2.1 with $\alpha = 2$, taking as generators all finite unions of intervals in $[0,1]$ with rational endpoints. Let P be a law on $\mathcal{H}_1$ which is 0 on countable sets and 1 on sets with countable complement (on any uncountable set, such a P is clearly defined and countably additive on a σ-algebra; under the continuum hypothesis we can take $\mathcal{H}_1 = [0,1]$ and P = Lebesgue measure or any other nonatomic law).

Now, let $P_1 = \delta_x$, $P_1' = \delta_y$ and $P_2' = (\delta_y + \delta_z)/2$ where x, y and z are coordinates in $(\mathcal{H}_1)^3$ with law P^3 (so x, y and z are i.i.d. P). Then $\sup_{A\in\mathcal{C}}(P_1-P_1')(A) = 1$ if and only if $x < y$. Now $\iint 1_{x<y}\, dP(x)dP(y) = 0$ but $\iint 1_{x<y}\, dP(y)dP(x) = 1$. Thus by Fubini's theorem, $T := \{\langle x,y\rangle : x < y\}$ must not be measurable, even for the completion of $P \times P$, and $\sup_{A\in\mathcal{C}}(P_1-P_1')(A)$ is non-measurable. (This "ordinal triangle" T also shows why some joint measurability is needed in Fubini's theorem).

Similarly, let $B := \{\langle x,y,z\rangle : \|P_1-P_2'\|_{\mathcal{C}} = 1\}$. Then $\langle x,y,z\rangle \in B$ if and only if either $x < \min(y,z)$ or $\max(y,z) < x$. So $\int 1_B(x,y,z)dP(x) = 1$ for all y, z and $\int 1_B(x,y,z)dP(y) = 1_{x<z}$, so $\iint 1_B(x,y,z)dP(y)dP(x) = 0$ for all z. Thus again by Fubini's theorem, B and $\|P_1-P_2'\|_{\mathcal{C}}$ must not be measurable even for the completion of the product measure $P \times P \times P$. Thus in Theorem 10.3.2 the Suslin assumption on $\mathcal{F}$ cannot simply be removed.

10.3.4 Effros Borel structure. Let us return to positive results. Let (X,d) be a separable metric space and $\mathcal{F}_0$ the collection of all non-empty closed subsets of X. Then $\mathcal{F}_0$ is admissible by Theorem 10.2.1, since the topology of X has a countable base. We may metrize the topology by some d which makes (X,d) totally bounded (e.g. Kelley, 1955, p. 125). If X is Polish, and h_d is the Hausdorff metric on $\mathcal{F}_0$ defined by d, then $(\mathcal{F}_0,h_d)$ is also Polish (Effros, 1965), hence a Suslin space, and the Borel σ-algebra of h_d is admissible (since $\{\langle x,F\rangle : x \in F\}$ is closed in the product of two separable metric spaces (X,d) and $(\mathcal{F}_0,h_d)$, its complement is a countable union of products of open sets). Reasonable subcollections of $\mathcal{F}_0$ will also be Suslin, hence admissible Suslin.

The Borel σ-algebra of h_d, called the <u>Effros</u> Borel structure on $\mathcal{F}_0$, is also generated by the sets $\{F: F \subset H\}$, $H \in \mathcal{F}_0$. Indeed, such sets are closed, hence Borel, for h_d. Let A be a countable dense set in X. Let $\mathcal{H}$ be the collection of all finite unions of closed balls with rational radii and centers in A. Then for any $F_1 \neq F_2$ in $\mathcal{F}_0$, there is some $H \in \mathcal{H}$ with $F_1 \subset H$, $F_2 \not\subset H$ or $F_2 \subset H$, $F_1 \not\subset H$. Thus the σ-algebra generated by the $\{F: F \subset H\}$ is the Borel σ-algebra of h_d by the Kuratowski isomorphism theorem (Parthasarathy, 1967, Cor. 3.3, p. 22).

On the other hand for U open, $\{F \in \mathcal{F}_0: F \subset U\}$ may not be a Borel or even Suslin set for h_d, e.g. in a separable infinite-dimensional Hilbert space X (Christensen, 1971, 1974).

Next, if $\mathcal{U}_0$ is the collection of open subsets of X other than X itself (which, like the empty closed set, may be adjoined as an additional discrete point if desired), then the Borel structure on $\mathcal{U}_0$ defined by the Effros Borel structure of the complements makes $\mathcal{U}_0$ admissible Suslin.

<u>Notes.</u> Cohn (1980, p. 261) defines an analytic set as a subset of a Polish space which is a continuous image of some Polish space. Cohn (1980, p. 288) calls a measurable space $(Y, \mathcal{S})$ <u>analytic</u> iff it is measurably isomorphic to some analytic set with relative Borel σ-algebra. Then it is clearly Suslin as defined above. Conversely, a Suslin measurable space is analytic (Cohn, 1980, Cor. 8.6.4, Prop. 8.6.5). Schwartz (1973, p. 96) calls a topological space <u>Suslin</u> iff it is Hausdorff and the continuous image of a Polish space. Then a measurable space $(Y, \mathcal{S})$ is analytic (or Suslin) iff $\mathcal{S}$ is the Borel σ-algebra of some Suslin topology (Cohn, 1980, Cor. 8.6.12, for "if"; "only if" follows from his definitions (cf. also Sion, 1960).

The older terminology "P$\in$ -Suslin" (Dudley, 1978; Kuelbs and Dudley, 1980) or "image P$\in$ -Suslin" (Dudley, 1981b) is now outdated since the measurability of open collections of subsets with respect to pseudometrics defined by P turns out to follow from admissibility (Theorem 10.2.3). Also, partly because of the extension from sets to functions, "$\in$" (membership) has been replaced by the more explicit "admissible".

Theorem 10.3.1 is an extension of results of Lusin and Sierpiński, and follows from a result of Sion (1960, Cor. 4.5); Cohn (1980, Theorem 8.5.3) is an expository reference. Example 10.3.3, with further properties, appeared earlier in Durst and Dudley (1981, Prop. 2.2). Regarding the Effros Borel structure as used here see Effros (1965), Christensen (1971, 1974) and Dudley (1978, sec. 4).

Chapter 11. Limit theorems on Vapnik-Červonenkis classes and using Kolčinskii-Pollard entropy.

11.1 Kolčinskii-Pollard entropy and Glivenko-Cantelli theorems.

For central limit theorems over Vapnik-Červonenkis and certain related classes the best sufficient conditions known up to now are those of Pollard (1982a), using the following form of "entropy" or "capacity".

Let $(X,\mathcal{A})$ be a measurable space and $\mathcal{F} \subset \mathcal{L}^0(X,\mathcal{A})$, the space of all real-valued measurable functions on X. Let $F_{\mathcal{F}}(x) := \sup\{|f(x)|: f \in \mathcal{F}\}$. Let $F \in \mathcal{L}^0(X,\mathcal{A})$ with $F \geq F_{\mathcal{F}}$ (only classes $\mathcal{F}$ for which such a finite valued F exists will be considered).

Let Γ be the set of all laws on X of the form $n^{-1}\sum_{j=1}^{n} \delta_{x(j)}$ for some $x(j) \in X$, $j = 1,\ldots,n$, and $n = 1,2,\ldots$, where the $x(j)$ need not be distinct. For $\delta > 0$, $0 < p < \infty$, and $\gamma \in \Gamma$ let

(11.1.1)
$$D_F^{(p)}(\delta,\gamma,\mathcal{F}) := \sup\{m: \text{for some } f_1,\ldots,f_m \in \mathcal{F}, \text{ and all } i \neq j,\ \int |f_i - f_j|^p d\gamma > \delta^p \int F^p d\gamma\}.$$

Let

$$D_F^{(p)}(\delta,\mathcal{F}) := \sup_{\gamma\in\Gamma} D_F^{(p)}(\delta,\gamma,\mathcal{F}).$$

Let $\mathcal{G} := \{f/F: f \in \mathcal{F}\}$, where $0/0$ is replaced by 0. Then $|g(x)| \leq 1$ for all $g \in \mathcal{G}$ and $x \in X$. Given F, p, and $\gamma \in \Gamma$ let $Q(B) := \int_B F^p d\gamma/\gamma(F^p)$, if $\gamma(F^p) > 0$. Then $Q := Q_\gamma$ is a law and for $1 \leq p \leq \infty$,

$$D_F^{(p)}(\delta,\gamma,\mathcal{F}) = D(\delta,\mathcal{G},d_{p,Q}),$$

where

$$d_{p,Q}(f,g) := (\int |f-g|^p dQ)^{1/p}.$$

The next few results will connect $D_F^{(p)}(\delta,\mathcal{F})$ with other ways of measuring the size of certain classes $\mathcal{F}$. First we have Vapnik-Červonenkis classes $\mathcal{C}$ of sets with $S(\mathcal{C}) < +\infty$ (Chap.9):

11.1.2 **Theorem.** If $\mathcal{C} \subset \mathcal{A}$, $S(\mathcal{C}) < +\infty$, $1 \leq p \leq \infty$, $F \in \mathcal{L}^p(X,\mathcal{A},P)$, $F \geq 0$, and $\mathcal{F} := \{F1_A: A \in \mathcal{C}\}$ then for any $w > S(\mathcal{C})$ there is a $K < \infty$ such that

$$D_F^{(p)}(\delta,\mathcal{F}) \leq K\delta^{-pw}, \quad 0 < \delta \leq 1.$$

Proof. Let $\gamma \in \Gamma$ and let G be the smallest set with $\gamma(G) = 1$. We may assume $F(x) > 0$ for some $x \in G$ since otherwise for any $A \geq 1$

$$D_F^{(p)}(\delta,\gamma,\mathcal{F}) = 1 \leq A\delta^{-wp}, \quad 0 < \delta \leq 1.$$

If $C(1),\ldots,C(m) \in \mathcal{C}$ are such that

$$d_{p,\gamma}(F1_{C(i)}, F1_{C(j)})^p > \delta^p \int F^p d\gamma, \quad i \neq j,$$

with m maximal, then for $Q = Q_\gamma$

$$Q(C(i)\Delta C(j)) = \int_{C(i)\Delta C(j)} F^p d\gamma / \int F^p d\gamma > \delta^p.$$

Then by maximality and Theorem 9.3.1,

$$D_F^{(p)}(\delta,\gamma,\mathcal{F}) \leq m \leq D(\delta^p,\mathcal{C},d_Q) \leq K(w,\mathcal{C})\delta^{-pw}. \qquad \square$$

Next, here is a kind of converse to Theorem 11.1.2:

11.1.3 **Proposition.** Suppose $1 \leq p < \infty$, $\mathcal{F} = \{1_B: B \in \mathcal{C}\}$ for some collection $\mathcal{C}$ of sets, $F \equiv 1$, and for some δ with $0 < \delta^p < 1/2$, $D_F^{(p)}(\delta,\mathcal{F}) < \infty$. Then $S(\mathcal{C}) < \infty$.

Proof. Suppose $\mathcal{C}$ shatters a set G with card $G = n = 2^m$ for some positive integer m. Let γ have mass $1/n$ at each point of G. Then G has m subsets $A(i)$, $i = 1,\ldots,m$, independent for γ, with $\gamma(A(i)) = 1/2$, $i = 1,\ldots,m$. Then for $i \neq k$, $\int |1_{A(i)} - 1_{A(k)}|^p d\gamma = 1/2$. So for each j in (11.1.1) we can take $f_j = 1_{A(j)}$. Thus

$$D_F^{(p)}(\delta,\mathcal{F}) \geq D_F^{(p)}(\delta,\gamma,\mathcal{F}) \geq m.$$

For large m this is impossible, so $S(\mathcal{C}) < \infty$. $\square$

Next let us consider families of functions of the form $\mathcal{F} = \{Fg: g \in \mathcal{G}\}$ where $\mathcal{G}$ is a family of functions totally bounded in the supremum norm

$$||g||_{sup} := \sup_{x \in X} |g(x)|,$$

with the associated metric $d_{sup}(g,h) := ||g-h||_{sup}$ and with $||g||_{sup} \leq 1$ for all $g \in \mathcal{G}$.

11.1.4 **Proposition.** For $1 \leq p < \infty$ and $0 < \varepsilon \leq 1$ then for any such $\mathcal{F}$,

$$D_F^{(p)}(\varepsilon,\mathcal{F}) \leq D(\varepsilon,\mathcal{G},d_{sup}).$$

Proof. If $d_{sup}(g,h) \leq \delta$ then for all x, $|Fg-Fh|^p(x) \leq \delta^p F(x)^p$, so the result follows from the definition (11.1.1). □

Next we come to the Kolčinskii-Pollard method of symmetrization of empirical measures. Given $n = 1,2,\ldots$, let $x_1,\ldots,x_{2n}$ be coordinates on $(X^{2n}, \mathcal{A}^{2n}, P^{2n})$, hence i.i.d. P. Let $\sigma(1),\ldots,\sigma(n)$ be random variables independent of each other and the x_i with $Pr(\sigma(i) = 2i) = Pr(\sigma(i) = 2i-1) = \frac{1}{2}$, $i = 1,\ldots,n$. Let $\tau(i) = 2i$ if $\sigma(i) = 2i-1$ and $\tau(i) = 2i-1$ if $\sigma(i) = 2i$. Let $x(i) := x_i$. Then the $x(\sigma(j))$ are i.i.d. P.

Let

$$P'_n := n^{-1}\textstyle\sum_{j=1}^n \delta_{x(\sigma(j))}, \quad P''_n := n^{-1}\sum_{j=1}^n \delta_{x(\tau(j))},$$

$$\nu'_n := n^{1/2}(P'_n - P), \quad \nu''_n := n^{1/2}(P''_n - P),$$

$$P^0_n := P'_n - P''_n, \quad \nu^0_n := n^{1/2}P^0_n.$$

Note that $P''_n = 2P_{2n} - P'_n$ and that ν'_n and ν''_n are two independent copies of ν_n. Then we have:

11.1.5 Symmetrization Lemma. Let $\zeta > 0$ and $\mathcal{F} \subset \mathcal{L}^2(X, \mathcal{A}, P)$ with $\int |f|^2 dP \leq \zeta^2$ for all $f \in \mathcal{F}$. Assume $\mathcal{F}$ is image admissible Suslin via some $(Y, \mathcal{S}, T)$. Then for any $\eta > 0$

$$Pr\{||\nu^0_n||_{\mathcal{F}} > \eta\} \geq (1 - \zeta^2\eta^{-2})Pr\{||\nu_n||_{\mathcal{F}} > 2\eta\}.$$

Proof. Note that the given events are measurable by Theorem 10.3.2. For $x = \langle x_1,\ldots,x_{2n}\rangle$ let $H := \{\langle x,f\rangle : |\nu''_n(f)| > 2\eta,\ f \in \mathcal{F}\}$. Then by admissibility, H is a product measurable subset of $X^{2n} \times \mathcal{F}$. The Suslin properties imply (Theorem 10.3.1) that there is a universally measurable selector h such that whenever $\langle x,f\rangle \in H$ for some $f \in \mathcal{F}$, $h(x) \in \mathcal{F}$ and $\langle x,h(x)\rangle \in H$. Let $x'_\tau := \langle x_{\tau(1)},\ldots,x_{\tau(n)}\rangle$. Then for some function h_1, $h(x) = h_1(x'_\tau)$ where $h_1(y)$ is defined if and only if $y \in J$ for some Suslin set $J \subset X^n$ (by Theorem 10.3.1). Let $\mathfrak{I}_n$ be the smallest σ-algebra with respect to which $x'_\tau(\cdot)$ is measurable. Then on the set where $x'_\tau \in J$,

$$Pr(||\nu^0_n||_{\mathcal{F}} > \eta | \mathfrak{I}_n) \geq Pr(|\nu'_n(h_1(x'_\tau)(\cdot)| \leq \eta | \mathfrak{I}_n).$$

Given $\mathfrak{I}_n$, i.e. given x'_τ, $h_1(x'_\tau)(\cdot)$ is a fixed function $f \in \mathcal{F}$ with

$\int|f|^2dP \leq \zeta^2$. Then since ν_n' is independent of $\mathfrak{J}_n$, we can apply Chebyshev's inequality to obtain

$$\Pr(|\nu_n'(f)| \leq \eta) \geq 1-(\zeta/\eta)^2.$$

Integrating gives

$$\Pr\{\|\nu_n^0\|_{\mathcal{F}} > \eta\} \geq (1-(\zeta/\eta)^2)\Pr\{\|\nu_n''\|_{\mathcal{F}} > 2\eta\}$$

which, since ν_n'' is a copy of ν_n, gives the result. □

Here is a law of large numbers (generalized Glivenko-Cantelli theorem):

11.1.6 Theorem. Let $(X,\mathcal{A},P)$ be a probability space, $F \in \mathcal{L}^1(X,\mathcal{A},P)$, and $\mathcal{G}$ a collection of measurable functions on X with $|g(x)| \leq 1$ for all $x \in X$, $g \in \mathcal{G}$. Let $\mathcal{F} := \{Fg\colon g \in \mathcal{G}\}$. Suppose $\mathcal{F}$ is image admissible Suslin via $(Y,\mathcal{S},T)$. Assume that

(11.1.7) $$D_F^{(1)}(\delta,\mathcal{F}) < \infty \quad \text{for all } \delta > 0.$$

Then

$$\lim_{n\to\infty}\|P_n-P\|_{\mathcal{F}} = 0 \text{ a.s.}$$

Proof. Again, $\|P_n-P\|_{\mathcal{F}}$ is measurable by Theorem 10.3.2.

Let $\mathcal{S}_n$ be the smallest σ-algebra for which all $P_j(f)$ are measurable for $j \geq n$ and $f \in \mathcal{F}$. By convexity, for all $n \geq 1$

$$E\|P_n-P\|_{\mathcal{F}} \leq E\|P_1-P\|_{\mathcal{F}} \leq 2P(F) < \infty.$$

For any $f \in \mathcal{F}$ we have

$$E((P_n-P)(f)|\mathcal{S}_{n+1}) = (P_{n+1}-P)(f)$$

(e.g. Ash, 1972, p. 310). Thus

$$E(\|P_n-P\|_{\mathcal{F}}|\mathcal{S}_{n+1}) \geq \sup_{f\in\mathcal{F}}|E((P_n-P)(f)|\mathcal{S}_{n+1})| = \|P_{n+1}-P\|_{\mathcal{F}}.$$

Hence $\{\|P_n-P\|_{\mathcal{F}},\mathcal{S}_n\}_{n\geq1}$ is a reversed submartingale. Being nonnegative, it converges almost surely and in $\mathcal{L}^1$ (Doob, 1953, Theorem 4.22, p. 329; or Hunt, 1966, p. 36, or Meyer, 1972, p. 37). Its limit is a.s. equal to a constant K by the Hewitt-Savage 0-1 law (Ash, 1972, p. 279). It remains to prove $K = 0$, using the main hypothesis (11.1.7). For this we need only show $\|P_n-P\|_{\mathcal{F}} \to 0$ in probability.

Given $\varepsilon > 0$, take M large enough so that $P(F1_{F>M}) < \varepsilon/4$. Then

$$||(P_n-P)1_{F>M}||_{\mathcal{F}} \leq ||P_n 1_{F>M}||_{\mathcal{F}} + P(F1_{F>M}) \leq (P_n+P)(F1_{F>M}) \to 2P(F1_{F>M}) < \varepsilon/2$$

almost surely.

Replacing each $f \in \mathcal{F}$ by $f1_{F\leq M}$ takes $\mathcal{F}$ onto another class of functions which is still image admissible Suslin since

$$\langle x,y\rangle \to T(y)(x)1_{F(x)\leq M}$$

is jointly measurable while the Suslin spaces $(X,\mathcal{Q})$ and $(Y,\mathcal{S})$ are unchanged.

So we may assume $F \leq M$.

Next apply Lemma 11.1.5 with

$$\zeta = M \quad \text{and} \quad \eta = n^{1/2}\varepsilon/4 > 2M \quad \text{for } n \text{ large.}$$

Then it is enough to show that $||P_n^0||_{\mathcal{F}} = ||P_n'-P_n''||_{\mathcal{F}} \to 0$ in probability.

In the definition of $D_F^{(1)}(\delta,\gamma,\mathcal{F})$ take $\delta = \varepsilon/(9M)$ and $\gamma = P_{2n}$. Choose functions $f_1,\ldots,f_m$ to satisfy (11.1.1), with

$$m = D_F^{(1)}(\delta,\gamma,\mathcal{F}) \leq D_F^{(1)}(\delta,\mathcal{F}).$$

One can choose the $f_j := f_{jn}^{(x)}$ to depend measurably on $x := \langle x_1,\ldots,x_{2n}\rangle$ as follows. For each $k \geq 1$, the measurable space $(Y,\mathcal{S})^k$ is Suslin. Let

$$B_k := \{\langle x,y_1,\ldots,y_k\rangle: P_{2n}(|T(y_i)-T(y_j)|) > \delta P_{2n}(F),\ i \neq j\}.$$

Then B_k is product measurable by image admissibility. Let $y^{(k)} := \langle y_1,\ldots,y_k\rangle$ and let A_k denote the projection $\{x: \langle x,y^{(k)}\rangle \in B_k$ for some $y^{(k)}\}$, $k \geq 2$; $A_1 := X^{2n}$. Then by selection (Theorem 10.3.1), A_k is a Suslin set and there is a universally measurable function η_k from A_k into Y^k such that $\langle x,\eta_k(x)\rangle \in B_k$ for all $x \in A_k$. Let $\eta(x) := \eta_k(x)$ if and only if $x \in A_k \backslash A_{k+1}$, $k \geq 1$. For each x let $k(x)$ be the unique k such that $x \in A_k \backslash A_{k+1}$. Thus we have the following:

11.1.8 <u>Lemma</u>. For each $x \in X^{2n}$, and $P_{2n} := (2n)^{-1}\sum_{j\leq 2n}\delta_{x_j}$, $D_F^{(1)}(\delta,P_{2n},\mathcal{F}) = k(x)$ where $k(\cdot)$ is universally measurable and for $k(x) = k$, η_k is a universally measurable function from X^{2n} into Y^k such that the functions f_i in (11.1.1) can be taken as $T((\eta_k(x))_i)$, $i = 1,\ldots,k$.

Now for all $f \in \mathcal{F}$ and

$$m = D_F^{(1)}(\delta,\gamma,\mathcal{F}) \quad (\gamma := P_{2n}),\ \min_{j\leq m}\gamma(|f-f_j|) \leq \delta P_{2n}(F).$$

For each j at which the minimum occurs,

$$|P_n^0(f) - P_n^0(f_j)| = |P_n'(f-f_j) - P_n''(f-f_j)|$$

$$\leq (P_n' + P_n'')(|f-f_j|) = 2P_{2n}(|f-f_j|) \leq 2\delta P_{2n}(F).$$

As $n \to \infty$, almost surely

$$2\delta P_{2n}(F) \to 2\delta P(F) < \varepsilon/4.$$

Since m remains bounded as $n \to \infty$,

$$\Pr\{\max_{j\leq m}|P_n^0(f_j)| > \varepsilon/4\}$$

$$\leq m\cdot\sup\{\Pr\{|P_n^0(f)| > \varepsilon/4\}: |f| \leq M\}$$

$$\leq m\cdot 2M^2(4/\varepsilon)^2/n < \varepsilon/4$$

for n large by Chebyshev's inequality. So for n large enough

$$\Pr\{\|P_n^0\|_{\mathcal{F}} > \varepsilon\} < \varepsilon, \text{ proving 11.1.6.} \quad \square$$

11.1.9 Corollary. Let $(X,\mathcal{A})$ be a Suslin measurable space and $\mathcal{C} \subset \mathcal{A}$ where $S(\mathcal{C}) < \infty$ and $\mathcal{C}$ is image admissible Suslin. Then for any probability law P on $\mathcal{A}$, we have

$$\lim_{n\to\infty} \sup_{A\in\mathcal{C}} |(P_n-P)(A)| = 0 \text{ a.s.}$$

Proof. In Theorem 11.1.6, take $F \equiv 1$ and apply Theorem 11.1.2 with $p = 1$. $\square$

Notes. Kolčinskii (1981a) defined $D_F^{(p)}(\delta,\gamma,\mathcal{F})$ when $F \equiv 1$ and $\gamma = P_n$. Pollard (1982a) independently defined it for general F, for $p = 2$, and for distinct $x(j)$. Both, in fact, used the minimal cardinality of an ε-net (see Sec. 6.0 above). Theorem 11.1.2 is due to Pollard (1982a, proof of Theorem 9). I do not know references for Props. 11.1.3 or 11.1.4. The symmetrization method is also due independently to Kolčinskii (1981a) and Pollard (1982a). Lemma 11.1.5 is adapted from Pollard (1982a, Lemma 11), who proves it for a countable class $\mathcal{F}$, thus avoiding measurability difficulties. From the countable case one can infer the result if there is a countable subset $\mathcal{H} \subset \mathcal{F}$ such that $\|\nu_n\|_{\mathcal{H}} = \|\nu_n\|_{\mathcal{F}}$ almost surely. Thus it suffices for ν_n on $\mathcal{F}$ to be separable in the sense of Doob. For most classes $\mathcal{F}$ arising in applications the empirical process is naturally separable. But the collection $\mathcal{C}$ of all singletons in $[0,1]$, for P = Lebesgue measure, appears as a rather regular collection for which the empirical process is not separable. In such a case it may appear unnatural to use a separable modification of the process, so that e.g. $P_1(\{x\}) = 0$ for all x, even $x = x_1$!

Theorem 11.1.6 extends Pollard (1982a, Theorem 12), and Wolfowitz (1954) for half-spaces in $\mathbb{R}^d$.

11.2 Vapnik-Červonenkis-Steele laws of large numbers.

Let $(X, \mathcal{A}, P)$ be a probability space and $\mathcal{C} \subset \mathcal{A}$. Let $\{x_n\}_{n \geq 1}$ be coordinates in $(X^\infty, \mathcal{A}^\infty, P^\infty)$, so that x_j are i.i.d. P.

For certain classes $\mathcal{C}$ with $S(\mathcal{C}) = +\infty$ one will have $\Delta^{\mathcal{C}}(\{x_1, \ldots, x_n\}) < 2^n$ with P^n-probability converging to 1. For such classes, with sufficient measurability properties, a law of large numbers will still hold. Here a main result is a as follows (on uniformly bounded classes of functions see 11.2.14 below):

11.2.1 Theorem. If $\mathcal{C}$ is image admissible Suslin the following are equivalent:

a) $||P_n - P||_{\mathcal{C}} \to 0$ a.s. as $n \to \infty$;

b) $||P_n - P||_{\mathcal{C}} \to 0$ in probability as $n \to \infty$;

c) $\lim_{n\to\infty} n^{-1} E \log \Delta^{\mathcal{C}}(\{x_1, \ldots, x_n\}) = 0$.

To begin the proof, here is a

Definition (J. M. Steele). For a finite set $F \subset X$ and collection $\mathcal{C} \subset 2^X$, let
$k^{\mathcal{C}}(F) := \max\{m\colon$ for some $G \subset F$, card $G = m$ and $\Delta^{\mathcal{C}}(G) = 2^m\}$.

11.2.2 Lemma. For any image admissible Suslin class $\mathcal{C}$ via $(Y, \mathcal{S}, T)$, $\Delta^{\mathcal{C}}(\{x_1, \ldots, x_n\})$ and $k^{\mathcal{C}}(\{x_1, \ldots, x_n\})$ are universally measurable.

Proof. For any decomposition of $\{1, \ldots, n\}$ into subsets, there is a measurable set of ordered n-tuples $\langle x_1, \ldots, x_n \rangle$ such that $x_i = x_j$ if and only if i and j belong to the same subset. It is enough to prove the Lemma on each such measurable set; specifically, the set on which the x_i are all different.

For each set $J \subset \{1, \ldots, n\}$, and $x := \langle x_1, \ldots, x_n \rangle$, $U(J) := \{\langle x, y \rangle\colon x_j \in T(y)$ if and only if $j \in J\}$ is product measurable by image admissibility. Thus its projection $\Pi U(J)$ into X^n is universally measurable (Suslin, see 10.3.1). Now

(11.2.3)
$$\Delta^{\mathcal{C}}(\{x_1, \ldots, x_n\}) = \sum_J 1_{\Pi U(J)}(x) \text{ and}$$
$$\{k^{\mathcal{C}}(\{x_1, \ldots, x_n\}) \geq m\} = \bigcup_{G \subset \{1, \ldots, n\}, \text{card } G = m} \bigcap_{H \subset G} \Pi U(H). \quad \square$$

The main step in Steele's proof uses Kingman's (1968, 1973) subadditive ergodic theorem. To state it, here is some terminology. Let $\mathbb{N}$ denote the set of nonnegative integers. A <u>subadditive process</u> is a doubly indexed set $\{x_{mn}\}_{0\leq m<n<\infty}$ of real random variables, $m, n \in \mathbb{N}$, $m < n$, such that

(11.2.4) $$x_{kn} \leq x_{km} + x_{mn} \quad \text{whenever} \quad k < m < n.$$

Let $x_{nn} := 0$ for all $n \in \mathbb{N}$. If instead of (11.2.4),

(11.2.5) $$x_{kn} \geq x_{km} + x_{mn}, \quad k < m < n,$$

then $\{x_{mn}\}_{0\leq m<n}$ is called <u>superadditive</u>.

A process which is both subadditive and superadditive is called <u>additive</u> and can clearly be written as

$$x_{kn} = \sum_{k<j\leq n} x_j, \quad x_j := x_{j-1,j},$$

i.e. one has just partial sums of a sequence of random variables.

A subadditive process $\{x_{mn}\}_{0\leq m<n}$ will be called stationary if the joint distribution of the x_{mn} is the same as that of the x_{m+1}, x_{n+1} (and hence as that of $x_{m+k,n+k}$ for each $k \in \mathbb{N}$).

Another useful hypothesis for such processes is:

(11.2.6) For each $n \in \mathbb{N}$, $E|x_{0n}| < +\infty$, and $\gamma := \inf_{n\geq 1} Ex_{0n}/n > -\infty$.

11.2.7 <u>Theorem</u>. (Kingman's subadditive ergodic theorem). Let $\{x_{mn}\}_{0\leq m<n}$ be a stationary subadditive process satisfying (11.2.6). Then as $n \to \infty$, x_{0n}/n converges a.s. and in $\mathcal{L}^1$ to a random variable y with $Ey = \gamma$. Let $\mathcal{S}$ be the σ-algebra of measurable events defined from the process $\{x_{mn}\}_{0\leq m<n}$ and invariant under $\{x_{mn}\} \to \{x_{m+1,n+1}\}$. Then $y = \lim_{n\to\infty} n^{-1}E(x_{0n}|\mathcal{S})$. Thus if $\mathcal{S}$ is trivial, $y = \gamma$ a.s.

Theorem 11.2.7 is proved in Kingman (1968). (No proof will be given here.) To indicate how 11.2.7 is used to prove 11.2.1 here is a result due essentially to Steele (1978):

11.2.8 <u>Theorem</u>. Let $X_1, X_2, \ldots$ be a stationary sequence of random variables with values in X and law P. Let $\mathcal{C}$ be image admissible Suslin, $\emptyset \neq \mathcal{C} \subset \mathcal{A}$. Then each of the following is a stationary subadditive process satisfying (11.2.6):

a) $D^{\mathcal{C}}_{mn} := \sup_{A\in\mathcal{C}} \left|\sum_{m<i\leq n} (1_A(X_i) - P(A))\right|;$

b) $\log\Delta^{\mathcal{C}}_{mn} := \log\Delta^{\mathcal{C}}(\{X_{m+1},\ldots,X_n\})$;

c) $k^{\mathcal{C}}_{mn} := k^{\mathcal{C}}(\{X_{m+1},\ldots,X_n\})$.

Proof. We have measurability by 10.3.2 in a) and by 11.2.2 in b) and c). Stationarity follows from that of $\{X_j\}$. Subadditivity is clear in a) and not difficult for b) and c). All three processes are nonnegative: in b), $\mathcal{C}$ non-empty implies $\Delta^{\mathcal{C}} \geq 1$, so (11.2.6) holds. □

Now let us prove Theorem 11.2.1. By Theorems 11.2.7 and 11.2.8, we have almost sure limits

$$\lim_{n\to\infty} D^{\mathcal{C}}_{0n}/n := c_1,$$

(11.2.9) $$\lim_{n\to\infty} (\log\Delta^{\mathcal{C}}_{0n})/n := c_2,$$

$$\lim_{n\to\infty} k^{\mathcal{C}}_{0n}/n := c_3.$$

Thus in Theorem 11.2.1, a) and b) are equivalent (as also shown in the proof of Theorem 11.1.6 via a reversed submartingale). It remains to prove b) equivalent to c) (Theorem 4 of Vapnik and Červonenkis, 1971).

To show c) implies b), given $0 < \varepsilon < 1$, first apply Pollard's symmetrization lemma 11.1.5 with $\zeta = 1$ and $\eta = \varepsilon n^{1/2}/2$ for n large enough so that $\eta \geq 2$, giving

$$\Pr\{||P_n - P||_{\mathcal{C}} > \varepsilon\} \leq 2\Pr\{||P'_n - P''_n||_{\mathcal{C}} > \varepsilon/2\}.$$

Thus it suffices to show $||P'_n - P''_n||_{\mathcal{C}} \to 0$ in probability (these variables are universally measurable by 10.3.2).

Let $\mathcal{A}^n$ be the smallest σ-algebra for which $x_1,\ldots,x_n$ are measurable. For any fixed set $A \in \mathcal{A}$, the conditional probability

$$\Pr_{A,\varepsilon,n} := \Pr\{|(P'_n - P''_n)(A)| > \varepsilon|\mathcal{A}^{2n}\}$$

$$= \Pr\{|\textstyle\sum_{j=1}^n s_j(\delta_{x_{2j}} - \delta_{x_{2j-1}})(A)| > n\varepsilon\}$$

where $s_j := 2 1_{\{\sigma(j)=2j\}} - 1 = \pm 1$ with probability $1/2$ each, independently of each other (Rademacher variables) and of the x_i. Thus by Hoeffding's inequality (2.2.4) with $a_j = -1$, 0 or 1, $\Pr_{A,\varepsilon,n} \leq \exp(-n\varepsilon^2/2)$.

For some n_0, $E\log\Delta^{\mathcal{C}}_{0n} < 2n\varepsilon^4$, $n \geq n_0$.

Then by Markov's inequality,

$$\Pr(\log\Delta^{\mathcal{C}}_{0n} > 2n\varepsilon^3) > \varepsilon.$$

Given $\mathcal{A}^{2n}$, the event $|(P_n'-P_n'')(A)| > \varepsilon$ is the same for any two sets A having the same intersection with $\{x_1,\ldots,x_{2n}\}$. Thus

$$\Pr\{||P_n'-P_n''||_{\mathcal{C}} > \varepsilon|\mathcal{A}^{2n}\} \leq \Delta_{0,2n}^{\mathcal{C}}\exp(-n\varepsilon^2/2).$$

Hence, for $\varepsilon < 1/8$ and $n \geq n_0$ large enough so that $\exp(-n\varepsilon^2/4) < \varepsilon$, we have

$$\Pr(||P_n'-P_n''||_{\mathcal{C}} > \varepsilon) < \varepsilon + \exp(2n\varepsilon^3 - n\varepsilon^2/2)$$

$$< \varepsilon + \exp(-n\varepsilon^2/4) < 2\varepsilon,$$

so c) implies b).

Now let us show b) implies c). By (11.2.9) we have

$$1 \geq n^{-1}\log\Delta_{0n}^{\mathcal{C}} \to c \quad \text{a.s.}$$

for some constant $c := c_2 \geq 0$. Thus $n^{-1}E\log\Delta_{0n}^{\mathcal{C}} \to c$ and we want to prove $c = 0$. Suppose $c > 0$. Given $\varepsilon > 0$, for n large enough

$$\Pr\{(2n)^{-1}\log\Delta_{0,2n}^{\mathcal{C}} > c/2\}$$

$$= \Pr\{\Delta_{0,2n}^{\mathcal{C}} > e^{nc}\} > 1-\varepsilon.$$

Next, to symmetrize,

$$\Pr\{||P_n'-P_n''||_{\mathcal{C}} > 2\varepsilon\}$$

$$\leq \Pr\{||P_n'-P||_{\mathcal{C}} > \varepsilon\} + \Pr\{||P_n''-P||_{\mathcal{C}} > \varepsilon\}$$

$$= 2\Pr\{||P_n'-P||_{\mathcal{C}} > \varepsilon\}.$$

So it will suffice to prove $||P_n'-P_n''||_{\mathcal{C}} \not\to 0$ in probability.

If $2 \leq k := [\alpha n]$ where $0 < \alpha < 1/2$, then by the inequality (2.2.9) we have

$$_{2n}C_{\leq k} \leq (2ne/k)^k \leq (3e/\alpha)^{\alpha n}.$$

As $\alpha \downarrow 0$, $(3e/\alpha)^{\alpha} \to 1$. Thus for α small enough, $(3e/\alpha)^{\alpha} < e^c$. Choose and fix such an $\alpha > 0$. Then for n large enough,

(11.2.11) $$n \geq 2/\alpha \quad \text{and} \quad (3e/\alpha)^{\alpha n} < e^{nc}.$$

Hence by Sauer's theorem (9.1.2), if $\Delta_{0,2n}^{\mathcal{C}} > e^{nc}$, then $k_{0,2n}^{\mathcal{C}} \geq [\alpha n]$. Fix n satisfying (11.2.11). Let $k := [\alpha n]$.

Now on an event U with $\Pr(U) > 1-\varepsilon$, there is a subset T of the indices $\{1,\ldots,2n\}$ such that card $T = k$, $\mathcal{C}$ shatters $\{x_i: i \in T\}$, and $x_i \neq x_j$ for $i \neq j$ in T. If there is more than one such T, let us select each of the possible T's with equal probability, using for this a random variable Y independent of x_j

and σ_j, $1 \leq j \leq 2n$. Then since x_j are i.i.d., T is uniformly distributed over its $\binom{2n}{k}$ possible values. For any distinct $j_i \in \{1,\ldots,2n\}$, $N := 2n$, we have, where the following equations are conditional on U,

$$\Pr(j_1 \in T) = k/N,$$

(11.2.12)
$$\Pr(j_1, j_2 \in T) = k(k-1)/N(N-1),$$

$$\Pr(j_i \in T,\ i = 1,2,3,4) = \binom{k}{4}/\binom{N}{4}.$$

Let M_n be the number of values of $j \leq n$ such that both $2j-1$ and $2j$ are in T. Then from (11.2.12),

$$EM_n = k(k-1)/2(N-1) = \alpha^2 n/4 + \mathcal{O}(1),\quad n \longrightarrow \infty;$$

$$EM_n^2 = k(k-1)/2(N-1) + n(n-1)\binom{k}{4}/\binom{N}{4},$$

and a bit of algebra gives

$$\sigma^2(M_n) = EM_n^2 - (EM_n)^2 = \mathcal{O}(n),\quad n \longrightarrow \infty.$$

Thus for $0 < \delta < \alpha^2/4$, by Chebyshev's inequality, $\Pr(M_n \geq \delta n) > 1-2\varepsilon$ for n large. On the event $U \cap \{M_n \geq \delta n\}$, let us make a measurable selection (10.3.1) of a sequence J of $[\delta n]$ values of i such that $J' := \bigcup_{i\in J}\{2i-1, 2i\} \subset T$. Here M_n and J are independent of the $\sigma(j)$. Now measurably select a set $A = A(\omega) = T(y(\omega)) \in \mathcal{C}$ such that $\{j \in J' : x_j \in A\} = \{\sigma(i) : i \in J\}$. Then

$$\sum_{i\in J} (\delta_{x_{\sigma(i)}} - \delta_{x_{\tau(i)}})(A) = [n\delta].$$

Here $y(\cdot)$ is measurable for the σ-algebra $\mathcal{B}_J$ generated by all the x_j, by Y, and by $\sigma(i)$ for $i \in J$. Conditional on $\mathcal{B}_J$,

$$\sum_{i\notin J} (\delta_{x_{\sigma(i)}} - \delta_{x_{\tau(i)}})(A) = \sum_{i\notin J} s_i a_i$$

where a_i are $\mathcal{B}_J$-measurable functions with values -1, 0 and 1, and s_i have values ± 1 with probability 1/2 each, independently of each other and of $\mathcal{B}_J$. Thus by Chebyshev's inequality

$$\Pr\{\textstyle\sum_{i\notin J} s_i a_i > n\delta/3 \,|\, \mathcal{B}_J\} \leq 9/(n\delta^2)$$

on the event where J is defined. Thus for n large

$$\Pr((P_n' - P_n'')(A(\omega)) > \delta/3) > 1 - 3\varepsilon$$

and $\|P_n' - P_n''\|_{\mathcal{C}} \not\to 0$ in probability. □

11.2.13 Theorem. In (11.2.9), if any c_i is 0, all three are 0.

Proof. In the last proof we saw that $c_1 = 0$ if and only if $c_2 = 0$ and that if $c_2 > 0$ then $c_3 > 0$. On the other hand if $c_3 > 0$, then for some $\delta > 0$, $k^{\mathcal{C}}_{0n} \geq \delta n$ for n large enough a.s., and then $\Delta^{\mathcal{C}}_{0n} \geq 2^{\delta n}$. Thus $c_2 \geq c_3 \log 2 > 0$. □

Remark. Blum's theorem, the special case of Theorem 6.1.5 for classes of sets, is (using 6.1.4): $N_I(\varepsilon, \mathcal{C}, P) < \infty$ for all $\varepsilon > 0$ implies $\|P_n - P\|^*_{\mathcal{C}} \to 0$ a.s. One might try to deduce this from Theorem 11.2.1, although it would seem any such proof must be longer than the original proof.

To extend Theorem 11.2.1 to classes of functions one can first extend (11.1.1) to $p = \infty$ as follows. For any law P on X we have the space $\mathcal{L}^\infty(X,P)$ of essentially bounded, measurable functions with seminorm

$$\|f\|_{\infty,P} := \inf\{M: P(|f| > M) = 0\} \leq \|f\|_{\sup}.$$

Then for F bounded and measurable let

$$D_F^{(\infty)}(\delta,\gamma,\mathcal{F}) := \sup\{m: \text{for some } f_1,\ldots,f_m \in \mathcal{F}, \text{ and all } i \neq j, \ \|f_i - f_j\|_{\infty,\gamma} > \delta\|F\|_{\infty,\gamma}\}.$$

Results of Vapnik and Červonenkis (1981) then support the following.

11.2.14 Conjecture. If $\|f\|_{\sup} \leq 1 := F$ for all $f \in \mathcal{F}$ and $\mathcal{F}$ is image admissible Suslin then the following are equivalent, as $n \to \infty$:

a) $\|P_n - P\|_{\mathcal{F}} \to 0$ a.s.;

b) $\|P_n - P\|_{\mathcal{F}} \to 0$ in probability;

c) $n^{-1} E \log D_1^{(\infty)}(\varepsilon, P_n, \mathcal{F}) \to 0$ for each $\varepsilon > 0$.

Here are a few remarks on this conjecture. Let the theory of subadditive processes be applied as in the rest of this section. Then equivalence of a) and b) should follow as before. Vapnik and Červonenkis (1981, main Theorem) state equivalence of b) and c) (it seems to me prudent to add a measurability hypothesis as I have). Parts of their proof can apparently be shortened as for 11.2.1 here.

The Conjecture, if true, directly implies 11.2.1. As a characterization, it should in principle imply 6.1.5 (at least under measurability conditions) and 11.1.6, but I do not know how long the proofs would be.

Notes. In Theorem 11.2.1, Vapnik and Červonenkis (1971) proved equivalence of b) and c). The proof here is apparently new, as the Kolčinskii-Pollard techniques allow some simplification of the Vapnik-Červonenkis proof. Steele (1978) proved equivalence of a) and b) using Kingman's theorem and proved (11.2.8) and (11.2.9). However, Steele's assumption that $||P_n-P||_{\mathcal{C}}$ is measurable has been strengthened to "image admissible Suslin". Some such strengthening seems needed in the proof.

11.3 Pollard's central limit theorem. Using D. Pollard's kind of entropy and law of large numbers (Sec. 11.1 above) will let us prove the following:

11.3.1 Theorem. Let $(X,\mathcal{A},P)$ be a probability space and $\mathcal{F} \subset \mathcal{L}^2(X,\mathcal{A},P)$. Let $\mathcal{F}$ be image admissible Suslin via $(Y,\mathcal{S},T)$ and have an envelope function $F_{\mathcal{F}}^* \leqslant F \in \mathcal{L}^2(X,\mathcal{A},P)$. Suppose that

(11.3.2) $$\int_0^1 (\log D_F^{(2)}(x,\mathcal{F}))^{1/2} dx < \infty.$$

Then $\mathcal{F}$ is a functional Donsker class for P.

Before we prove the theorem, here is a consequence:

11.3.3 Corollary (Jain and Marcus, 1975a). Let (K,d) be a compact metric space. Let $C(K)$ be the space of continuous real functions on K with supremum norm. Let $X_1, X_2, \ldots$ be i.i.d. random variables in $C(K)$. Suppose $EX_1(t) = 0$ and $EX_1(t)^2 < \infty$ for all $t \in K$. Assume that for some random variable M with $EM^2 < \infty$,

$$|X_1(s) - X_1(t)|(\omega) \leqslant M(\omega)d(s,t) \quad \text{for all } \omega \text{ and } s, t \in K.$$

Suppose that

(11.3.4) $$\int_0^1 (\log D(\varepsilon,K,d))^{1/2} d\varepsilon < \infty.$$

Then the central limit theorem holds, i.e. in $C(K)$, $\mathcal{L}(n^{-1/2}(X_1+\ldots+X_n))$ converges to some Gaussian law.

Proof. For a real-valued function h on K let

$$||h||_L := \sup_{s\neq t}|h(s)-h(t)|/d(s,t), \quad ||h||_{sup} := \sup_t |h(t)|,$$

$$||h||_{BL} := ||h||_L + ||h||_{sup}, \quad BL(K) := \{h \in C(K): ||h||_{BL} < \infty\}.$$

To apply Theorem 11.3.1, take as probability space $X = BL(K)$, $\mathcal{A}$ = σ-algebra induced by the Borel sets of $|| \,||_{sup}$ (or equivalently by evaluations at points of K). Let $P = \mathcal{L}(X_1)$, $F(h) := ||h||_{BL}$, $h \in X$. Then for any $s \in K$,

$$E||X_1||_{\sup}^2 \leq 2E(X_1(s)^2 + M(\omega)^2 \sup_t d(s,t)^2) < \infty,$$

and

$$E||X_1||_L^2 \leq EM(\omega)^2 < \infty, \quad \text{so} \quad EF^2 < \infty$$

(note that F is measurable, $F \in \mathcal{L}^2(X,\mathcal{A},P)$).

Let $\mathcal{G}$ be the collection of functions δ_t/F: $h \to h(t)/F(h)$, $h \in X$, where we replace $h(t)/F(h)$ by 0 if $h \equiv 0$ in $BL(K)$, and t runs through K. Then $|g(h)| \leq 1$ for all $g \in \mathcal{G}$ and $h \in X$. Let

$$\mathcal{F} := \{Fg: g \in \mathcal{G}\} = \{\delta_t := (h \to h(t)): t \in K\}.$$

For any $s,t \in K$ and $h \in X$,

$$|(\delta_s/F - \delta_t/F)(h)| \leq d(s,t).$$

Then by Prop. 11.1.4, for $0 < x \leq 1$,

$$D_F^{(2)}(x,\mathcal{F}) \leq D(x,\mathcal{G},d_{\sup}) \leq D(x,K,d).$$

Thus (11.3.2) holds and Theorem 11.3.1 applies to give that $\mathcal{F}$ is a functional Donsker class. Since $\mathcal{F}$ is the set of evaluations at points of K, uniform convergence over $\mathcal{F}$ (as in the definition of functional Donsker class) implies uniform convergence of functions on K. Since $BL(K) \subset C(K)$ which is complete for uniform convergence, the limiting Gaussian process G_P for our functional Donsker class must also have sample functions in $C(K)$ (almost surely). Since $C(K)$ is separable, the laws $\mathcal{L}(n^{-1/2}(X_1 + \ldots + X_n))$ are defined on all Borel sets of $C(K)$ and converge to $\mathcal{L}(G_P)$. □

Remark. The metric d may not be an "original", "given", "usual" metric on K but may be chosen, perhaps as a function $d = f(e)$ of the original metric e, where $f(x)$ may approach 0 slowly as $x \downarrow 0$, e.g. $f(x) = x^\varepsilon$ for $\varepsilon > 0$ or $f(x) = 1/\max(|\log x|,2)$. Thus one can increase the possibilities for obtaining the Lipschitz property of X_1 with respect to d, so long as (11.3.4) still holds.

Now to prove Theorem 11.3.1 we first have:

11.3.5 Lemma. Let $(X,\mathcal{A},P)$ be a probability space, $F \in \mathcal{L}^2(X,\mathcal{A},P)$ and $\mathcal{F} \subset \mathcal{L}^2(X,\mathcal{A},P)$ with $|f(x)| \leq F(x)$ for all $f \in \mathcal{F}$ and $x \in X$. Let $H := 4F^2$ and $\mathcal{H} := \{(f-g)^2: f,g \in \mathcal{F}\}$. Then $0 \leq \varphi(x) \leq H(x)$ for all $\varphi \in \mathcal{H}$ and $x \in X$, and for any $\delta > 0$,

$$D_H^{(1)}(4\delta,\mathcal{H}) \leq D_F^{(2)}(\delta,\mathcal{F})^2.$$

<u>Proof</u>. Clearly $0 \leq \varphi \leq H$, $\varphi \in \mathcal{H}$. Given any $\gamma \in \Gamma$, choose $m \leq D_F^{(2)}(\delta,\mathcal{F})$ and $f_1,\ldots,f_m \in \mathcal{F}$ such that (11.1.1) holds. For any $f, g \in \mathcal{F}$, take i and j such that

$$\max(\gamma((f-f_i)^2),\ \gamma((g-f_j)^2)) \leq \delta^2\gamma(F^2).$$

Then by the Cauchy-Bunyakovsky inequality,

$$\begin{aligned}\gamma((f-g)^2-(f_i-f_j)^2) &= \gamma((f-g-f_i+f_j)(f-g+f_i-f_j))\\ &\leq \gamma((f-f_i-(g-f_j))^2)^{1/2}4\gamma(F^2)^{1/2}\\ &\leq 8\delta\gamma(F^2) = 2\delta\gamma(H).\end{aligned}$$

Thus letting $h_{k(i,j)} := f_i-f_j$ where $k(i,j) := mi-m+j$, $i,j = 1,\ldots,m$, we get an approximation of all functions in $\mathcal{H}$, in the $\mathcal{L}^1(\gamma)$ norm, within $2\delta\gamma(H)$, by functions h_k^2, $k = 1,\ldots,m^2$, which implies the Lemma. □

Lemma 11.3.5 gives in particular that if $D_F^{(2)}(\delta,\mathcal{F}) < \infty$ for all $\delta > 0$ then $D_H^{(1)}(\varepsilon,\mathcal{H}) < \infty$ for all $\varepsilon > 0$. Thus hypothesis (11.3.1) lets us apply Theorem 11.1.6, with $\mathcal{F}$ there $= \mathcal{H}$.

11.3.6 <u>Proposition</u>. If $D_F^{(2)}(\delta,\mathcal{F}) < \infty$ for all $\delta > 0$ and $F \in \mathcal{L}^2(X,\mathcal{A},P)$, then $\mathcal{F}$ is totally bounded in $\mathcal{L}^2(X,\mathcal{A},P)$.

<u>Proof</u>. We may assume $\int F^2dP > 0$, as otherwise $\mathcal{F}$ is quite totally bounded. By Lemma 11.3.5 and Theorem 11.1.6 we have

$$\sup\{|(P_n-P)((f-g)^2)|:\ f,g \in \mathcal{F}\} \to 0 \text{ a.s., } n \to \infty.$$

Also, $\int F^2dP_{2n} \to \int F^2dP$ a.s., $n \to \infty$. Given $\varepsilon > 0$ take n_0 large enough and a value of P_{2n}, $n \geq n_0$, such that $\int F^2dP_{2n} < 2\int F^2dP$ and

$$\sup\{|(P_{2n}-P)((f-g)^2)|:\ f,g \in \mathcal{F}\} < \varepsilon/2.$$

Take $0 < \delta < (\varepsilon/(4P(F^2)))^{1/2}$ and choose $f_1,\ldots,f_m \in \mathcal{F}$ to satisfy (11.1.1) for $p = 2$ and $\gamma = P_{2n}$. Then for each $f \in \mathcal{F}$ we have for some j

$$\int(f-f_j)^2dP < \frac{\varepsilon}{2} + \int(f-f_j)^2dP_{2n} \leq \frac{\varepsilon}{2} + \delta^2\int F^2dP_{2n} < \varepsilon. \qquad \square$$

The total boundedness gives us condition 4.1.10(a); it remains to check the asymptotic equicontinuity condition 4.1.10(b). For this, given $\varepsilon > 0$, let us apply the symmetrization lemma 11.1.5 to sets $\mathcal{F}_{j,\delta}$ of the form

$$\{f-f_j: f \in \mathcal{F},\ \int (f-f_j)^2 dP < \delta^2\}$$

with $\delta > 0$ small enough, $\eta = \varepsilon/2$, and $\zeta = \varepsilon/4$. By Theorem 10.2.2 with $p = 2$, $\{y \in Y: T(y)-f_j \in \mathcal{F}_{j,\delta}\} \in \mathcal{S}$, and since a measurable subset of a Suslin space is Suslin with its relative Borel structure (Cohn, 1980, Props. 8.2.2, 8.2.3), $\mathcal{F}_{j,\delta}$ is image admissible Suslin. Thus 11.1.5 applies and we need only check 4.1.10(b) for the symmetrized ν_n^0 in place of ν_n. To do this we may prove 4.1.10(b) conditionally on P_{2n} for P_{2n} in a set with probability converging to 1 as $n \to \infty$. So by Lemma 11.3.5 and Theorem 11.1.6 again, $\mathcal{F}_{j,\delta}$ can be replaced (once the conditioning is justified) by

$$\mathcal{F}_{j,\delta,n} := \{f-f_j: f \in \mathcal{F},\ \int (f-f_j)^2 dP_{2n} < \delta^2\}$$

which, for each fixed P_{2n}, is image admissible Suslin.

Let $\mathcal{A}_{2n}$ be the smallest σ-algebra making $x = (x_1,\dots,x_{2n})$ measurable as in Pollard's symmetrization (11.1.5). Then given $\varepsilon, \eta > 0$ it will be enough to prove

(11.3.7) $\quad \Pr\{\Pr[\sup\{|\nu_n^0(f-g)|: f,g \in \mathcal{F},\ \int (f-g)^2 dP_{2n} < \delta^2\} > 3\eta \mid \mathcal{A}_{2n}] > 3\varepsilon\}$

$< 3\varepsilon$ for n large enough.

Given $\mathcal{A}_{2n}$, i.e. given x, let $\|f\|_{2n} := (P_{2n}(f^2))^{1/2}$. Let $\delta_i := 2^{-i}$, $i = 1,2,\dots$. Choose finite subsets $\mathcal{F}(1,x), \mathcal{F}(2,x),\dots$ of $\mathcal{F}$ such that for all i, and $f \in \mathcal{F}$,

(11.3.8) $$\min\{\|f-g\|_{2n}: g \in \mathcal{F}(i,x)\} \leq \delta_i \|F\|_{2n},$$

with $\mathrm{card}(\mathcal{F}(i,x)) \leq D_F^{(2)}(\delta_i, \mathcal{F})$.

We can write $\mathcal{F}(i,x) = \{g_{i,1}^{(x)},\dots,g_{i,k(i,x)}^{(x)}\}$ where by Lemma 11.1.8, we have $g_{ir}^{(x)} = T(y_{ir}(x))$, $k(i,\cdot)$ and $y_{ir}(\cdot)$ being universally measurable in x, where $y_{ir}(x)$ is defined iff $r \leq k(i,x)$.

For each $f \in \mathcal{F}$, let $f_i := g := g_{ir} \in \mathcal{F}(i,x)$ achieve the minimum in (11.3.8), with r minimal in case of a tie. Let $\bar{\mathcal{A}}^k$ denote the σ-algebra of universally measurable sets for laws defined on $\mathcal{A}^k$. For each $f = T(y) \in \mathcal{F}$ and i, we have $f_i = T(y)_i = g_{ir(x,y,i)}^{(x)}$ where $r(\cdot,\cdot,i)$ is $\bar{\mathcal{A}}^{2n} \times \mathcal{S}$ measurable. Thus $\langle u,x,y\rangle \to g_{ir(x,y,i)}^{(x)}(u)$ is $\bar{\mathcal{A}}^{2n+1} \times \mathcal{S}$ measurable.

Hence $\nu_n^0(T(y) - T(y)_i)$ is $\bar{\mathcal{A}}^{2n} \times \mathcal{S}$ measurable and thus equals some $\mathcal{A}^{2n} \times \mathcal{S}$ measurable function $G(x,y)$ for $x \notin V$ where $P^{2n}(V) = 0$. Then by the selection theorem 10.3.1, $\sup_y G(x,y)$ is universally measurable in x. Thus for each j,

(11.3.9) $$\sup_y |\nu_n^0(T(y) - T(y)_j)| \text{ is } P^{2n}\text{-measurable in } x.$$

Now $||f_i - f||_{2n} \to 0$ as $i \to \infty$ by (11.3.8), and for any fixed r, $f - f_r = \sum_{r<j<\infty} f_j - f_{j-1}$ pointwise on $S = \{x_1, \ldots, x_{2n}\}$.

Let $H_j := \log D_F^{(2)}(2^{-j}, \mathcal{F})$, $j = 1,2,\ldots$. The integral condition (11.3.2) is equivalent to $\sum_j 2^{-j} H_j^{1/2} < \infty$. For all x, card $\mathcal{F}(j,x) \leq \exp(H_j)$. Let

$$\eta_j := \max(j\delta_j, (576P(F^2)\delta_j^2 H_j)^{1/2}) > 0.$$

Then

(11.3.10) $$\sum_{j\geq 1} \eta_j < \infty,$$

(11.3.11) $$\eta_j^2 \geq 576P(F^2)\delta_j^2 H_j, \text{ and}$$

(11.3.12) $$\sum_{j\geq 1} \exp(-\eta_j^2/(288\delta_j^2 P(F^2)))$$

$$\leq \sum_{j\geq 1} \exp(-j^2/(288P(F^2))) < \infty.$$

Then

(11.3.13) $$\Pr\{\sup_{f\in\mathcal{F}} |\nu_n^0(f - f_r)| > \sum_{j>r} \eta_j | \mathcal{A}_{2n}\}$$
$$\leq \sum_{j>r} \exp(H_j)\exp(H_{j-1}) \sup_{f\in\mathcal{F}} \Pr\{|\nu_n^0(f_j - f_{j-1})| > \eta_j | \mathcal{A}_{2n}\}.$$

For a fixed j and f let

$$z_i := (f_j - f_{j-1})(x_{2i}) - (f_j - f_{j-1})(x_{2i-1}).$$

Then

$$\nu_n^0(f_j - f_{j-1}) = n^{-1/2} \sum_{i=1}^n (-1)^{e(i)} z_i$$

where $e(i) := 1_{\{\sigma(i)=2i-1\}}$ are random variables taking values 0 and 1 with probability 1/2 each, independently of each other and the z_i. Then by an inequality of Hoeffding (2.2.4 above)

$$\Pr\{n^{-1/2} |\sum_{i=1}^n (-1)^{e(i)} z_i| > \eta_j\}$$

$$\leq 2\exp(-\tfrac{1}{2} n\eta_j^2 / \sum_{i=1}^n z_i^2).$$

Now

$$\sum_{i=1}^n z_i^2 \leq 4n\int (f_j - f_{j-1})^2 dP_{2n} \leq 4n(||f - f_j||_{2n} + ||f - f_{j-1}||_{2n})^2$$
$$\leq 4n||F||_{2n}^2 (\delta_j + \delta_{j-1})^2$$

(by (11.3.8) and the few lines after it) $\leq 72n\delta_j^2 P(F^2)$ on

$B_n := \{||F||_{2n}^2 \leq 2P(F^2)\}$.

Then the last sum in (11.3.13) is less than

$$\textstyle\sum_{j>r} \exp(2H_j) 2 \exp(-\eta_j^2/(144\delta_j^2 P(F^2)))$$

$$\leq 2\textstyle\sum_{j>r} \exp(-\eta_j^2/(288\delta_j^2 P(F^2))) \quad \text{by (11.3.11)}$$

$< \varepsilon$ for r large enough by (11.3.12).

If r is also large enough so that $\sum_{j>r} \eta_j < \eta$ we obtain almost surely on B_n that, by (11.3.13),

(11.3.14) $$\Pr\{\sup_{f\in\mathcal{F}} |\nu_n^0(f-f_r)| > \eta | \mathcal{A}_{2n}\} < \varepsilon.$$

Next, if $||f-g||_{2n}^2 < \delta_r^2 P(F^2)$ and $||F||_{2n}^2 \leq 4P(F^2)$ then by choice of f_r (after 11.3.8)

$$\begin{aligned}||f_r-g_r||_{2n} &\leq ||f_r-f||_{2n} + ||f-g||_{2n} + ||g-g_r||_{2n}\\ &< \delta_r P(F^2)^{1/2} + 2\delta_r ||F||_{2n}\\ &\leq 5\,\delta_r (P(F^2))^{1/2}.\end{aligned}$$

Then by the same Hoeffding inequality (2.2.4) and with measurability as in (11.3.9), if $\delta^2 \leq \delta_r^2 P(F^2)$,

$$\Pr\{\sup\{|\nu_n^0(f_r-g_r)|: ||f-g||_{2n} < \delta,\ f,g\in\mathcal{F}\} > \eta | \mathcal{A}_{2n}\}$$

$$\leq (\text{card}\,\mathcal{F}(r,x))^2 2\exp(-\eta^2/8\sup||f_r-g_r||_{2n}^2)$$

(11.3.15) $$\leq 2\exp(2H_r - \eta^2/(200\delta_r^2 P(F^2)))$$

$$\leq 2\exp(-\eta^2/(400\delta_r^2 P(F^2))) \quad \text{if } \eta^2 \geq 800 H_r \delta_r^2 P(F^2)$$

$< \varepsilon$ for r large enough, on B_n,

noting that $H_r\delta_r^2 \to 0$ as $r \to \infty$ (see just before (11.3.10)). Clearly $\Pr(B_n) \to 1$ as $n \to \infty$. Now

$$\sup\{|\nu_n^0(f-g)|: f,g\in\mathcal{F},\ ||f-g||_{2n} < \delta\}$$

$$\leq 2\sup\{|\nu_n^0(f-f_r)| + |\nu_n^0(f_r-g_r)|: f,g\in\mathcal{F},\ ||f-g||_{2n} < \delta\}.$$

Thus from (11.3.14) and (11.3.15) we get (11.3.7) for $\delta = \delta_r P(F^2)^{1/2}$ and r large enough. The event in (11.3.7) is measurable as follows: as $Y \times Y$ is Suslin, and $\langle x,y,z\rangle \mapsto \int (T(y) - T(z))^2 dP_{2n}$ is jointly measurable by admissibility, so the supremum over a measurable subset is universally measurable as in Sec. 10.3. □

Now let us recall the classes of functions $\mathcal{G}_{\alpha,K,d}$ treated in Sec. 7.1, which are not functional Donsker classes if $\alpha < d/2$ by Theorem 8.1.1 for P = Lebesgue measure λ^d on the cube I^d. Thus the following is sharp:

11.3.16 Corollary. If $\alpha > d/2$ then $\mathcal{G}_{\alpha,K,d}$ is a functional Donsker class for any $K < \infty$, $d = 1,2,\ldots$, and any law P on I^d.

Proof. By Theorem 7.1.1 we have

$$\log D(\varepsilon, \mathcal{G}_{\alpha,K,d}, d_{sup}) \asymp \varepsilon^{-d/\alpha}, \quad \varepsilon \downarrow 0.$$

Next apply Prop. 11.1.4 with $p = 2$ and $F \equiv K$ to get

$$D_F^{(2)}(\varepsilon, \mathcal{G}_{\alpha,K,d}) \leq D(\varepsilon, \mathcal{G}_{\alpha,K,d}, d_{sup}).$$

Thus Theorem 11.3.1 gives the result. □

For $\alpha = d/2$, Theorem 8.1.1 still implies that $\mathcal{G}_{\alpha,K,d}$ is not a functional Donsker class for λ^d, using an additional result of J. Feldman (1971) on Gaussian processes: if a class $\mathcal{F}$ is G_PBUC, then for all $\varepsilon > 0$,

$$\Pr\{\sup_{f\in\mathcal{F}} |G_P(f)| < \varepsilon\} > 0$$

for a suitable version of G_P.

Kolčinskii (1981a) gives the following theorem: let $\mathcal{F}$ be a uniformly bounded class of measurable functions, say $|f| \leq F \equiv 1$ for all $f \in \mathcal{F}$. It is assumed that $\mathcal{F}$ satisfies suitable measurability conditions (see his paper for details). Suppose that for the $L_2(P)$ metric $d_{2,P}$ we have

i) $\int_0^1 (\log D(x, \mathcal{F}, d_{2,P}))^{1/2} dx < \infty$.

Suppose that for Kolčinskii's entropy (defined in Sec. 11.1) with respect to the empirical P_n we have for every $\varepsilon > 0$

ii) $\Pr\{u|\log u|^{1/2} \log D_1^{(1)}(\varepsilon u, P_{n(u)}, \mathcal{F}) \geq \varepsilon\} \to 0$

as $u \downarrow 0$, where $n(u) := [u^{-2}]$.

Then $\mathcal{F}$ satisfies the central limit theorem [is a functional Donsker class].

Note that i) implies $\mathcal{F}$ is a G_PBUC class (Dudley, 1973). Also, with $F \equiv 1$ here, $D_1^{(p)}(x, P_{n(u)}, \mathcal{F}) = D(x, \mathcal{F}, d_{p,P_{n(u)}})$.

Even when restricted to cases where both apply, with $F \equiv 1$ and $D_1^{(1)}$ replaced by its supremum over all P_n, it appears that neither Kolčinskii's theorem nor Pollard's (11.3.1) implies the other. As it seems less clear that Kolčinskii's result has reached a final form, I will not reproduce its proof, but future research seems likely to aim for conditions which, like his and those in Sec. 11.2, may not be uniform over all P_n.

Next, let us note that so far not all functional Donsker classes can be obtained directly from the results given above (in Chaps. 5-6 and this section). For a law P on $\mathbb{R}$ let $\mathcal{F} = \{x \to e^{itx}: |t| \leq 1/2\}$. Marcus (1981) in effect proved that the following are equivalent:

a) $\mathcal{F}$ is a functional Donsker class for P;

b) $\mathcal{F}$ is a G_PBUC class;

c) $\int_0^1 (\log D(x, \mathcal{F}, d_{2,P}))^{1/2} dx < \infty$,

where the main new step was to show that b) or c) implies a) in this case. This implication cannot follow from Theorem 11.3.1 since here b) and c) do depend on P. Here Theorem 6.4.1 is weaker than 6.2.1, which is hampered from applying by its x^2 and bracketing. Kolčinskii's theorem has its additional hypothesis ii). So it is a challenge to see if Marcus' theorem follows from some more general result which would also imply 6.2.1, 11.3.1 and/or Kolčinskii's theorem.

For relations of a)-c) to the tail behavior of P at $\pm\infty$ (under some conditions) see S. Csörgo (1981). Also on empirical characteristic functions see Ledoux (1982), Marcus and Philipp (1982).

In a special "lacunary" case, Marcus' theorem can be improved as follows:

11.3.17 <u>Theorem.</u> If P on $\mathbb{R}$ has $P\{2^m\} = p_m$, $m = 1,2,\ldots$, with $\sum_{m=1}^{\infty} p_m = 1$, and $\{G_m\}$ are i.i.d. $N(0,1)$, then a), b) and c) above are also equivalent to each of:

d) the series $\sum_{m\geq 1} (p_m(1-p_m))^{1/2} G_m(\omega)\exp(i2^m t)$ converges absolutely and uniformly on $\mathbb{R}$ a.s.;

e) $\sum_m p_m^{1/2} < \infty$.

Proof. Since the series in d) represents $G_P(f_t)$ where $f_t(x) = e^{itx}$, $f_t \in \mathcal{F}$, clearly d) implies b). The converse follows from a theorem of Feldman (1971).

Clearly e) implies d). The converse follows from a theorem of Paley and Zygmund, see Kahane (1968, p.77, Theorem 1). □

Comparing this last result with the Borisov theorem 6.3.1 above, one may ask if the strong analogy is more than superficial.

Notes. Pollard (1982a) proves Theorem 11.3.1 when the empirical processes ν_n are stochastically separable in the sense of Doob, as is usually true ab initio in cases of interest and can always be obtained by modifications of the process which may, occasionally, appear unnatural (see the Notes at the end of Sec. 11.1). The current "image admissible Suslin" formulation is being presented here for the first time. The proof, based on Pollard's, is somewhat different, not only as regards measurability, but notably in that Proposition 11.3.6 extends a more specific result of Pollard. Dudley (1981b) puts a stronger moment condition on $F^*_{\mathcal{F}}$ than in 11.3.1.

The implication from Pollard's theorem to that of Jain and Marcus is new as far as I know.

11.4 Necessary conditions for limit theorems.

Theorems 11.1.2 and 11.3.1 imply that every class $\mathcal{C} \subset \mathcal{A}$ with $S(\mathcal{C}) < +\infty$, and which is image admissible Suslin, is a functional Donsker class, for an arbitrary law P on $\mathcal{A}$. In this section it will be shown that to obtain, for all P, such a central limit theorem (or even the existence of the suitable Gaussian limit process G_P), the condition $S(\mathcal{C}) < +\infty$ is necessary. Then it will be noted that some measurability, beyond that of $||P_n - P||_{\mathcal{C}}$, is needed to obtain even a law of large numbers for $S(\mathcal{C}) < +\infty$ (11.1.8).

11.4.1 Theorem. Let $(X, \mathcal{A})$ be a measurable space and $\mathcal{C} \subset \mathcal{A}$. Suppose that for all laws P on $\mathcal{A}$, $\{1_A: A \in \mathcal{C}\}$ is a G_PBUC class (as defined in Sec. 2.1). Then $S(\mathcal{C}) < \infty$.

Proof. Suppose $S(\mathcal{C}) = +\infty$. Then for each $n \geq 1$, $\mathcal{C}$ shatters some set F_n with card $F_n = 4^n$. Let $G_n := F_n \setminus \bigcup_{j<n} F_j$. Then the sets G_n are disjoint, $\mathrm{card}(G_n) > 2^n$, and $\mathcal{C}$ shatters G_n. Take $E_n \subset G_n$ with $\mathrm{card}(E_n) = 2^n$. Then the E_n are disjoint and shattered by $\mathcal{C}$. Some countable subset $\mathcal{D} \subset \mathcal{C}$ shatters every E_n.

Let P be the law on $\bigcup_{j=1}^{\infty} E_n$ with $P(\{x\}) = 6/(\pi^2 n^2 2^n)$ for each $x \in E_n$.

Given n, for the process W_P (as in Sec. 2.1), for each $C \in \mathcal{D}$, $W_P(C) = W_P(C \cap E_n) + W_P(C \setminus E_n)$. For $0 < K < \infty$, define the events

$$\mathcal{E}_1 := \{|W_P(B)| \leq 2K \text{ for all } B \subset E_n\},$$

$$\mathcal{E}_2 := \{|W_P(B)| > 2K \text{ for some } B \subset E_n, \text{ and for all such } B \text{ and all } C \in \mathcal{D} \text{ with } C \cap E_n = B, |W_P(C \setminus E_n)| > K\}.$$

Then $\{|W_P(C)| < K \text{ for all } C \in \mathcal{D}\} \subset \mathcal{E}_1 \cup \mathcal{E}_2$.

Let $S_n := \sum\{|W_P(\{x\})| : x \in E_n\}$. Then $\sup\{|W_P(B)| : B \subset E_n\} \geq S_n/2$, so $\mathcal{E}_1 \subset \{S_n \leq 4K\}$. For each $x \in E_n$, $W_P(\{x\})$ has a Gaussian law with mean 0 and variance $6/(\pi^2 n^2 2^n) := \sigma_n^2$, so $E|W_P(\{x\})| = (2/\pi)^{1/2} \sigma_n$ and $\mathrm{var}(|W_P(\{x\})|) = \sigma_n^2(1 - \frac{2}{\pi})$. Thus $ES_n = 2^n (2/\pi)^{1/2} \sigma_n$. Since W_P has independent values on disjoint sets, $\mathrm{var}(S_n) = (6/\pi^2 n^2)(1 - \frac{2}{\pi})$. For n large, $ES_n \geq 4K$, and then by Chebyshev's inequality

$$\Pr\{S_n \leq 4K\} \leq \Pr\{|S_n - ES_n| \geq ES_n - 4K\} < \frac{1}{n^2 (ES_n - 4K)^2}$$

$$\leq 1/((2^n \cdot \frac{12}{\pi^3})^{1/2} - 4Kn)^2 := f(n,K) \to 0 \quad \text{as} \quad n \to \infty.$$

Turning to $\mathcal{E}_2$, let $t(n) := 2^{2^n}$. Let the subsets of E_n be $B_1, \ldots, B_{t(n)}$. Let $M_0 := \emptyset$ and recursively for $j \geq 1$, $M_j := \{|W_P(B_j)| > 2K\} \setminus \bigcup_{0 \leq i < j} M_i$. Let

$$D_j := M_j \cap \{\text{for all } C \in \mathcal{D} \text{ such that } C \cap E_n = B_j, |W_P(C \setminus E_n)| > K\}.$$

For any set A, $W_P(A)$ has a Gaussian distribution with mean 0 and variance $P(A)$. Let $\Phi(x) := (2\pi)^{-1/2} \int_{-\infty}^{x} \exp(-t^2/2)\,dt$. Then since W_P has independent values on disjoint sets,

$$\Pr(D_j) = \Pr(M_j)\Pr(\text{for all } C \in \mathcal{D} \text{ with } C \cap E_n = B_j, |W_P(C \setminus E_n)| > K\}$$

$$\leq \Pr(M_j) \cdot 2\Phi(-K).$$

Now $\mathcal{E}_2 \subset \bigcup_{1 \leq j \leq t(n)} D_j$, so

$$\Pr(\mathcal{E}_2) \leq \sum_{1\leq j\leq t(n)} \Pr(M_j)\cdot 2\Phi(-K)$$
$$= 2\Phi(-K)\Pr(|W_P(B)| > 2K \text{ for some } B \subset E_n) \leq 2\Phi(-K).$$

Hence

$$\Pr(|W_P(C)| < K \text{ for all } C \in \mathcal{D}) \leq f(n,K) + 2\Phi(-K).$$

If we let $n \to +\infty$, then $K \to +\infty$, we see that W_P is a.s. unbounded on $\mathcal{D} \subset \mathcal{C}$, and apply (2.1.2). □

11.4.2. Now let us recall example 10.3.3, where $\mathcal{C}$ is the collection of all countable initial segments of an uncountable well-ordered set $(X,<)$ and P is a continuous law on some σ-algebra $\mathcal{A}$ containing all countable subsets of X. Then $S(\mathcal{C}) = 1$ but $\sup_{A\in\mathcal{C}}|(P_n-P)(A)| \equiv 1$ for all n. Thus the latter random variable is measurable. For this class the weak law of large numbers, hence the strong law and central limit theorem, all fail as badly as possible. This shows that in Theorem 11.1.6 and Cor. 11.1.8, the "image admissible Suslin" condition cannot simply be removed, nor replaced by simple measurability of random variables appearing in the statements of the results. Further, for all $A \in \mathcal{C}$, $1_A = 0$ a.s. (P), strengthening Remark 6.1.6B in one sense: vanishing a.s. (P) even with $S(\mathcal{C}) = 1$ does not imply a law of large numbers. Of course, central limit theorems such as the functional Donsker property also fail for this class $\mathcal{C}$.

11.4.3 <u>Remark.</u> If X is a countably infinite set and $\mathcal{A} = 2^X$, then for an arbitrary law P on $\mathcal{A}$, $\lim_{n\to\infty} \sup_{A\in\mathcal{A}}|(P_n-P)(A)| = 0$ a.s., but $S(\mathcal{A}) = +\infty$, so the hypothesis of 11.4.1 cannot be weakened to a law of large numbers for all P.

<u>Note.</u> The results of this section are from Durst and Dudley (1981).

Chapter 12 Non-parametric statistics and speeds of convergence.

This chapter has the character of a summary of some topics which, for lack of time, could not be covered more thoroughly.

12.1 Non-parametric statistics.

Suppose a statistician has observed $x_1,\ldots,x_n$ i.i.d. P and has no other information about P. To make inferences about the unknown law P, all the useful information in the observations is contained in the empirical law P_n (Sec. 10.1). For any measurable set A, we have $P_n(A) \to P(A)$ a.s. as $n \to \infty$. For approximation of the unknown $P(A)$ to be useful, it is helpful if, first, we can approximate $P(A)$ by $P_n(A)$ uniformly for A in some non-trivial family $\mathcal{C}$ of measurable sets; and, second, if we can bound the speed of convergence $||P_n-P||_{\mathcal{C}} \to 0$.

If $\mathcal{C}$ is a functional Donsker class for P, then $||P_n-P||_{\mathcal{C}}$ is of order $\mathcal{O}_p(1/n^{1/2})$ (where $U_n = \mathcal{O}_p(a_n)$ means: for any $\varepsilon > 0$ there is a $K < \infty$ such that $\Pr(U_n > Ka_n) < \varepsilon$ for all n large enough). This rate of convergence is, up to constant multiples, as good as can be obtained for one set A with $0 < P(A) < 1$.

For $\mathcal{C}$ to be a functional Donsker class for all laws P on σ-algebras $\mathcal{A} \supset \mathcal{C}$, it is sufficient that $S(\mathcal{C}) < \infty$ and $\mathcal{C}$ be image admissible Suslin (by Theorems 11.1.2 and 11.3.1). On the other hand it is necessary that $S(\mathcal{C}) < \infty$ and $\mathcal{C}$ satisfy some further condition (Sec. 11.4). In this sense, Vapnik-Červonenkis classes $\mathcal{C}$ (those with $S(\mathcal{C}) < \infty$), under suitable measurability conditions, are the natural classes of events for non-parametric statistics. Further, there are inequalities of the form

$$(12.1.1) \qquad \Pr\{||\nu_n||_{\mathcal{C}} > M\} \leq K(S(\mathcal{C}),\delta)\exp(-(2-\delta)M^2)$$

(Alexander, 1982), where the $\delta > 0$ cannot be replaced by 0 except in special cases, if $\mathcal{C}$ contains sets A with $P(A) \approx 1/2$. Still the values of the constants $K(S(\mathcal{C}),\delta)$ might be quite large. More optimal constants would help actual applications.

Devroye (1982) proves the inequality

$$(12.1.2) \qquad \Pr\{||\nu_n||_{\mathcal{C}} > M\} \leq 4m^{\mathcal{C}}(n^2)\exp(-2M^2 + 4Mn^{-1/2} + 4M^2n^{-1}),$$

sometimes improving on the Vapnik-Červonenkis (1971) inequality

(12.1.3) $\Pr\{\|\nu_n\|_{\mathcal{C}} > M\} \leq 4m^{\mathcal{C}}(2n)\exp(-M^2/8)$.

The latter two inequalities (unlike 12.1.1) become useless as $n \to \infty$ for fixed M, but it may be that inequalities like (12.1.2) will be more useful statistically than those like (12.1.1).

Given two independent empirical laws P_m and Q_n, one may ask whether the unknown laws P and Q are the same, rejecting the hypothesis $P = Q$ if $\|P_m - Q_n\|_{\mathcal{C}}$ is too large. How large is "too large" remains to be clarified in general. For concrete classes $\mathcal{C}$, Monte Carlo experiments should indicate better bounds than those obtainable from (12.1.1) - (12.1.3).

In the classical case $X = \mathbb{R}$, $\mathcal{C} = \{]-\infty, x]: x \in \mathbb{R}\}$, the quantities $\|P_n - P\|_{\mathcal{C}}$, $\|P_m - Q_n\|_{\mathcal{C}}$ and some related ones are called Kolmogorov-Smirnov statistics. Here the distribution of $\|P_n - P\|_{\mathcal{C}}$ is the same for all nonatomic laws P, and stochastically smaller for all laws having atoms. This helpful property does not carry over to general classes $\mathcal{C}$. It does hold, however, for suitable $\mathcal{C}$ and P restricted to some subsets; for example, if P is a Gaussian law on $\mathbb{R}^d$ with unknown mean and covariance matrix, and $\mathcal{C}$ the class of all half-spaces, or of ellipsoids in $\mathbb{R}^d$.

Specific methods in non-parametric statistics (hypothesis testing, especially the chi-squared test; classification or pattern recognition; density estimation, and clustering) have been briefly surveyed in this connection elsewhere (Dudley, 1982, Sec.5).

12.2 <u>Speeds of convergence in central limit theorems</u>. For a functional Donsker class $\mathfrak{F}$ one can ask how fast is the convergence to 0 in probability (2.1.3) or almost surely: in (2.1.4), when can $(nLLn)^{1/2}$ be replaced by a smaller sequence such as $n^{1/2-\varepsilon}$, $\varepsilon > 0$? Here are statements of results in that direction.

12.2.1 <u>Theorem</u>. Let $\mathfrak{F} \subset \mathcal{L}^2(X, \mathcal{A}, P)$ and assume for some $0 < \delta \leq 1$ that $E(F_{\mathfrak{F}}^*)^{2+\delta} < \infty$. Let $X_j := \delta_{x_j} - P$, $j = 1,2,\ldots$, on $\mathfrak{F}$. Suppose that for all $m \geq 1$ there is a linear map Λ_m, defined on the linear span S of $\{\delta_x - P: \mathfrak{F} \to \mathbb{R}, x \in X\}$, with values in a finite-dimensional vector space of functions on $\mathcal{F}$, and satisfying the following for some constants $C_i \geq 1$, $D > 2$ and $\beta > 0$:

(12.2.2) $\sup\{||\Lambda_m g||_{\mathcal{F}}: \quad m \geq 1, \quad ||g||_{\mathcal{F}} \leq 1\} < \infty;$

(12.2.3) for each $m \geq 1$ there is an $n_0(m) \leq C_1 m^D$ such that for all $n \geq n_0(m)$

$$\Pr^*\{n^{-1/2}||\sum_{j\leq n} X_j - \Lambda_m X_j||_{\mathcal{F}} \geq 1/m\} \leq 1/m;$$

(12.2.4) for each $m \geq 1$, $\Lambda_m X_1$ is a measurable random variable with $E\Lambda_m X_1 = 0$.

(12.2.5) $\dim \Lambda_m S \leq C_2 \exp(C_3 m^\beta)$, $m \geq 1$.

Then $\mathcal{F}$ is a functional Donsker class and in (2.1.4), $(nLLn)^{1/2}$ can be replaced by $n^{1/2}(\log n)^{-\theta}$ for any $\theta < 1/(2\beta)$.

If instead of (12.2.5), for some $\gamma \geq 1$, $C_4 < \infty$

(12.2.6) $\dim \Lambda_m S \leq C_4 m^\gamma$,

then in (2.1.4) one can replace $(nLLn)^{1/2}$ by $n^{1/2-\lambda}$ where $\lambda = \min(\delta, 4/(D-2))^2/(600\gamma)$.

Theorem 12.2.1 is proved in Dudley and Philipp (1982, Sec. 6). The proof, being rather long, will not be reproduced here. I only remark here that, in general, one needs to make one choice of Y_j to obtain a weak invariance principle (2.1.3), and a possibly different choice for the strong invariance principle (2.1.4). The rates of convergence in Theorem 12.2.1, though slow, are fast enough to allow Y_j to be chosen to satisfy both (2.1.3) and (2.1.4) at once.

Theorem 12.2.1 can be applied to give rates of convergence for families $\mathcal{C}$ of sets satisfying a metric entropy with inclusion condition (cf. Sec. 6.1 above) or $S(\mathcal{C}) < \infty$ (Dudley and Philipp, 1982, Sec. 7). Comparison with others' results in particular cases suggests that our rates are far from optimal.

The proof of Theorem 12.2.1 uses rates of convergence in the finite-dimensional central limit theorem due to Yurinskii (1977) and Dehling (1980).

INDEX OF NOTATION AND TERMINOLOGY

After each term is listed the section in which it is first defined. Greek letters are listed just after the corresponding Latin letters. Notations containing no letters (e.g. norm signs) are listed at the end, but if there are some letters, as in $||\cdot||_{\mathcal{F}}$, the first letter appearing is used. Most terms appearing in only one section are not listed here.

REFERENCES

The numbers at the end of each reference give the section(s) in which it is cited.

Alexander, Kenneth (1982). Some limit theorems and inequalities for weighted non-identically distributed empirical processes. Ph. D. thesis, Mass. Inst. of Tech. (12.1).

Ash, Robert B. (1972). Real Analysis and Probability. Academic Press. N.Y. (11.1).

Assouad, P. (1981). Sur les classes de Vapnik-Červonenkis. C.R. Acad. Sci. Paris 292 Sér. I 921-924. (9.1, 9.2, 9.3).

Assouad, P. (1983). Densité et dimension. Ann.Inst.Fourier(Grenoble)33 n°3 233-282. (9.1, 9.2, 9.3).

Aumann, R.J. (1961). Borel structures for function spaces. Illinois J. Math. 5 614-630. (10.2).

Bahvalov [Bakhvalov], N.S. (1959). On approximate calculation of multiple integrals (in Russian). Vestnik Moskov. Univ. Ser. Mat. Meh. Astron. Fiz. Khim. 1959 n°. 4, 3-18. (8.1).

Banach, Stefan, and K. Kuratowski (1929). Sur une généralisation du problème de la mesure. Fund. Math. 14 127-131. (3.1).

Bennett, George W. (1962). Probability inequalities for the sum of bounded random variables. J. Amer. Statist. Assoc. 57, 33-45. (2.2).

Berkes, István, and Walter Philipp (1977). An almost sure invariance principle for the empirical distribution function of mixing random variables. Z. Wahrsch. verw. Geb. 41,115-137. (1.2).

Bernštein, Sergei N. (1924). Ob odnom vidoizmenenii neravenstva Chebysheva i o pogreshnosti formuly Laplasa (in Russian). Uchen. Zapiski Nauchn.-issled. Kafedr Ukrainy, Otdel. Mat., vyp. 1, 38-48; reprinted in S. N. Bernštein, Sobranie Sochineniĭ [Collected Works], Tom IV, Teoriya Veroiatnostei, Matematicheskaya Statistika, Nauka, Moscow, 1964 pp. 71-79. (2.2).

Bernštein, Sergei N. (1927). Teoriya Veroiatnostei (in Russian). Moscow. (2.2).

Billingsley, Patrick (1968). Convergence of Probability Measures. Wiley, N.Y. (1.1).

Birgé, Lucien (1982). Personal communication. (7.2).

Blum, J.R. (1955). On the convergence of empiric distribution functions. Ann. Math. Statist. 26 527-529. (6.1).

Borisov, I.S. (1981 a). On the accuracy of the approximation of empirical fields. Theor. Probability Appls. 26 632-633 (English), 641-642 (Russian). (1.1).

Borisov, I.S. (1981 b). Some limit theorems for empirical distributions (in Russian). Abstracts of Reports, Third Vilnius Conf. Probability Th. Math. Statist. I, 71-72 (6.3).

Borovkov, A.A., and V.S. Korolyuk (1965). On the results of asymptotic analysis in problems with boundaries. Theor. Probability Appls. 10 236-246 (English), 255-266 (Russian). (1.1).

Bourbaki, N. (1956). Eléments de mathématique, Première partie, livre VI, Intégration, Chap. 5, Intégration des mesures. Hermann, Paris. (1.2).

Breiman, Leo (1968). Probability. Addison-Wesley, Reading, Mass. (1.1, 3.2).

Bronštein [Bronshtein] , E.M. (1976). ε-entropy of convex sets and functions. Siberian Math. J. 17 393-398 = Sibirskii Mat. Zh. 17 508-514. (7.3).

Chernoff, Herman (1952). A measure of asymptotic efficiency for tests of a hypothesis based on the sum of observations. Ann. Math. Statist. 23 493-507 (2.2).

Christensen, J.P.R. (1971). On some properties of Effros Borel structure on spaces of closed subsets. Math. Ann. 195 17-23. (10.3).

Christensen, J.P.R. (1974). Topology and Borel Structure. North-Holland, Amsterdam; American Elsevier, N.Y. (10.3).

Clements, G.F. (1963). Entropies of several sets of real valued functions. Pacific J. Math. 13 1085-1095. (7.1).

Cohn, Donald L. (1980). Measure theory. Birkhäuser, Boston, Basel. (10.3, 11.3).

Csörgő Sándor (1981). Limit behavior of the empirical characteristic function. Ann. Probability 9 130-144. (11.3).

Danzer, L., B. Grünbaum and V. L. Klee (1963). Helly's theorem and its relatives. Proc. Symp. Pure Math. (Amer. Math. Soc.) 7 101-180. (9.2).

Dehardt, J. (1971). Generalizations of the Glivenko-Cantelli theorem. Ann. Math. Statist. 42 2050-2055. (6.1).

Dehling, Herold (1983). Limit theorems for sums of weakly dependent Banach space valued random variables. Z. Wahrsch. verw. Geb. 63 393-342. (12.2).

Devroye, Luc (1982). Bounds for the uniform deviation of empirical measures. J. Multivariate Analysis 12 72-79. (12.1).

Donsker, Monroe D. (1951). An invariance principle for certain probability limit theorems. Mem. Amer. Math. Soc. 6. (1.1).

Donsker, Monroe D. (1952). Justification and extension of Doob's heuristic approach to the Kolmogorov-Smirnov theorems. Ann. Math. Statist. 23 277-281 (1.1).

Doob, J.L. (1953). Stochastic Processes. Wiley, N.Y. (11.1).

Dudley, R.M. (1966). Weak convergence of probabilities on nonseparable metric spaces and empirical measures on Euclidean spaces. Illinois J. Math. 10 109-126. (1.1, 4.1).

Dudley, R.M. (1967). The sizes of compact subsets of Hilbert space and continuity of Gaussian processes. J. Functional Analysis 1 290-330. (2.1, 4.1).

Dudley, R.M. (1973). Sample functions of the Gaussian process. Ann. Probability 1 66-103. (2.1, 11.3).

Dudley, R.M. (1974). Metric entropy of some classes of sets with differentiable boundaries. J. Approximation Th. 10 227-236, Correction 26 (1979) 192-193. (7.1, 7.3).

Dudley, R.M. (1978). Central limit theorems for empirical measures. Ann. Probability 6 899-929; Correction 7 (1979) 909-911. (4.1, 5.1, 6.2, 9.3, 10.3).

Dudley, R.M. (1981 a). Donsker classes of functions. Statistics and Related Topics (Proc. Symp. Ottawa, 1980), North-Holland, N.Y., 341-352. (4.1, 5.2, 6.4).

Dudley, R.M. (1981 b). Vapnik-Červonenkis Donsker classes of functions. Aspects statistiques et aspects physiques des processus gaussiens (Proc. Colloque C.N.R.S. St-Flour, 1980), C.N.R.S., Paris, 251-269, (10.3, 11.3).

Dudley, R.M. (1982). Some recent results on empirical processes. Probability in Banach spaces III (Proc. Conf. Tufts Univ., 1980), Lecture Notes in Math. (Springer) 860 107-123. (12.1).

Dudley, R.M. (1982). Empirical and Poisson processes on classes of sets or functions too large for central limit theorems. Z. Wahrsch. verw. Geb. 61 355-368. (8.1, 8.2, 8.3, 8.4).

Dudley, R.M., and Walter Philipp (1983). Invariance principles for sums of Banach space valued random elements and empirical processes. Z. Wahrschein. verw. Geb. 62 509-552. (1.1, 3.1, 3.2, 3.3, 4.1, 4.2, 12.1, 12.2).

Durst, Mark, and R.M. Dudley (1981). Empirical processes, Vapnik-Chervonenkis classes and Poisson processes. Prob. Math. Statist. (Wrocław) 1 n° 2, 109-115. (6.3, 10.3, 11.4).

Eames, W., and L.E. May (1967). Measurable cover functions. Canad. Math. Bull. 10 519-523 (3.1).

Effros, E.G. (1965). Convergence of closed subsets in a topological space. Proc. Amer. Math. Soc. 16 929-931. (10.3).

Eršov, M.P. (1975). The Choquet theorem and stochastic equations. Analysis Math. 1, 259-271. (1.2).

Evstigneev, I.V. (1977). "Markov times" for random fields. Theor. Probability Appls. 22 563-569 = Teor. Veroiatnost. i Primenen. 22 575-581 (8.3).

Feldman, J. (1971). Sets of boundedness and continuity for the canonical normal process. Proc. Sixth Berkeley Symp. Math. Statist. Prob. 2 357-367. Univ. Calif. Press. (11.3).

Feller, W. (1968, 1971). An Introduction to Probability Theory and its Applications. Vol. I, 3d ed. (2.2, 9.1); Vol. II, 2d ed. (8.3). Wiley, N.Y.

Freedman, David (1966). On two equivalence relations between measures. Ann. Math. Statist. 37 686-689. (10.2).

Freedman, David (1971). Brownian Motion and Diffusion. Holden-Day, San Francisco. (1.1).

Gaenssler, P. (1983). Empirical processes. Institute of Mathematical Statistics, Lecture Notes - Monograph Series 3. (1.1).

Gaenssler, P. and W. Stute (1979). Empirical processes: a survey of results for independent and identically distributed random variables. Ann. Probability 7 193-243 (1.1, 6.1).

Giné M., Evarist (1974). On the central limit theorem for sample continuous processes. Ann. Probability 2 629-641. (2.2).

Gnedenko, B.V., and A.N. Kolmogorov (1949, 1954, 1968). Limit distributions for sums of independent random variables (transl. K.L. Chung). Addison-Wesley, Reading, Mass. (3.3).

Goodman, Victor, J. Kuelbs and J. Zinn (1981). Some results on the LIL in Banach space with applications to weighted empirical processes. Ann. Probability 9 713-752. (4.2).

Gruber, P.M. (1981). Approximation of convex bodies by polytopes. Comptes Rendus (Doklady) Acad. bulgar. Sci. 34 621-622. (7.3).

Gutmann, S. (1980). The empirical measure is a sufficient statistic. Manuscript (unpublished). (10.1).

Heinkel, B. (1979). Relation entre théorème central-limite et loi du logarithme itéré dans les espaces de Banach. Z. Wahrsch. verw. Geb. 49 211-220. (4.2).

Hoeffding, Wassily (1963). Probability inequalities for sums of bounded random variables. J. Amer. Statist. Assoc. 58 13-30. (2.2).

Hoffmann-Jørgensen, J. (1974). Sums of independent Banach space valued random variables. Studia Math. 52 159-186. (3.2).

Hunt, G.A. (1966). Martingales et processus de Markov. Dunod, Paris. (11.1).

Jain, Naresh, and Michael B. Marcus (1975 a). Central limit theorems for C(S)-valued random variables. J. Functional Analysis 19 216-231. (11.3)

Jain, Naresh, and Michael B. Marcus (1975 b). Integrability of infinite sums of independent vector-valued random variables. Trans. Amer. Math. Soc. 212 1-36. (3.2).

Kac, Mark (1949). On deviations between theoretical and empirical distributions. Proc. Nat. Acad. Sci. USA 35 252-257. (8.3).

Kahane, Jean-Pierre (1968). Some random series of functions. D.C. Heath, Lexington, Mass. (3.2, 11.3).

Kelley, John L. (1955). General Topology. Van Nostrand, Princeton. (10.3).

Kiefer, Jack (1972). Skorohod imbedding of multivariate rv's, and the sample df. Z. Wahrsch. verw. Geb. 24 1-35 (1.1).

Kingman, J.F.C. (1968). The ergodic theory of subadditive stochastic processes. J. Roy. Statist. Soc. B 30 499-510. (11.2).

Kingman, J.F.C. (1973). Subadditive ergodic theory (with discussion). Ann. Probability 1 883-909. (11.2).

Kirszbraun, M.D. (1934). Über die zusammenziehende und Lipschitzsche Transformationen. Fund Math. 22 77-108. (7.2).

Kolčinskii, V.I. (1981 a). On the central limit theorem for empirical measures. Theor. Probability Math. Statist. 24 71-82 = Teor. Verojatnost. i Mat. Statist. 24 63-75 (11.1, 11.3).

Kolčinskii, V.I. (1981 b). On the law of the iterated logarithm in Strassen's form for empirical measures. Ibid. 25, 40-47 (Russian), 43-49 (English). (4.2).

Kolmogorov, A. (1931). Eine Verallgemeinerung des Laplace-Liapounoffschen Satzes. Izv. Akad. Nauk SSSR Otdel. Mat. Estest. Nauk, VII. Ser., n° 7, 959-962. (Zentralblatt 3, 1932, p. 357). (1.1).

Kolmogorov, A. (1933). Über die Grenzwertsätze der Wahrscheinlichkeitsrechnung. Izv. Akad. Nauk SSSR (Bull. Acad. Sci. URSS). VII. Ser., n° 3, 363-372. (1.1).

Kolmogorov, A. (1955). Bounds for the minimal number of elements of an ε-net in various classes of functions and their applications to the question of representability of functions of several variables by superpositions of functions of fewer variables (in Russian). Uspekhi Mat. Nauk (N.S.) 10 n° 1 (63) 192-194. (7.1).

Kolmogorov, A.N., and V.M. Tikhomirov (1959). ε-entropy and ε-capacity of sets in function spaces. Amer. Math. Soc. Transls. (Ser. 2) 17 (1961) 277-364 = Uspekhi Mat. Nauk 14 vyp. 2(86), 3-86 (6.0, 7.1).

Komlós, J., P. Major and G. Tusnády (1975). An approximation of partial sums of independent RV's, and the sample DF. I. Z. Wahrsch. verw. Geb. 32 111-131. (1.1).

Kuelbs, James D. (1977). Kolmogorov's law of the iterated logarithm for Banach space valued random variables. Illinois J. Math. 21 784-800. (3.2).

Kuelbs, James D. and R.M. Dudley (1980). Log log laws for empirical measures. Ann. Probability 8 405-418. (4.2, 10.3).

Kuelbs, James D. and Joel Zinn (1979). Some stability results for vector valued random variables. Ann. Probability 7 75-84. (4.2).

Kuratowski, Kazimierz (1966). Topology I, 5th ed., transl. from the French by J. Jaworowski. Academic Press, N.Y.; PWN, Warszawa. (10.2).

Ledoux, Michel (1982). Loi du logarithme itéré dans C(S) et fonction caractéristique empirique. Z. Wahrsch. verw. Geb. 60 425-435. (11.3).

Lorentz, G.G. (1966). Metric entropy and approximation. Bull. Amer. Math. Soc. 72 903-937. (7.1).

Major, Péter (1976). Approximation of partial sums of i.i.d. rv's when the summands have only two moments. Z. Wahrsch. verw. Geb. 35 221-229. (1.1).

Marcus, Michael B. (1981). Weak convergence of the empirical characteristic function. Ann. Probability 9 194-201. (11.3).

Marcus, Michael B. and Walter Philipp (1982). Almost sure invariance principles for sums of B-valued random variables with applications to random Fourier series and the empirical characteristic process. Trans. Amer. Math. Soc. 269 67-90. (11.1).

McShane, E. J. (1934). Extension of range of functions. Bull. Amer. Math. Soc. 40 837-842. (7.2).

Meyer, Paul-André (1972). Martingales and Stochastic Integrals I. Lecture Notes in Math. (Springer) 284. (11.1).

Mourier, Edith (1951). Lois des grands nombres et théorie ergodique. C.R. Acad. Sci. Paris 232 923-925. (6.1).

Mourier, Edith (1953). Eléments aléatoires dans un espace de Banach. Ann. Inst. H. Poincaré 13 161-244. (6.1).

Nagaev, S.V. (1970). On the speed of convergence in a boundary problem, I, II. Theor. Probability Appls. 15 163-186, 403-429 = Teor. Veroiatnost. i Primenen. 15 179-199, 419-441. (1.1).

Natanson, I.P. (1957). Theory of Functions of a Real Variable, 2d ed., transl. by L. F. Boron, Vol. II, 1961. Ungar, N.Y. (10.2).

Neveu, J. (1977). Processus ponctuels. Ecole d'été de probabilités de Saint-Flour VI, 1976, Lecture Notes in Math. (Springer) 598 249-447. (10.1).

Okamoto, Masashi (1958). Some inequalities relating to the partial sum of binomial probabilities. Ann. Inst. Statist. Math. 10 29-35. (2.2).

Parthasarathy, K.R. (1967). Probability Measures on Metric Spaces. Academic Press, N.Y. (1.2).

Philipp, Walter (1979). Almost sure invariance principles for sums of B-valued random variables. Probability in Banach Spaces II (Proc. Conf. Oberwolfach, 1978), Lecture Notes in Math. (Springer) 709 171-193.

Philipp, Walter (1980). Weak and L^p-invariance principles for sums of B-valued random variables. Ann. Probability 8 68-82. Correction (to appear). (1.1).

Pisier, G. (1975). Le théorème de la limite centrale et la loi du logarithme itéré dans les espaces de Banach (suite et fin). Sém. Maurey-Schwartz 1975-76, Exposé IV, Ecole Polytechnique, Palaiseau. (4.2).

Pollard, David B. (1982 a). A central limit theorem for empirical processes. J. Austral. Math. Soc. Ser. A, 33 235-248. (11.1, 11.3).

Pollard, David B. (1982 b). Convergence of Stochastic Processes. (Preprint, 61 pp.) (1.1).

Pyke, R. (1968). The weak convergence of the empirical process with random sample size. Proc. Cambr. Philos. Soc. 64 155-160. (8.3).

Radon, J. (1921). Mengen konvexer Körper, die einen gemeinsamen Punkt enthalten. Math. Ann. 83 113-115. (9.2).

Rao, B.V. (1971). Borel structures for function spaces. Colloq. Math. 23 33-38. (10.2).

Rogers, C.A. (1964). Packing and Covering. Cambridge University Press, N.Y. (7.2).

Sauer, N. (1972). On the density of families of sets. J. Combin. Th. A 13 145-147. (9.1).

Schmidt, Wolfgang M. (1975). Irregularities of distribution IX. Acta Arith. 27 385-396. (8.1).

Schwartz, Laurent (1973). Radon measures on arbitrary topological spaces and cylindrical measures. Tata Institute; Oxford Univ. Press, London. (10.3).

Shelah, S. (1972). A combinatorial problem: stability and order for models and theories in infinitary languages. Pacific J. Math. 41 247-261. (9.1).

Shortt, Rae M. (1982). Existence of laws with given marginals and specified support. Ph. D. Thesis, Math., Mass. Inst. of Tech. (1.2).

Sion, M. (1960). On uniformization of sets in topological spaces. Trans. Amer. Math. Soc. 96 237-245. (10.3).

Skorohod, A.V. (1976). On a representation of random variables. Theor. Probability Appls. 21 628-632 (English), 645-648 (Russian). (1.2).

Sonis, M.G. (1966). Certain measurable subspaces of the space of all sequences with a Gaussian measure (in Russian). Uspehi Mat. Nauk 21 n° 5 (131) 277-279. (2.1).

Steele, J. Michael (1978). Empirical discrepancies and subadditive processes. Ann. Probability 6 118-127. (11.2).

Strassen, Volker (1964). An invariance principle for the law of the iterated logarithm. Z. Wahrsch. verw. Geb. 3 211-226. (1.1).

Ulam, Stanisław (1930). Zur Masstheorie in der allgemeinen Mengenlehre. Fund. Math. 16 140-150. (3.1).

Uspensky, J.V. (1937). Introduction to Mathematical Probability. McGraw-Hill, N.Y. (2.2).

Vapnik, V.N., and A. Ya. Červonenkis (1968). Uniform convergence of frequencies of occurrence of events to their probabilities. Doklady Akad. Nauk SSSR 181 781-783 (Russian) = Soviet Math. Doklady 9 915-918 (English). (9.1).

Vapnik, V.N., and A. Ya. Červonenkis (1971). On the uniform convergence of relative frequencies of events to their probabilities. Theor. Probability Appls. 16 264-280 = Teor. Verojatnost. i Primenen. 16 264-279. (9.1, 11.2, 12.1).

Vapnik, V.N., and A. Ya. Červonenkis (1974). Teoriya Raspoznavaniya Obrazov; Statisticheskie problemy obucheniya [Theory of Pattern Recognition; Statistical problems of learning; in Russian]. Nauka, Moscow. German ed.: Theorie der Zeichenerkennung, by W. N. Wapnik and A.J. Tscherwonenkis, transl. by K.G. Stöckel and B. Schneider, ed. S. Unger and K. Fritzsch. Akademie-Verlag, Berlin, 1979 (Elektronisches Rechnen und Regeln, Sonderband). (9.1, 12.1).

Vapnik, V.N., and A. Ya. Červonenkis (1981). Necessary and sufficient conditions for the uniform convergence of means to their expectations. Theor. Probability Appls. 26 532-553 (English), 543-563 (Russian). (11.2).

Vorob'ev, N.N. (1962). Consistent families of measures and their extensions. Theor. Probability Appls. 7 147-163 (English), 153-169 (Russian). (1.2).

Vulikh, B.Z. (1961). Introduction to the Theory of Partially Ordered Spaces (transl. by L.F. Boron, 1967). Wolters-Noordhoff, Groningen. (3.1).

Wenocur, R.S., and R.M. Dudley (1981). Some special Vapnik-Červonenkis classes. Discrete Math. 33 313-318. (9.2).

Wolfowitz, J. (1954). Generalization of the theorem of Glivenko-Cantelli. Ann. Math. Statist. 25 131-138. (11.1).

Wright, F.T. (1981). The empirical discrepancy over lower layers and a related law of large numbers. Ann. Probability 9 323-329. (7.2).

Yukich, J. (1982). Convergence of empirical probability measures. Ph. D. thesis, Math., Mass. Inst. of Tech. (6.4).

Yurinskii [Jurinskii] , V.V. (1977). On the error of the Gaussian approximation for convolutions. Theor. Probability Appls. 22 236-247 (English), 242-253 (Russian). (12.2).

STOCHASTIC DIFFERENTIAL EQUATIONS

AND

STOCHASTIC FLOWS OF DIFFEOMORPHISMS

PAR H. KUNITA

INTRODUCTION

This course presents recent results on the stochastic flow of diffeomorphisms generated by a stochastic differential equation and develop a differential geometric analysis for the stochastic flow of diffeomorphims. Main tools are stochastic integrals and differentials of Itô's.

We begin with recalling the relationship between a vector field and a (deterministic) flow of diffeomorphisms, which is basic to the differential geometry. For a moment we restrict the attention to Euclidean space R^d. Let $X(x) = (X^1(x),\dots,X^d(x))$ be a Lipschitz continuous R^d valued function (or vector field on R^d), and let $\phi_t(x)$ be the solution of the ordinary differential equation starting from $x \in R^d$ at time 0.

$$\frac{d\phi_t}{dt} = X(\phi_t), \qquad \phi_0 = x.$$

It is easily checked that the solution has the following properties.

(a) For each t, the map $\phi_t \; ; R^d \longrightarrow R^d$ is a homeomorphisms.

(b) $\phi_s \circ \phi_t = \phi_{s+t}$ for each s,t,

(c) $(t,x) \longrightarrow \phi_t(x)$ gives a continuous map from $R \times R^d$ onto R^d.

This $\{\phi_t\}$ is called a flow of homeomorphisms generated by the vector field X.

A central subject of this course is to establish the similar relationship between a stochastic differential equation (SDE) and a stochastic flow of diffeomorphisms under a reasonable mild condition. Let $X_0(t,x)$, $\dots,X_m(t,x)$, $t \in [0,a]$, $x \in R^d$ be R^d valued functions, continuous in (t,x) and Lipschitz continuous in x. Let $B_t = (B_t^1,\dots,B_t^m)$ be an m-dimensional Brownian motion. Consider an SDE on R^d:

$$d\xi_t = X_0(t,\xi_t)dt + \sum_{k=1}^{m} X_k(t,\xi_t)dB_t^k .$$

A rigorous definition, an existence and the uniqueness of the solution will be given at Chapter II. Let $\xi_{s,t}(x)$, $t \in [s,a]$ be the solution with the initial condition $\lim_{t \downarrow s} \xi_{s,t}(x) = x$ a.s. The problem we will be concerned is that whether the map $\xi_{s,t}$; $R^d \longrightarrow R^d$ induces a stochastic flow of homeomorphisms of R^d or not. The answer is yes but the verification is by no means simple, compared with the deterministic case. Perhaps, a reason is that the backward or the backward-forward calculus of stochastic differential equations is not so easy as the calculus of the ordinary differential equation.

Approaches to the above problem could be summarized by the following three types, so far:

(a) We switch the equation on R^d (or on manifold) to that on a Hilbert manifold consisting of diffeomorphisms (or homeomorphisms) and solve the equation directly to get a stochastic flow of diffeomorphisms. The approach is specially efficient if the underlying manifold is compact. See Elworthy [7] and Baxendale [1].

(b) We approximate to the equation a sequence of ordinary differential equations, replacing the Brownian paths by piecewise smooth curves. Then each approximating equation generates a flow of diffeomorphisms and the limit of the approximating flows will be the desired one. See Malliavin [30], Ikeda-Watanabe [13] and Bismut [2].

(c) We get several L^p estimates for quantities like $\xi_{s,t}(x) - \xi_{s',t'}(x')$ making use of stochastic calculus, and then apply Kolmogorov's criterion of the continuity of the random field to show the homeomorphic property of $\xi_{s,t}$. See Kunita [23], [25] [26] and Meyer [36].

In this course we will adopt the approach (c). An advantage of the last approach might be that the stochastic flow of hemeomorphisms can be obtained under a quite mild assumption on vector fields. In fact, global

Lipschitz conditions to vector fields $X_0, \ldots, X_m$ are sufficients. In case that global Lipschitz condition is not assumed to vector fields, we will obtain a necessary and sufficient condition for this, assuming some smoothness conditions to vector fields. Main results are found in Theorems 4.3, 4.4. 6.1 and 9.3 in Chapter II.

This course consists of three chapters. Chapter I deals with the stochastic calculus connected with continuous semimartingales, which will furnish a basic tool for various stochastic analysis in Chapters II and III. We will present a quick expositon of stochastic integrals and differentials, restricting our attention to continuous processes. The prerequisites are some classical theorems on martingales such as Doob's inequality and Doob's optional sampling theorem, while we will not require Meyer's decomposition theorem of supermartingales. Instead, we will discuss the quadratic variation of continuous (local) martingales in details in four beginning sections.

Stochastic integrals by continuous semimartingales are defined in a standard manner as Kunita-Watanabe [28] or Meyer [34]. Then we will establish a differential rule for the composition of two continuous semimartingales. It is a generalization of Itô's formula which states the differential rule for change of variables. The formula is particularly useful for the analysis of the stochastic flow of diffeomorphisms.

In Chapter II, we prove that a stochastic differential equation generates a flow of homeomorphisms or diffeomorphisms in various situations. The contents can be divided into three parts. The first part (Sections 1-4) discusses the case that the equation is defined on a Euclidean sapce, whose coefficients are globally Lipschitz continuous. In addition to the homeomorphic property of the map $\xi_{s,t} ; R^d \longrightarrow R^d$, the smoothness of $\xi_{s,t}(x)$ will be studied: The solution $\xi_{s,t}(x)$ is k-th continuously differentiable and its k-th derivatives are locally Hölder continuous of order β less than α for any $s < t$ a.s. if coefficients $X_0, \ldots, X_m$ are k-th continuously differentiable and their k-th derivatives are Hölder continuous of order α.

The second part (Sections 5-7) deals with SDE on R^d whose coefficients are locally Lipschitz continuous. Generally, if the solution is (strictly) conservative, the map $\xi_{s,t}$; $R^d \longrightarrow R^d$ is an injection a.s. We will obtain a necessary and sufficient condition that the map $\xi_{s,t}$ is a surjection, assuming that coefficients are C^2 (C^1 and their derivatives are locally Lipschitz continuous). The condition will be stated in terms of the adjoint equation.

The third part (Sections 8 and 9) studies SDE on manifold. We will consider Stratonovich SDE rather than Itô SDE, since the former is more adapted to the differential geometry. We will again obtain a necessary and sufficient condition that the solution defines a stochastic flow of homeomorphisms a.s. The result is very close to the case of Euclidean space.

Chapter III concerns the stochastic differential geometry related to stochastic flow of diffeomorphisms. The flow $\xi_{s,t}$ acts naturally on tensor fields. In Section 3 we will obtian Itô's formula for $\xi_{s,t}$ acting on vector fields, where the Lie derivatives play an important role. As an application, we decompose the equation to two simpler ones and get the solution as the composition of solutions corresponding to these two equations. An aim of the decomposition is to represent the solution by means of Brownian motion explicitly. The complexity of the representation depends on the structure of the Lie algebra generated by coefficients $X_0, \ldots, X_m$ of the SDE. The case that Lie algebra is solvable will be studied in details in Section 3.

Itô's formula for $\xi_{s,t}$ acting on general tensor fields will be obtained at Section 4. The formula can be used in some case to determine the possible diffeomorphisms that the flow $\xi_{s,t}$ can take. It will be carried out in several examples in Section 5. Stochastic parallel displacement of tensor field will give us another Itô's formula, where the covariant derivatives play an important role. See Section 6.

Another subject of Chapter III is the backward calculus of stochastic flow. In Section 1, a differential rule of $\xi_{s,t}$ for the backward variables will be obtained. The formula is helpful for solving some type of the second order parabolic partial differential equation. See Section 7.

Acknowledgement. It is my pleasure to thank Professor Hennequin and the École d'Été for giving me the opportunity to present the lecture. I received many valuable comments from audiences, which encouraged me to rewrite the first version of this note. Finally I wish to express my hearty gratitude to Miss Setsuko Okabe for her surperb typing.

CHAPTER I

STOCHASTIC CALCULUS FOR CONTINUOUS SEMIMARTINGALES

1. PRELIMINARIES

Let $(\Omega,\underline{\underline{F}},P)$ be a complete probability space equipped with a family of sub σ-fields $\{\underline{\underline{F}}_t, t\in[0,a]\}$ with following properties, where a is a finite positive constant:

(i) Each $\underline{\underline{F}}_t$ contains all null sets of $\underline{\underline{F}}$.

(ii) $\{\underline{\underline{F}}_t\}$ is increasing, i.e., $\underline{\underline{F}}_t \supset \underline{\underline{F}}_s$ if $t \geq s$.

(iii) $\{\underline{\underline{F}}_t\}$ is right continuous, i.e., $\bigcap_{\varepsilon>0}\underline{\underline{F}}_{t+\varepsilon} = \underline{\underline{F}}_t$ for any $t < a$.

The probability space $(\Omega,\underline{\underline{F}},P;\underline{\underline{F}}_t)$ will be fixed throughout this chapter.

Let X_t, $t\in[0,a]$ be a stochastic process with values in $R = (-\infty,\infty)$. We will assume, unless otherwise mentioned, that it is $(\underline{\underline{F}}_t)$-adapted, i.e., X_t is $\underline{\underline{F}}_t$-measurable for any $t\in[0,a]$. The process X_t is called continuous if $X_t(\omega)$ is a continuous function of t for almost all ω.

Let $\underline{\underline{L}}_c$ be the linear space consisting of all continuous stochastic processes. We introduce the metric ρ by

$$\rho(X-Y) = \rho(X,Y) = E\left[\frac{\sup_t |X_t - Y_t|^2}{1+\sup_t |X_t - Y_t|^2}\right]^{\frac{1}{2}} .$$

It is equivalent to the topology of the uniform convergence in probability: A sequence $\{X^n\}$ of $\underline{\underline{L}}_c$ is a Cauchy sequence if and only if for any $\varepsilon > 0$,

$$P(\sup_t |X_t^n - X_t^m| > \varepsilon) \xrightarrow[n,m \to \infty]{} 0.$$

Obviously $\underline{\underline{L}}_c$ is a complete metric space.

We introduce the norm $\| \quad \|$ by $\|X\| = E[\sup_t |X_t|^2]^{\frac{1}{2}}$ and denote by $\underline{\underline{L}}_c^2$ the set of all elements in $\underline{\underline{L}}_c$ with finite norms. We may say that the topology of $\underline{\underline{L}}_c^2$ is the uniform convergence in L^2. Since $\rho(X) \leq \|X\|$, the topology by $\| \quad \|$ is stronger than that by ρ. It is easy to see that $\underline{\underline{L}}_c^2$ is a dense subset of $\underline{\underline{L}}_c$.

1.1. Definition. Let X_t, $t \in [0,a]$ be a continuous $(\underline{\underline{F}}_t)$-adapted process.

(i) It is called a martingale if $E|X_t| < \infty$ for any t and satisfies $E[X_t|\underline{\underline{F}}_s] = X_s$ for any $t > s$.

(ii) It is called a local martingale if there is an increasing sequence of stopping times [1] $\{T_n\}$ such that $T_n \uparrow \infty$ and each stopped process $X_t^{T_n} \equiv X_{t \wedge T_n}$ is a martingale.

(iii) It is called an increasing process if $X_t(\omega)$ is an increasing function of t a.s.

(iv) It is called a process of bounded variation if it is written as the difference of two increasing processes.

(v) It is called a semimartingale if it is written as the sum of a local martingale and a process of bounded variation.

We will quote two famous results of Doob's concerning martingales without giving proofs.

1.2. Theorem Let X_t, $t \in [0,a]$ be a martingale.

---

1) A random variable T is called a stopping time if it takes values in $[0,a] \cup \{\infty\}$ and satisfies $\{\omega | T(\omega) \leq t\} \in \underline{\underline{F}}_t$ for any $t \in [0,a]$.

(i) <u>Optional sampling theorem.</u> Let S and T be stopping times with values in $[0,a]$. Then X_S is integrable and satisfies $E[X_S|\underline{\underline{F}}_T] = X_{S\wedge T}$.[1)]

(ii) <u>Inequality.</u>[2)] Suppose $E[|X_a|^p] < \infty$ with $p > 1$. Then $E[\sup_s |X_s|^p] \leq q^p E[|X_a|^p]$ where q is the conjugate of p.

Remark. Let S be a stopping time. If X_t is a martingale, the stopped process X^S is also a martingale. In fact, by Doob's optional sampling theorem, we have for $t \geq s$ $E[X_t^S|\underline{\underline{F}}_s] = X_{t\wedge S\wedge s} = X_{S\wedge s} = X_s^S$. Similarly if X is a local martingale, the stopped process X^S is a local martingale.

A martingale is a local martingale, obviously. The following theorem gives us a criterion that a local martingale is a martingale.

1.3. <u>Theorem.</u> Let X_t be a continuous local martingale.

(i) If $E[\sup_t |X_t|] < \infty$, then X is a martingale.

(ii) Let $p > 1$. Then X is an L^p-martingale if and only if $E[\sup_t |X_t|^p] < \infty$.

<u>Proof.</u> Let $T_n \uparrow \infty$ be a sequence of stopping times such that each stopped process X^{T_n} is a martingale. Then for each t, $X_t^{T_n}$ converges to X_t a.s. and further $|X_t^{T_n} - X_t| \leq 2\sup_s |X_s|$. Hence if $\sup_s |X_s|$ is integrable, $E|X_t^{T_n} - X_t|$ tends to 0. This implies that X_t is a martingale, because $E[X_t|\underline{\underline{F}}_s] = \lim_{n\to\infty} E[X_t^{T_n}|\underline{\underline{F}}_s] = \lim_{n\to\infty} X_s^{T_n} = X_s$. The second assertion (ii) is immediate from Doob's inequality and the first assertion.

---

1) $\underline{\underline{F}}_T = \{A \in \underline{\underline{F}}_a \; ; \; A \cap \{T \leq t\} \in \underline{\underline{F}}_t$ holds for any $t \in [0,a]\}$.

2) The inequality is valid to positive submartingales, too.

Remark. Let X be a local martingale. Then there is an increasing sequence of stopping times $S_k \uparrow \infty$ such that each stopped process X^{S_k} is a bounded martingale. In fact, define S_k by

$$S_k = \inf\{t > 0 \; ; \; |X_t| \geq k\} \qquad (= \infty \text{ if } \{\dots\} = \phi).$$

Then $S_k \uparrow \infty$ and it holds $\sup_t |X_t^{S_k}| \leq k$, so that each X^{S_k} is a martingale.

Let $\underline{\underline{M}}_c$ be the set of all square integrable martingales X_t with $X_0 = 0$. Because of Doob's inequality, the norm $\|X\|$ is finite for any X of $\underline{\underline{M}}_c$. Hence $\underline{\underline{M}}_c$ is a subset of $\underline{\underline{L}}_c^2$. We denote by $\underline{\underline{M}}_c^{loc}$ the set of all continuous local martingales X_t such that $X_0 = 0$. It is a subset of $\underline{\underline{L}}_c$.

1.4. Theorem. $\underline{\underline{M}}_c$ is a closed subspace of $\underline{\underline{L}}_c^2$. $\underline{\underline{M}}_c^{loc}$ is a closed subspace of $\underline{\underline{L}}_c$. Furthermore, $\underline{\underline{M}}_c$ is dense in $\underline{\underline{M}}_c^{loc}$.

Proof. The first assertion is obvious. Let $\{X^n\}$ be a Cauchy sequence in $\underline{\underline{M}}_c^{loc}$ converging to X of $\underline{\underline{L}}_c$ by the topology ρ. Choosing a subsequence if necessary, we may assume that X^n converges to X uniformly a.s. Set $A_t = \sup_n \sup_s |X_s^n|$. Then it is a continuous increasing process. Define for $k=1,2,\dots$, $T_k = \inf\{t > 0 \; ; \; A_t \geq k\}$ $(= \infty$ if $\{\dots\} = \phi)$. Then it holds $\sup_t |X_t^{n,T_k}| \leq k$ for all n. Therefore, for each k, $\{X_t^{n,T_k}\}$ is a sequence of martingales converging to $X_t^{T_k}$ boundedly. Therefore X^{T_k} is a martingale for each k, proving that X is an element of $\underline{\underline{M}}_c^{loc}$. The last assertion is immediate from the remark after Theorem 1.3.

2. QUADRATIC VARIATIONS OF CONTINUOUS SEMIMARTINGLES

This section is devoted to the study of the quadratic variation of a continuous stochastic process $X_t, t \in [0,a]$. Let Δ be a partition of the interval $[0,a]$: $\Delta = \{0 = t_0 < \dots < t_n = a\}$ and let $|\Delta| = \max(t_{i+1} - t_i)$. Associated with the partition Δ, we define a continuous process $\langle X\rangle_t^\Delta$ as

$$\langle X\rangle_t^\Delta = \sum_{i=0}^{k-1} (X_{t_{i+1}} - X_{t_i})^2 + (X_t - X_{t_k})^2,$$

where k is the number such that $t_k \le t < t_{k+1}$. We call it the quadratic variation of X_t associated with the partition Δ.

Now let $\{\Delta_m\}$ be a sequence of partitions such that $|\Delta_m| \to 0$. If the limit of $\langle X\rangle_t^{\Delta_m}$ exists in probability and it is independent of the choice of sequences $\{\Delta_m\}$ a.s., it is called the quadratic variation of X_t and is denoted by $\langle X\rangle_t$.

The quadratic variation is not well defined to any continuous stochastic process. We will see in the sequel that a natural class of processes where quadratic variations are well defined is that of continuous semimartingales.

We begin the discussion with a process of bounded variation.

2.1. Lemma. Let X be a continuous process of bounded variation. Then the quadratic variation exists and equals 0 a.s.

Proof. Let $|X|_t(\omega)$ be the total variation of the function $X_s(\omega)$, $0 \le s \le t$. Then it holds

$$\langle X\rangle_t^\Delta \le \Big(\sum_{j=0}^{k-1}|X_{t_{j+1}}-X_{t_j}|+|X_t-X_{t_k}|\Big)\max_i|X_{t_{i+1}}-X_{t_i}|$$

$$\le |X|_t \max_i |X_{t_{i+1}}-X_{t_i}|$$

The right hand side converges to 0 as $|\Delta| \to 0$ a.s.

We next consider the quadratic variation of a bounded continuous martingale.

2.2. <u>Theorem.</u> Let M be a bounded continuous martingale. Let $\{\Delta_n\}$ be a sequence of partitions such that $|\Delta_n| \to 0$. Then $\langle M\rangle_t^{\Delta_n}$, $t \in [0,a]$ converges uniformly to a continuous increasing process $\langle M\rangle_t$ in L^2-sense, i.e.,

$$\lim_{n\to\infty} E[\sup_t |\langle M\rangle_t^{\Delta_n} - \langle M\rangle_t|^2] = 0.$$

Before the proof, we prepare two lemmas for a bounded continuous martingale.

2.3. <u>Lemma.</u> For any $t > s$, it holds

$$E[\langle M\rangle_t^\Delta|\underline{\underline{F}}_s] - \langle M\rangle_s^\Delta = E[(M_t-M_s)^2|\underline{\underline{F}}_s] = E[M_t^2|\underline{\underline{F}}_s] - M_s^2.$$

In particular, $M_t^2 - \langle M\rangle_t^\Delta$ is a continuous martingale.

<u>Proof.</u> Choose t_ℓ and t_k of Δ such that $t_\ell \le s < t_{\ell+1}$ and $t_k \le t < t_{k+1}$. Then a simple computation yields

$$\langle M\rangle_t^\Delta - \langle M\rangle_s^\Delta = (M_t-M_{t_k})^2 + \dots + (M_{t_{\ell+1}}-M_s)^2 + 2(M_{t_{\ell+1}}-M_s)(M_s-M_{t_\ell}).$$

Take the conditional expectation relative to $\underline{\underline{F}}_s$ to each of the above. Clearly the conditional expectation of the last member is 0. Note the conditional orthogonality property such as

$$E[(M_t-M_{t_k})(M_{t_k}-M_{t_{k-1}})|\underline{\underline{F}}_s] = E[E[M_t-M_{t_k}|\underline{\underline{F}}_{t_k}](M_{t_k}-M_{t_{k-1}})|\underline{\underline{F}}_s]$$

$$= 0 .$$

Then we get

$$E[\langle M\rangle_t^{\Delta}|\underline{\underline{F}}_s] - \langle M\rangle_s^{\Delta} = E[(M_t-M_s)^2|\underline{\underline{F}}_s]$$

$$= E[M_t^2 - 2M_tM_s + M_s^2|\underline{\underline{F}}_s]$$

$$= E[M_t^2|\underline{\underline{F}}_s] - M_s^2.$$

This proves the lemma.

2.4. Lemma. It holds $\lim_{n,m\to\infty} E[|\langle M\rangle_a^{\Delta_n} - \langle M\rangle_a^{\Delta_m}|^2] = 0.$

Proof. Given two partitions Δ, Δ' of $[0,a]$, we denote by $\Delta\cup\Delta' = \{0 = s_0 < \ldots < s_N = a\}$ the joint partition of Δ and Δ'. We will write $\langle M\rangle_t^{\Delta}$ as A_t^{Δ} for convenience. Consider the quadratic variation of the process $A_t^{\Delta} - A_t^{\Delta'}$ associated with the partition $\Delta\cup\Delta'$. Since $A_t^{\Delta} - A_t^{\Delta'}$ is a martingale, we have from Lemma 2.3

$$E[|A_a^{\Delta} - A_a^{\Delta'}|^2] = E[\langle A^{\Delta} - A^{\Delta'}\rangle_a^{\Delta\cup\Delta'}].$$

It holds

$$\langle A^{\Delta} - A^{\Delta'}\rangle_a^{\Delta\cup\Delta'} \leq 2\{\langle A^{\Delta}\rangle_a^{\Delta\cup\Delta'} + \langle A^{\Delta'}\rangle_a^{\Delta\cup\Delta'}\}.$$

It is sufficient to prove that $E[\langle A^{\Delta}\rangle_a^{\Delta\cup\Delta'}]$ converges to 0 as $|\Delta| + |\Delta'| \to 0$. Given s_k, s_{k+1} of $\Delta\cup\Delta'$, choose t_ℓ of Δ such that $t_\ell \leq s_k < s_{k+1} \leq t_{\ell+1}$. Then, since $A_{s_{k+1}}^{\Delta} - A_{s_k}^{\Delta} = (M_{s_{k+1}} - M_{t_\ell})^2 - (M_{s_k} - M_{t_\ell})^2$, we get

$$A_{s_{k+1}}^{\Delta} - A_{s_k}^{\Delta} = (M_{s_{k+1}} - M_{s_k})\{(M_{s_{k+1}} - M_{s_k}) + 2(M_{s_k} - M_{t_\ell})\}.$$

Therefore,

$$\langle A^{\Delta}\rangle_a^{\Delta\cup\Delta'} \le \sup_k |M_{s_{k+1}} + M_{s_k} - 2M_{t_\ell}|^2 \bullet A_a^{\Delta\cup\Delta'} .$$

By Schwarz's inequality

$$E[\langle A^{\Delta}\rangle_a^{\Delta\cup\Delta'}] \le E[\sup_k |M_{s_{k+1}} + M_{s_k} - 2M_{t_\ell}|^4]^{\frac{1}{2}} \cdot E[(A_a^{\Delta\cup\Delta'})^2]^{\frac{1}{2}} .$$

The first member of the right hand side converges to 0 as $|\Delta| + |\Delta'| \to 0$. We will prove that the second member is dominated by a constant independent of partitions: then $E[\langle A^{\Delta}\rangle_a^{\Delta\cup\Delta'}]$ would tend to 0 and the lemma will follow.

Observe the relation

$$(A_a^{\Delta})^2 = 2\sum_{k=1}^{n} (A_a^{\Delta} - A_{t_k}^{\Delta})(A_{t_k}^{\Delta} - A_{t_{k-1}}^{\Delta}) + \sum_{k=1}^{n} (M_{t_k} - M_{t_{k-1}})^4 .$$

From Lemma 2.3, it holds $E[(A_a^{\Delta} - A_{t_k}^{\Delta})|\underline{F}_{t_k}] = E[(M_a - M_{t_k})^2|\underline{F}_{t_k}]$.

Therefore,

$$E[(A_a^{\Delta})^2] = 2\sum_{k=1}^{n} E[(M_a - M_{t_k})^2(A_{t_k}^{\Delta} - A_{t_{k-1}}^{\Delta})] + \sum_{k=1}^{n} E[(M_{t_k} - M_{t_{k-1}})^4]$$

$$\le E[\{2\sup_k |M_a - M_{t_k}|^2 + \sup_k |M_{t_k} - M_{t_{k-1}}|^2\} A_a^{\Delta}] .$$

Since $|M_t|$ is dominated by a constant, say C, the last member of the above is dominated by $12C^2 E[A_a^{\Delta}] \le 48C^4$.

<u>Proof</u> of Theorem 2.2. By Doob's inequality for martingales, it holds

$$E[\sup_t |\langle M\rangle_t^{\Delta_n} - \langle M\rangle_t^{\Delta_m}|^2] \le 4E[|\langle M\rangle_a^{\Delta_n} - \langle M\rangle_a^{\Delta_m}|^2] .$$

Choosing a subsequence if necessary, $\langle M\rangle_t^{\Delta_n}$ converges uniformly a.s. Denote the limit as $\langle M\rangle_t$. It is a continuous process. We will prove that $\langle M\rangle_t$ is increasing in t a.s. Taking joint partitions if necessary, we may and do assume that Δ_{n+1} is a refined partition of Δ_n for each n and the set $\bigcup_n \Delta_n$ is dense in $[0,T]$. Let $s < t$ be two points in $\bigcup_n \Delta_n$. There is a natural number n_0 such that $s,t \in \Delta_{n_0}$. Then $\langle M\rangle_t^{\Delta_n} \geq \langle M\rangle_s^{\Delta_n}$ is satisfied for all $n \geq n_0$. Therefore $\langle M\rangle_t \geq \langle M\rangle_s$ is satisfied. The inequality is then satisfied for any real numbers $s < t$, since $\langle M\rangle_t$ is continuous in t a.s.

We next consider the quadratic variation of a local martingale. This time the quadratic variations associated with partitions do not converge in L^2 in general, but they converge in probability. In fact, we have the following theorem.

2.5. <u>Theorem</u>. Let M_t be a continuous local martingale. Then there is a continuous increasing process $\langle M\rangle_t$ such that $\langle M\rangle_t^{\Delta}$ converges uniformly to $\langle M\rangle_t$ in probability.

<u>Proof.</u> Let $\{T_n\}$ be a sequence of stopping times such that $T_n \uparrow \infty$ and each stopped process $M_t^n \equiv M_t^{T_n}$ is a bounded martingale (See Remark after Theorem 1.3). Then it holds $\langle M^n\rangle_{t\wedge T_m} = \langle M^m\rangle_t$ if $m \leq n$, because $\langle M^n\rangle_{t\wedge T_m}^{\Delta} = \langle M^m\rangle_t^{\Delta}$ is satisfied. Hence we can define a continuous increasing process $\langle M\rangle_t$ such that $\langle M\rangle_{t\wedge T_n} = \langle M^n\rangle_t$ holds for $t < T_n$. This $\langle M\rangle_t$ satisfies

$$\rho(\langle M\rangle^{\Delta} - \langle M\rangle)^2 \leq \rho(\langle M\rangle^{\Delta,T_n} - \langle M\rangle^{T_n})^2 + P(T_n < a).$$

For any $\varepsilon > 0$ choose n so large as $P(T_n < a) < \varepsilon$ and let $|\Delta|$ tend to 0. Then we have $\lim_{|\Delta|\to 0} \rho(\langle M\rangle^{\Delta} - \langle M\rangle) < \varepsilon$. The proof is complete.

Remark. Let M_t be a continuous local martingale and let T be a stopping time. Then it holds $<M^T>_t = <M>_t^T$ for all t a.s. In fact, it is easy to see that $<M^T>_t^\Delta = (<M>^\Delta)_t^T$ holds for any partition Δ. Letting $|\Delta|$ tend to 0, we get the desired relation.

2.6. Corollary. $M_t^2 - <M>_t$ is a local martingale if M_t is a continuous local martingale.

Proof. Suppose that M_t is a bounded martingale. Then $M_t^2 - <M>_t^\Delta$ is an L^2-martingale by Lemma 2.3. It converges to $M_t^2 - <M>_t$ in L^2-sense as $|\Delta| \to 0$ by Theorem 2.2. Therefore, the limit $M_t^2 - <M>_t$ is an L^2-martingale. Now let M_t be an arbitrary continuous local martingale and let $\{S_k\}$ be a sequence of stopping times such that $S_k \uparrow \infty$ and each stopped process $M_t^{S_k}$ is a bounded martingale. Then $(M^2 - <M>)_t^{S_k} = (M_t^{S_k})^2 - <M^{S_k}>_t$ is a martingale for each k. This proves that $M_t^2 - <M>_t$ is a local martingale.

2.7. Corollary. An element M of $\underline{\underline{M}}_c^{loc}$ belongs to $\underline{\underline{M}}_c$ if and only if $<M>_a$ is integrable. In this case, $M_t^2 - <M>_t$ is a martingale.

Proof. Let $M_t \in \underline{\underline{M}}_c^{loc}$ and $\{S_k\}$ be a sequence of stopping times mentioned above. Suppose that $<M>_a$ is integrable. Then

$$E[\sup_t |M_t|^2] = E[\lim_{k\to\infty} \sup_t |M_t^{S_k}|^2] \le \underline{\lim}_{k\to\infty} E[\sup_t |M_t^{S_k}|^2]$$

$$\le 4\underline{\lim}_{k\to\infty} E[|M_a^{S_k}|^2] = 4 \lim_{k\to\infty} E[<M>_a^{S_k}] = 4E[<M>_a] \quad \infty.$$

Therefore M_t is an L^2-martingale. Furthermore, $(M_t^{S_k})^2 - <M>_t^{S_k}$ is dominated by an integrable random variable $\sup_t |M_t|^2 + <M>_a$ for each t and k. Hence the sequence of martingales $(M_t^{S_k})^2 - \langle M\rangle_t^{S_k}$, k=1, 2,... converges to $M_t^2 - \langle M\rangle_t$ in L^1-sense. This proves that $M_t^2 - \langle M\rangle_t$ is a martingale.

Conversely if M_t is an L^2-martingale, then

$$E[\sup_t |M_t|^2] \geq E[\sup_t |M_t^{S_k}|^2] \geq E[|M_a^{S_k}|^2] = E[\langle M\rangle_a^{S_k}].$$

Therefore $E[\langle M\rangle_a] \leq E[\sup_t |M_t|^2] < \infty$. The proof is complete.

The following characterization of the quadratic variation is sometimes useful for finding the quadratic variation of a given local martingale, explicitly.

2.8. Theorem. Let M_t be a continuous local martingale. A continuous increasing process A_t satisfying $A_0 = 0$ coincides with the quadratic variation of M_t if and only if $M_t^2 - A_t$ is a local martingale.

Proof. "Only if" is clear from Corollary 2.6. Suppose that $M_t^2 - A_t$ is a local martingale. Then $\langle M\rangle_t - A_t$ is a continuous local martingale of bounded variation, whose quadratic variation is 0. Therefore by Corollary 2.7, $\langle M\rangle_t - A_t$ is an L^2-martingale and $(\langle M\rangle_t - A_t)^2$ is an L^1-martingale. This proves $E[(\langle M\rangle_t - A_t)^2] = 0$, and we get $\langle M\rangle_t = A_t$.

Remark. Corollary 2.7 indicates that the submartingale M_t^2 is decomposed into the sum of martingale $N_t = M_t^2 - \langle M\rangle_t$ and increasing process $\langle M\rangle_t$. The decomposition is known as the Doob-Meyer decomposition of the submartingale. Note that we did not use the decomposition theorem for the proof of Theorem 2.2. If one knows the theorem and apply it, then one can prove the theorem more easily. See Meyer [34] for this direction.

We will finally consider the quadratic variation of a continuous semimartingale. Let X_t be a continuous semimartingale and let $X_t = M_t + A_t$ be the decomposition to the local martingale M_t and a process of bounded variation A_t. The quadratic variation $\langle X\rangle_t^\Delta$ associated with the partition Δ satisfies

$$|\langle X\rangle_t^\Delta - \langle M\rangle_t^\Delta - \langle A\rangle_t^\Delta| \leq 2\{\langle M\rangle_t^\Delta \langle A\rangle_t^\Delta\}^{\frac{1}{2}}.$$

$\langle M\rangle_t^\Delta$ converges uniformly to $\langle M\rangle_t$ in probability and $\langle A\rangle_t^\Delta$ converges uniformly to 0 a.s. Therefore $\langle X\rangle_t^\Delta$ converges uniformly to $\langle M\rangle_t$ in probability. We then have the following theorem.

2.9. <u>Theorem.</u> Let X_t be a continuous semimartingale. Then $\langle X\rangle_t^\Delta$ converges uniformly to $\langle M\rangle_t$ in probability as $|\Delta| \to 0$, where M_t is the local martingale part of X_t.

3. CONTINUITY OF QUADRATIC VARIATIONS IN $\underline{\underline{M}}_c$ AND $\underline{\underline{M}}_c^{loc}$ -

Quadratic variations are continuous in the space $\underline{\underline{M}}_c$ and $\underline{\underline{M}}_c^{loc}$ in their topologies.

3.1. <u>Theorem.</u> (1) Let $\{M^n\}$ be a sequence in $\underline{\underline{M}}_c$. It converges to M of $\underline{\underline{M}}_c$ if and only if $\{\langle M^n - M\rangle_a\}$ converges to 0 in L^1-norm.

(2) Let $\{M^n\}$ be a sequence in $\underline{\underline{M}}_c^{loc}$. It converges to M of $\underline{\underline{M}}_c^{loc}$ if and only if $\{\langle M^n - M\rangle_a\}$ converges to 0 in probability.

<u>Proof.</u> The first assertion (1) is immediate from the relation

$$\|M^n - M\|^2 \geq E[\,|M_a^n - M_a|^2] = E[\langle M^n - M\rangle_a] \geq \frac{1}{4}\|M^n - M\|^2.$$

Suppose next that $\{M^n\}$ of $\underline{\underline{M}}_c^{loc}$ converges to M of $\underline{\underline{M}}_c^{loc}$. If $\{\langle M^n - M\rangle_a\}$ does not converge to 0 in probability, there are $\varepsilon > 0$ and a subsequence $\{M^{n_i}\}$ such that $\overline{\lim_{i\to\infty}} P(\langle M^{n_i} - M\rangle_a > \varepsilon) > 0$. Choose a subsequence of $\{M^{n_i}\}$ denoted by $\{M^{n_i'}\}$ converging to M uniformly a.s. Then there is a sequence of stopping times $\{T_k\} \uparrow \infty$ such that stopped processes $M^{n_i'}_{t\wedge T_k}$ are in $\underline{\underline{M}}_c$ and converge to $M_{t\wedge T_k}$ in the space $\underline{\underline{M}}_c$ as $n_i' \to \infty$. See the proof of Theorem 1.3. Then $\{\langle M^{n_i'} - M\rangle_{a\wedge T_k}\}$ converges to 0 in L^1-norm. Since it is valid for any T_k, $\{\langle M^{n_i'} - M\rangle_a\}$ converges to 0 in probability. This is a contradiction. We have thus shown that $\langle M^n - M\rangle_a$ converges to 0 in probability.

Conversely suppose that $\{\langle M^n - M\rangle_a\}$ converges to 0 in probability. If $\{M^n\}$ does not converge to M in $\underline{\underline{M}}_c^{loc}$, there are $\varepsilon > 0$ and a subsequence $\{M^{n_i}\}$ such that $\overline{\lim_{i\to\infty}} P(\sup_t |M_t^{n_i} - M_t| > \varepsilon) > 0$. Choose a subsequence of $\{M^{n_i}\}$ denoted by $\{M^{n_i'}\}$ such that $\langle M^{n_i'} - M\rangle_a$ converges to 0 a.s. Then $\{\langle M^{n_i'}\rangle_t\}$ converges uniformly a.s., because

$$\sup_t |\langle M^{n_i'}\rangle_t^{\frac{1}{2}} - \langle M\rangle_t^{\frac{1}{2}}| \leq \langle M^{n_i'} - M\rangle_a^{\frac{1}{2}}.$$

Now set $B_t = \sup_i \langle M^{n_i'}\rangle_t$. It is a continuous increasing process. For a positive integer k, define $S_k = \inf\{t > 0 ; B_t \geq k\}$ ($= \infty$ if $\{\dots\} = \phi$). Then $\langle M^{n_i'}\rangle_{t\wedge S_k} \leq k$, so that we have

$$\langle M^{n_i'} - M\rangle_{t\wedge S_k} \leq 2(\langle M^{n_i'}\rangle_{t\wedge S_k} + \langle M\rangle_{t\wedge S_k}) \leq 2k$$

for all n_i', t. Therefore $\langle M^{n_i'} - M\rangle_{t\wedge S_k}$ converges to 0 in L^1-norm. Consequently, $\{M^{n_i'}_{t\wedge S_k},\ i = 1, 2, \dots\}$ is in $\underline{\underline{M}}_c$ and converges to $M_{t\wedge S_k}$ in $\underline{\underline{M}}_c$ by the assertion (1). Hence $\{M_t^{n_i'}\}$ converges to M_t uniformly in probability. This is a contradiction. We have thus shown that $\{M^n\}$ converges to M in $\underline{\underline{M}}_c^{loc}$.

3.2. <u>Theorem.</u> (1) Let $\{M^n\}$ be a sequence in $\underline{\underline{M}}_c$ converging to M of $\underline{\underline{M}}_c$. Then it holds

$$\sup_\Delta E[\sup_t \langle M^n - M\rangle_t^\Delta] \xrightarrow[n\to\infty]{} 0.$$

(2) Let $\{M^n\}$ be a sequence in $\underline{\underline{M}}_c^{loc}$ converging to M of $\underline{\underline{M}}_c^{loc}$. Then it holds for any $\varepsilon > 0$

$$\sup_\Delta P[\sup_t \langle M^n - M\rangle_t^\Delta > \varepsilon] \xrightarrow[n\to\infty]{} 0.$$

<u>Proof.</u> The first assertion is obvious since

$$E[\sup_t \langle M^n - M\rangle_t^\Delta] \le 17E[|M_a^n - M_a|^2].$$

For the proof of (2), suppose on the contrary that the assertion is not valid. Then for some $\varepsilon > 0$ there is a sequence $n_1 < n_2 < \dots$ and $\Delta_{n_1}, \Delta_{n_2}, \dots$ such that $\varlimsup_{i\to\infty} P(\sup_t \langle M^{n_i} - M\rangle_t^{\Delta_{n_i}} > \varepsilon) > 0$. Choose a subsequence denoted by $\{M^{n_i'}\}$ converging to M uniformly a.s. There is an increasing sequence of stopping times $\{T_k\} \uparrow \infty$ such that $M_{t\wedge T_k}^{n_i'}$ converges to $M_{t\wedge T_k}$ in $\underline{\underline{M}}_c$ as $n_i' \to \infty$ for each k. Then $E[\langle M^{n_i'} - M\rangle_{a\wedge T_k}] \longrightarrow 0$ as $n_i' \to \infty$ by Theorem 3.1. Therefore,

$$\begin{aligned}
P(\sup_t \langle M^{n_i'} - M\rangle_t^{\Delta_{n_i'}} > \varepsilon) &\le P(\sup_t \langle M^{n_i'} - M\rangle_{t\wedge T_k}^{\Delta_{n_i'}} > \varepsilon,\ T_k > a) + P(T_k \le a) \\
&\le \frac{1}{\varepsilon} E[\sup_t \langle M^{n_i'} - M\rangle_{t\wedge T_k}^{\Delta_{n_i'}}] + P(T_k \le a) \\
&\le \frac{4}{\varepsilon} E[\langle M^{n_i'} - M\rangle_{a\wedge T_k}] + P(T_k \le a) \\
&\longrightarrow 0,\quad n_i' \longrightarrow \infty,\ k \longrightarrow \infty.
\end{aligned}$$

This is a contradiction. The proof is complete.

4. JOINT QUADRATIC VARIATIONS.

Let M and N be elements of $\underline{\underline{M}}_c^{loc}$. The joint quadratic variation of M, N associated with the partition $\Delta = \{0 = t_0 < \dots < t_n = a\}$ is defined by

$$\langle M,N\rangle_t^\Delta = \sum_{i=0}^{k-1} (M_{t_{i+1}} - M_{t_i})(N_{t_{i+1}} - N_{t_i}) + (M_t - M_{t_k})(N_t - N_{t_k}),$$

where k is the number such that $t_k \le t < t_{k+1}$.

4.1. Theorem. $\langle M,N\rangle^\Delta$ converges uniformly to a continuous process of bounded variation $\langle M,N\rangle$ in probability as $|\Delta| \to 0$.

Proof is immediate from

$$\langle M,N\rangle_t^\Delta = \frac{1}{4}\{\langle M+N\rangle_t^\Delta - \langle M-N\rangle_t^\Delta\}$$

and Theorem 2.5.

Remark. $\langle M,M\rangle = \langle M\rangle$.

The following is immediate from Theorem 2.8.

4.2. Corollary. Given M, N of $\underline{\underline{M}}_c^{loc}$, a continuous process of bounded variation A[1] coincides with the joint quadratic variation $\langle M,N\rangle$ if and only if $MN - A$ is a local martingale.

4.3. Theorem. Joint quadratic variations have the following properties.

(i) bilinear: $\langle aM^1 + bM^2, N\rangle = a\langle M^1,N\rangle + b\langle M^2,N\rangle$ holds for any

---

1) $A_0 = 0$ a.s. is assumed.

M^1, M^2, N of $\underline{\underline{M}}_c^{loc}$ and real numbers a, b.

(ii) <u>symmetric:</u> $\langle M,N\rangle = \langle N,M\rangle$ for any M, N of $\underline{\underline{M}}_c^{loc}$.

(iii) <u>positive definite:</u> $\langle M\rangle_t - \langle M\rangle_s \geq 0$ holds for any $t \geq s$ and the equality holds a.s. if and only if $M_r = M_s$ holds for all $r \in [s,t]$ a.s.

(iv) <u>Schwarz's inequality:</u>

$$|\langle M,N\rangle_t - \langle M,N\rangle_s| \leq (\langle M\rangle_t - \langle M\rangle_s)^{\frac{1}{2}}(\langle N\rangle_t - \langle N\rangle_s)^{\frac{1}{2}}.$$

(v) <u>extended Schwarz's inequality:</u> Let f_u, g_u, $u \in [0,a]$ be processes measurable with respect to the smallest σ-field on $[0,a]\times\Omega$ for which all continuous stochastic processses are measurable. Suppose

$$\int_0^t |f_u|^2 d\langle M\rangle_u < \infty, \quad \int_0^t |g_u|^2 d\langle N\rangle_u < \infty.$$

Then

$$\left|\int_0^t f_u g_u d\langle M,N\rangle_u\right| \leq \left(\int_0^t |f_u|^2 d\langle M\rangle_u\right)^{\frac{1}{2}}\left(\int_0^t |g_u|^2 d\langle N\rangle_u\right)^{\frac{1}{2}}.$$

<u>Proof.</u> Properties (i) - (iii) are obvious. The property (iv) follows immediately from (i) - (iii). We will prove (v). Suppose that f_u and g_u are step functions: there are a partition $\Delta = \{0 = t_0 < \ldots < t_n = t\}$ of $[0,t]$ and bounded random variables f_i and g_i such that $f_u = f_i$, $g_u = g_i$ if $t_i \leq u < t_{i+1}$. Then

$$\left|\int_0^t f_u g_u d\langle M,N\rangle_u\right|$$

$$= \left|\sum_{i=0}^{n-1} f_i g_i [\langle M,N\rangle_{t_{i+1}} - \langle M,N\rangle_{t_i}]\right|$$

$$\leq \sum_{i=0}^{n-1} |f_i||g_i| [\langle M\rangle_{t_{i+1}} - \langle M\rangle_{t_i}]^{\frac{1}{2}} [\langle N\rangle_{t_{i+1}} - \langle N\rangle_{t_i}]^{\frac{1}{2}}$$

$$\leq \left\{\sum_{i=0}^{n-1} |f_i|^2 [\langle M\rangle_{t_{i+1}} - \langle M\rangle_{t_i}]\right\}^{\frac{1}{2}} \left\{\sum_{i=0}^{n-1} |g_i|^2 [\langle N\rangle_{t_{i+1}} - \langle N\rangle_{t_i}]\right\}^{\frac{1}{2}}$$

$$\leq \{\int_0^t |f_u|^2 d\langle M\rangle_u\}^{\frac{1}{2}} \{\int_0^t |g_u|^2 d\langle N\rangle_u\}^{\frac{1}{2}} .$$

Extensions to general f and g will be clear.

Joint quadratic variations are continuous in $\underline{\underline{M}}_c$ and $\underline{\underline{M}}_c^{loc}$ in their topologies.

4.4. Theorem. (1) Let $\{M^n\}$ be a sequence of $\underline{\underline{M}}_c$ converging to M. Then it holds for any N of $\underline{\underline{M}}_c$,

$$\lim_{n\to\infty} E[\sup_t |\langle M^n - M, N\rangle_t|] = 0 ,$$

$$\lim_{n\to\infty} \sup_\Delta E[\sup_t |\langle M^n - M, N\rangle_t^\Delta|] = 0.$$

(2) Let $\{M^n\}$ be a sequence of $\underline{\underline{M}}_c^{loc}$ converging to M. Then it holds for any $\varepsilon > 0$ and N of $\underline{\underline{M}}_c^{loc}$,

$$\lim_{n\to\infty} P(\sup_t |\langle M^n - M, N\rangle_t| > \varepsilon) = 0,$$

$$\lim_{n\to\infty} \sup_\Delta P(\sup_t |\langle M^n - M, N\rangle_t^\Delta| > \varepsilon) = 0.$$

Proof is immediate from Theorems 3.1 and 3.2.

Finally we will mention the joint quadratic variations of continuous semimartingales. Let X and Y be continuous semimartingales. The joint quadratic variation associated with the partition Δ is defined as before and is written as $\langle X,Y\rangle^\Delta$. The following theorem is immediate.

4.5. Theorem. $\langle X,Y\rangle^\Delta$ converges uniformly in prabability to a continuous process of bounded variation $\langle X,Y\rangle_t$. If M and N are local martingale parts of X and Y, respectively, then $\langle X,Y\rangle$ coincides with $\langle M,N\rangle$.

5. STOCHASTIC INTEGRALS.

Let M_t be a continuous local martingale and let f_t be a continuous $(\underline{\underline{F}}_t)$-adapted process. We will define the stochastic integral of f_t by the differential dM_t. Here, the differential does not mean a signed measure, since the sample function of a continuous local martingale is not of bounded variation, except a trivial martingale $M_t \equiv$ constant a.s. Nevertheless, the integral is well defined if the integrand f_t is $(\underline{\underline{F}}_t)$-adapted: Our discussion will be based on the properties of martingales, specially those of quadratic variations.

Let $\Delta = \{0 = t_0 < \ldots < t_n = a\}$ be a partition of the interval $[0,a]$. For any $t \in [0,a]$, choose t_k of Δ such that $t_k \leq t < t_{k+1}$ and define

$$(1) \quad L_t^{\Delta} = \sum_{i=0}^{k-1} f_{t_i}(M_{t_{i+1}} - M_{t_i}) + f_{t_k}(M_t - M_{t_k}).$$

It is easy to see that L_t^{Δ} is a continuous local martingale. The quadratic variation is computed directly as

$$(2) \quad \langle L^{\Delta} \rangle_t = \sum_{i=0}^{k-1} f_{t_i}^2(\langle M \rangle_{t_{i+1}} - \langle M \rangle_{t_i}) + f_{t_k}^2(\langle M \rangle_t - \langle M \rangle_{t_k})$$

$$= \int_0^t |f_s^{\Delta}|^2 d\langle M \rangle_s,$$

where f_s^{Δ} is a step process defined from f_s by $f_s^{\Delta} = f_{t_k}$ if $t_k \leq s < t_{k+1}$.

Let Δ' be another partition of $[0,a]$. We define $L_t^{\Delta'}$ similarly using the same f_s and M_s. Then it holds

$$\langle L^{\Delta} - L^{\Delta'} \rangle_t = \int_0^t |f_s^{\Delta} - f_s^{\Delta'}|^2 d\langle M \rangle_s.$$

Now let $\{\Delta_n\}$ be a sequence of partitions of $[0,a]$ such that $|\Delta_n| \to 0$. Then $\langle L^{\Delta_n} - L^{\Delta_m}\rangle_a$ converges to 0 in probability as $n,m \to \infty$. Hence $\{L^{\Delta_n}\}$ is a Cauchy sequence in $\underline{\underline{M}}_c^{loc}$ by Theorem 3.1. We denote the limit as L_t.

5.1. Definition. The above L_t is called the Itô integral of f_t by dM_t and is denoted by $\int_0^t f_s dM_s$.

The Itô integral can be defined to more general class of stochastic processes called predictable ones. Here the predictable σ-field is, by definition, the least σ-field on the product space $[0,a] \times \Omega$ for which all continuous $(\underline{\underline{F}}_t)$-adapted processes $f_t(\omega)$ are measurable. A predictable process is, by definition, a process measurable to the predictable σ-field. A continuous $(\underline{\underline{F}}_t)$-adapted process is predictable, obviously.

Now let M_t be a continuous local martingale and let $\langle M\rangle_t$ be the quadratic variation. We denote by $L^2(\langle M\rangle)$ the set of all predictable processes f_t such that $\int_0^a |f_s|^2 d\langle M\rangle_s < \infty$ a.s. Then the set of continuous $(\underline{\underline{F}}_t)$-adapted processes is dense in $L^2(\langle M\rangle)$, i.e., for any f of $L^2(\langle M\rangle)$, there is a sequence of continuous $(\underline{\underline{F}}_t)$-adapted processes f_t^n such that $\int_0^a |f_s^n - f_s|^2 d\langle M\rangle_s$ converges to 0 a.s. Then the sequence of stochastic integrals $\int_0^t f_s^n dM_s$, $n=1,2,\ldots$ forms a Cauchy sequence in $\underline{\underline{M}}_c^{loc}$. Denote the limit as $\int_0^t f_s dM_s$ and call it the Itô integral of f_t by dM_t.

5.2. Theorem. (i) Let $M \in \underline{\underline{M}}_c^{loc}$ amd $f \in L^2(\langle M\rangle)$. Then Itô integral satisfies the following relation

$$(3) \quad \left\langle \int f dM, N\right\rangle_t = \int_0^t f_s d\langle M,N\rangle_s, \qquad \forall N \in \underline{\underline{M}}_c^{loc}.$$

(ii) Conversely suppose that L of $\underline{\underline{M}}_c^{loc}$ satisfies

$$(4) \quad \langle L,N\rangle_t = \int_0^t f_s d\langle M,N\rangle_s, \qquad \forall N \in \underline{\underline{M}}_c^{loc}.$$

Then L is the Itô integral of f_t by dM_t, i.e., Itô integral is characterized as the unique element L in $\underline{\underline{M}}_c^{loc}$ satisfying (4).

Proof. If f is a step process, the relation (3) is direct from the computation of the joint quadratic variation. If f is an arbitrary one in $L^2(\langle M\rangle)$, we may choose a sequence of step processes f^n converging to f in $L^2(\langle M\rangle)$ a.s. Then, noting the continuity of the joint quadratic variation, we see that (3) is valid for this f. Now, for the proof of the second assertion, set $L'_t = \int_0^t f_s dM_s$. Then $\langle L-L',N\rangle_t = 0$ is satisfied for all N of $\underline{\underline{M}}_c^{loc}$ because of (3) and (4). Hence $\langle L-L'\rangle_t = 0$, proving $L_t = L'_t$. The proof is complete.

5.3. Corollary. It holds

$$\langle \int f dM\rangle_t = \int_0^t f_s^2 d\langle M\rangle_s.$$

We will list a few properties of Itô integrals.

5.4. Theorem. Let M be an element of $\underline{\underline{M}}_c^{loc}$.

(i) If f, g are in $L^2(\langle M\rangle)$ and a, b are constants, then af + bg is in $L^2(\langle M\rangle)$ and satisfies

$$\int_0^t (af_s + bg_s)dM_s = a\int_0^t f_s dM_s + b\int_0^t g_s dM_s.$$

(ii) Let $f \in L^2(\langle M\rangle)$ and $L_t = \int_0^t f_s dM_s$. Let g_s be a predictable process such that $\int_0^a f_s^2 g_s^2 d\langle M\rangle_s < \infty$ a.s. Then g is in $L^2(\langle L\rangle)$ and

(5) $\int_0^t g_s dL_s = \int_0^t g_s f_s dM_s$.

(iii) Let T be a stopping time. Then it holds

$$\int_0^{t\wedge T} f_s dM_s = \int_0^t f_s dM_s^T = \int_0^{t\wedge T} f_s dM_s^T.$$

5.5. <u>Definition.</u> Let X be a continuous semimartingale decomposed to the sum of a continuous local martingale M and a continuous process of bounded variation A. Let f be a predictable process such that $f \in L^2(\langle M\rangle)$ and $\int_0^a |f_s| d|A|_s < \infty$. Then the <u>Itô integral of f by</u> dX_t is defined as

$$\int_0^t f_s dX_s \equiv \int_0^t f_s dM_s + \int_0^t f_s dA_s.$$

We will define another stochastic integral by the differential $\circ dX_t$:

$$\int_0^t f_s \circ dX_s$$

$$= \lim_{|\Delta|\to 0} \{ \sum_{i=0}^{k-1} \tfrac{1}{2}(f_{t_{i+1}} + f_{t_i})(X_{t_{i+1}} - X_{t_i}) + \tfrac{1}{2}(f_t + f_{t_k})(X_t - X_{t_k})\}$$

5.6. <u>Definition.</u> If the above limit exists, it is called the <u>Stratonovich integral of</u> f <u>by</u> dX_s.

5.7. <u>Theorem.</u> If f is a continuous semimartingale, the Stratonovich integral is well defined and satisfies

$$\int_0^t f_s \circ dX_s = \int_0^t f_s dX_s + \tfrac{1}{2}\langle f, X\rangle_t.$$

<u>Proof</u> is immediate from the relation

$$\sum_{i=0}^{k-1} \tfrac{1}{2}(f_{t_{i+1}} + f_{t_i})(X_{t_{i+1}} - X_{t_i}) + \tfrac{1}{2}(f_t + f_{t_k})(X_t - X_{t_k})$$

$$= \sum_{i=0}^{k-1} f_{t_i}(X_{t_{i+1}} - X_{t_i}) + f_{t_k}(X_t - X_{t_k}) + \frac{1}{2}\langle f, X\rangle_t^{\Delta}.$$

6. STOCHASTIC INTEGRALS OF VECTOR VALUED PROCESSES

Let B a separable reflexive Banach space and let f_s be a B-valued process. Let M be a real valued continuous local martingale. In this and the next section, we will discuss the stochastic integral of the form $\int_0^t f_s dM_s$, which is to be a B-valued local martingale.

We begin with introducing conditional expectations for Banach space valued random variables. Let B be a separable reflexive Banach space with norm $\|\ \|$, and let $\underline{\underline{B}}$ be the topological Borel field of B. We denote by B' the dual space of B. Let $f(\omega)$ be a mapping from Ω into B. It is called a <u>B-valued random variable</u> if it is a measurable mapping from $(\Omega, \underline{\underline{F}})$ into $(B, \underline{\underline{B}})$. This is equivalent to saying that (f, Φ) is a real valued random variable for any Φ of B', where (,) is the canonical bilinear form on $B \times B'$.

Remark. If f is a B-valued random variable, then the norm $\|f\|$ is a real random variable. In fact, let $\underline{\underline{C}}$ be a countable dense subset of the unit ball $\{\Phi \in B' ; \|\Phi\| \leq 1\}$. Then it holds $\|f\| = \sup_{\Phi \in \underline{\underline{C}}} |(f, \Phi)|$, which is clearly measurable.

6.1. <u>Definition</u>. Let $\underline{\underline{G}}$ be a sub σ-field of $\underline{\underline{F}}$. A measurable mapping $g ; (\Omega, \underline{\underline{G}}) \longrightarrow (B, \underline{\underline{B}})$ is called a <u>conditional expectation of</u> f <u>with respect to</u> $\underline{\underline{G}}$ if

(1) $\quad (g, \Phi) = E[(f, \Phi) \mid \underline{\underline{G}}]$

is satisfied a.s. for any Φ of B'.

6.2. Lemma. Suppose that $\|f\|$ is integrable. Then a conditional expectation exists uniquely.

Proof. Suppose that there are two g_1, g_2 satisfying (1). The it holds $(g_1 - g_2, \Phi) = 0$ a.s. for any Φ of B'. This implies $\|g_1 - g_2\| = 0$ a.s. Hence the conditional expectation is at most unique.

We will prove the existence. Let $\underline{\underline{E}}$ be the linear space $\{\Sigma_{k=1}^{n} \alpha_k \Phi_k, n=1,2,\ldots\}$ where α_k are rational numbers and $\{\Phi_k\}$ is a dense subset of B'. Then for almost all ω, we can associate a linear form G_ω on $\underline{\underline{E}}$ such that

a) $G_\omega(\Phi)$ is $\underline{\underline{G}}$-measurable for any Φ of $\underline{\underline{E}}$,

b) $G_\omega(\Phi) = E[(f,\Phi)|\underline{\underline{G}}](\omega)$ a.s. for any Φ of $\underline{\underline{E}}$,

c) $|G_\omega(\Phi)| \le E[\|f\| | \underline{\underline{G}}](\omega) \|\Phi\|$ for all $\Phi \in \underline{\underline{E}}$ a.s.

Obviously the linear form G_ω can be extended uniquely to a continuous linear form on B'. Hence by Riesz's theorem, there is an element $g(\omega)$ of $B'' = B$ such that $G_\omega(\Phi) = (g(\omega), \Phi)$ holds for all Φ of $\underline{\underline{E}}$. Then it holds

$$(g(\omega), \Phi) = E[(f,\Phi)|\underline{\underline{G}}](\omega) \quad \text{a.s.} \quad \text{for any } \Phi.$$

Therefore $g(\omega)$ is a B-valued random variable and in fact the conditional expectation of f with respect to $\underline{\underline{G}}$.

6.3. Definition. The conditional expectation of f with respect to $\underline{\underline{G}}$ is denoted by $E[f|\underline{\underline{G}}]$.

Remark. For any $p \geq 1$, it holds $E[\|E[f|\underline{\underline{G}}]\|^p] \leq E[\|f\|^p]$. In fact, from (1) we have $|(E[f|\underline{\underline{G}}], \Phi)| \le E[\|f\| | \underline{\underline{G}}]$ for any $\Phi \in B'$ such that $\|\Phi\| \le 1$. This implies $\|E[f|\underline{\underline{G}}]\| \le E[\|f\| | \underline{\underline{G}}]$ and the assertion follows.

The following lemma corresponds to Doob's convergence theorem of a real L^p martingale.

6.4. Lemma. Let f be a B-valued random variable such that $E[\|f\|^p] < \infty$ for some $p \geq 1$. Let $\underline{\underline{G}}_n$, n=1,2,... be an increasing sequence of σ-fields such that $\bigvee_n \underline{\underline{G}}_n = \underline{\underline{F}}$. Set $f_n = E[f|\underline{\underline{G}}_n]$. Then it holds $E[\|f - f_n\|^p] \longrightarrow 0$ as $n \to \infty$.

Proof. Let $\underline{\underline{C}}_n$ be the set of all $\underline{\underline{G}}_n$-measurable B-valued random variables f such that $E[\|f\|^p] < \infty$. Set $\underline{\underline{C}} = \bigcup_n \underline{\underline{C}}_n$. The assertion is clearly true for any f of $\underline{\underline{C}}$. Let f be an arbitrary B-valued random variable such that $E[\|f\|^p] < \infty$. Then we may choose a sequence $\{g_m\}$ of $\underline{\underline{C}}$ converging to f in the sense $E[\|f - g_m\|^p] \longrightarrow 0$. Then

$$E[\|f - f_n\|^p]^{\frac{1}{p}} \leq E[\|f - g_m\|^p]^{\frac{1}{p}} + E[\|g_m - E[g_m|\underline{\underline{G}}_n]\|^p]^{\frac{1}{p}}$$
$$+ E[\|E[g_m - f|\underline{\underline{G}}_n]\|^p]^{\frac{1}{p}}.$$

Note the relation in the remark after Definition 6.3. Then the last member is dominated by $E[\|g_m - f\|^p]^{1/p}$. Therefore, making n tend to infinity, we have

$$\overline{\lim_{n\to\infty}} E[\|f - f_n\|^p]^{\frac{1}{p}} \leq 2E[\|f - g_m\|^p]^{\frac{1}{p}}.$$

The right hand side converges to 0 as $m \to \infty$. Thererfore we have $\lim_{n\to\infty} E[\|f - f_n\|^p]^{1/p} = 0$. The proof is complete.

B-valued martingale adapted to σ-fields $(\underline{\underline{F}}_t)$ is defined by means of B-valued conditional expectations. Let L_t, $t \in [0,a]$ be a B-valued,

$(\underline{\underline{F}}_t)$-adapted, measurable process. It is called a <u>martingale</u> if $E[\|L_t\|] < \infty$ for any t and $E[L_t | \underline{\underline{F}}_s] = L_s$ holds a.s. for any $t > s$. It is a <u>local martingale</u> if there is an increasing sequence of stopping times T_n such that $T_n \uparrow \infty$ and each stopped process $L_t^{T_n} = L_{t \wedge T_n}$ is a martingale. If L_t is a (local) martingale then (L_t, Φ) is a real valued (local) martingale for any Φ of B'.

6.5. <u>Definition.</u> Let M_t be a real continuous local martingale and let f_s be a B-valued predictable process such that

$$(2) \qquad \int_0^a \|f_s\|^2 d\langle M\rangle_s < \infty \qquad \text{a.s.}$$

A B-valued local martingale L_t is called the <u>stochastic integral of f_t by</u> dM_t if it satisfies

$$(3) \qquad (L_t, \Phi) = \int_0^t (f_s, \Phi) dM_s$$

for any Φ of B'. If it exists, we denote it by $\int_0^t f_s dM_s$.

It is obvious that stochastic integral is at most one. The existence is easily seen if f_t is a step process, i.e., $f_t = f_{t_i}$ holds for all $t \in (t_i, t_{i+1}]$, where $0 = t_0 < t_1 < \ldots < t_n = a$. In fact,

$$(4) \qquad L_t \equiv \sum_{i=0}^{k-1} f_{t_i}(M_{t_{i+1}} - M_{t_i}) + f_{t_k}(M_t - M_{t_k}), \quad t_k \le t < t_{k+1}$$

is a B-valued local martingale satisfying (3). However, it is not an easy problem to show in general the existence of stochastic integrals of any B-valued predictable processes satisfying (2). In this lecture, we will prove it in two cases. The first is the case that B is a Hilbert space, which will be discussed at the remainder of this section: The second is the case that B is a Sobolev space, which will be discussed at the next section.

In the sequel, we assume that B is a Hilbert space. Then stochastic integrals can be defined similarly as the case of real valued process. Key points are the following two lemmas.

6.6. Lemma. Suppose that the step process f_t satisfies $E[\int_0^a ||f_s||^2 d\langle M\rangle_s] < \infty$. Then L_t defined by (4) is a martingale and satisfies

(5) $$E[||L_t||^2] = E[\int_0^t ||f_s||^2 d\langle M\rangle_s], \qquad \forall t \in [0,a].$$

Proof. We will only prove the case that M_t is a square integrable martingale. From (4), it holds

$$E[||L_t||^2] = \sum_{i,j} E[(L_{t_{i+1}} - L_{t_i}, L_{t_{j+1}} - L_{t_j})]$$
$$= \sum_{i,j} E[(f_{t_i}, f_{t_j})(M_{t_{i+1}} - M_{t_i})(M_{t_{j+1}} - M_{t_j})].$$

If $i < j$, then

$$E[(f_{t_i}, f_{t_j})(M_{t_{i+1}} - M_{t_i})(M_{t_{j+1}} - M_{t_j})] = 0,$$

since $(f_{t_i}, f_{t_j})(M_{t_{i+1}} - M_{t_i})$ is $\underline{\underline{F}}_{t_j}$-adapted. If $i = j$, then it holds

$$E[||f_{t_i}||^2 (M_{t_{i+1}} - M_{t_i})^2] = E[||f_{t_i}||^2 (\langle M\rangle_{t_{i+1}} - \langle M\rangle_{t_i})],$$

since $E[(M_{t_{i+1}} - M_{t_i})^2 | \underline{\underline{F}}_{t_i}] = E[\langle M\rangle_{t_{i+1}} - \langle M\rangle_{t_i} | \underline{\underline{F}}_{t_i}]$ holds. Therefore we have

$$E[||L_t||^2] = \sum_i E[||f_{t_i}||^2 (\langle M\rangle_{t_{i+1}} - \langle M\rangle_{t_i})] = E[\int_0^t ||f_s||^2 d\langle M\rangle_s].$$

The proof is complete.

6.7. Lemma. Let f_t be a predictable B-valued process such that

$$(6) \qquad E\left[\int_0^a \|f_s\|^2 d\langle M\rangle_s\right] < \infty.$$

Then there is a sequence of B-valued step processes f_t^n such that

$$(7) \qquad E\left[\int_0^a \|f_s - f_s^n\|^2 d\langle M\rangle_s\right] \xrightarrow[n\to\infty]{} 0.$$

Proof. Let us define an increasing sequence of σ-fields on $[0,a]\times\Omega$ by

$$\underline{\underline{B}}_n = \sigma\{(\frac{ka}{2^n}, \frac{(k+1)a}{2^n}] \times A \; ; \; A \in \underline{\underline{F}}_{\frac{ka}{2^n}}, \; k=0,\dots,2^n\}.$$

Then it holds $\bigvee_n \underline{\underline{B}}_n = \underline{\underline{P}}$. Define a measure $\widetilde{P}$ on the predictable σ-field $\underline{\underline{P}}$:

$$\widetilde{P}(A) = E\left[\int_0^a I_A(s,\omega) d\langle M\rangle_s\right], \qquad A \in \underline{\underline{P}},$$

where I_A is the indicator function of the set A. Then it holds

$$\widetilde{E}[\|f\|^2] = E\left[\int_0^a \|f_s\|^2 d\langle M\rangle_s\right].$$

Consider a sequence of conditional expectations $f^n = \widetilde{E}[f|\underline{\underline{B}}_n]$. Then each f^n is a step process such that $f^n(t,\omega) = f^n(\frac{ka}{2^n},\omega)$ if $\frac{ka}{2^n} < t \le \frac{(k+1)a}{2^n}$. It converges to f in the sense $\widetilde{E}[\|f - f^n\|^2] \longrightarrow 0$ by Lemma 6.4. This shows (7). The proof is complete.

6.8. Theorem. For any predictable Hilbert space valued process f_t satisfying (2), the stochastic integral $\int_0^t f_s dM_s$ is well defined. Furthermore, it is a strongly continuous Hilbert space valued local martingale.

Proof. Suppose that f_t satisfies (6). Let f^n be a sequence of step processes converging to f in the sense of (7). Then by Lemma 6.6 and

Doob's inequality of positive submartingale, we have

$$E[\sup_{t\leq a}\|L_t^n - L_t^m\|^2] \leq 4E[\|L_a^n - L_a^m\|^2] = 4E[\int_0^a \|f_s^n - f_s^m\|^2 d\langle M\rangle_s] \longrightarrow 0$$

as $n,m \to \infty$. Therefore there is a strongly continuous B-valued process L_t such that $E[\sup_{t\leq a}\|L_t - L_t^n\|]$ tends to 0. This L_t satisfies (3) obviously.

Next suppose f_t satisfy (2). Define an increasing sequence of stopping times T_k by $\inf\{t>0 ; \int_0^t \|f_s\|^2 d\langle M\rangle_s \geq k\}$ (= ∞ if $\{\ldots\} = \phi$). For each k, there is a strongly continuous B-valued martingale $L_t^{(k)}$ such that $(L_t^{(k)},\Phi) = \int_0^{t\wedge T_k}(f_s,\Phi)dM_s$ holds for any Φ of B'. If $k<k'$, it holds $(L_t^{(k)},\Phi) = (L_{t\wedge T_k}^{(k')},\Phi) = \int_0^{t\wedge T_k}(f_s,\Phi)dM_s$. Therefore we have $L_t^{(k)} = L_{t\wedge T_k}^{(k')}$ if $k<k'$. Hence we may define a strongly continuous B-valued process L_t, $t\in[0,a]$ by $L_t = L_t^{(k)}$ if $t<T_k$. Then L_t satisfies (3) and is a strongly continuous B-valued local martingale. The proof is complete.

7. REGULARITY OF INTEGRALS WITH RESPECT TO PARAMETERS

Let $f_s(\lambda)$ be a real valued predictable process with parameter $\lambda\in\Lambda$ and let M_t be a continuous local martingale. If $\int_0^a |f_s(\lambda)|^2 d\langle M\rangle_s < \infty$ a.s. for any λ, Itô's stochastic integral $\int_0^t f_s(\lambda)dM_s$ is well defined except for a null set for each λ. However, the exceptional set may depend on the parameter λ. Therefore, in order to discuss the regularity of the integral with respect to the parameter, we have to choose a good modification of the integrals so that the exceptional set does not depend on λ. For this purpose, we shall consider that $f_s(\lambda)$ is a Sobolev space valued process and we shall define the integral as a Sobolev space valued local martingale.

Let us introduce some notations concerning Sobolev space. The parameter space Λ is assumed to be a bounded domain in R^d. Let $\lambda = (\lambda_1,\dots,\lambda_d) \in \Lambda$ and $k = (k_1,\dots,k_d)$ be a multi-index of non-negative integers. We denote by D^k the differential operator $(\frac{\partial}{\partial \lambda_1})^{k_1}$, $\dots (\frac{\partial}{\partial \lambda_d})^{k_d}$. Let p be a real number greater than 1 and m be a nonnegative integer. A Sobolev space of type p,m, denoted by W_p^m, is the set of all L_p functions ϕ on Λ such that derivatives $D^k\phi$, $|k|$ $(=k_1+\dots+k_d) \le m$ in the distributional sense are all L_p functions. For $\phi \in W_p^m$, we define the norm $\|\ \|_{p,m}$ by

$$\|\phi\|_{p,m} = | \sum_{|k|\le m} \int_\Lambda |D^k\phi(\lambda)|^p d\lambda |^{\frac{1}{p}} .$$

Then W_p^m is a separable reflexive Banach space.

Now let C_b^m be the set of all m-times continuously differentiable functions whose derivatives up to m are all bounded. For $\phi \in C_b^m$, we define the norm

$$\|\phi\|_{\infty,m} = \sum_{|k|\le m} \sup_\lambda |D^k\phi(\lambda)| .$$

Then C_b^m is a separable Banach space.

A fundamental result concerning Sobolev space is the following.

7.1. Sobolev's embedding theorem. Let ℓ be a nonnegative integer less than $m - \frac{d}{p}$. Then it holds $W_p^m \subset C_b^\ell$ and there is a positive constant $K_{p,m}^\ell$ such that

$$\|\phi\|_{\infty,\ell} \le K_{p,m}^\ell \|\phi\|_{p,m}, \qquad \forall \phi \in W_p^m .$$

In the following, we will fix p,m and omit it from the notation of the norm.

We shall now define the stochastic integral of W_p^m-valued process. If f_t is a predictable W_p^m-valued step process, the stochastic integral was defined by (4) in the previous section. In order to define the integral for more general class of f_t, we need a lemma analogous to Lemma 6.6.

7.2. Lemma. Let $p \geq 2$. There exists a positive constant C such that

(1) $$E[\|L_t\|^p] \leq CE[(\int_0^t \|f_s\|^p d\langle M\rangle_s)\langle M\rangle_t^{\frac{p}{2}-1}], \qquad \forall t \in [0,a]$$

holds for any step process f_t.

For the proof, we require Burkholder's inequality, which will be proved at the next section.

7.3. Theorem. (Burkholder's inequality). Let $p \geq 2$. Then there is a positive constant $C^{(p)}$ such that

(2) $$E[|M_t|^p] \leq C^{(p)} E[\langle M\rangle_t^{\frac{p}{2}}] \qquad \forall t \in [0,a]$$

holds for any $M \in \underline{\underline{M}}_c$ such that $E[|M_a|^p] < \infty$.

Proof of Lemma 7.2. It holds

$$\|L_t\|^p = \sum_{|k| \leq m} \int_\Lambda |\int_0^t D^k f_s(\lambda) dM_s|^p d\lambda.$$

By Burkholder's inequality and Hölder's inequality, we have

$$E[\|L_t\|^p] = \sum_{|k| \leq m} \int_\Lambda E[|\int_0^t D^k f_s(\lambda) dM_s|^p] d\lambda$$

$$\leq C^{(p)} \sum_{|k| \leq m} \int_\Lambda E[|\int_0^t |D^k f_s(\lambda)|^2 d\langle M\rangle_s|^{\frac{p}{2}}] d\lambda$$

$$\le C^{(p)} \sum_{|k|\le m} \int_\Lambda E[(\int_0^t |D^k f_s(\lambda)|^p d\langle M\rangle_s)\langle M\rangle_t^{\frac{p}{2}-1}]d\lambda$$

$$\le C^{(p)} E[(\int_0^t \|f_s\|^p d\langle M\rangle_s)\langle M\rangle_t^{\frac{p}{2}-1}].$$

This proves the lemma.

7.4. Lemma. Let $p \ge 2$. Let f_t be a predictable process such that

$$E[(\int_0^a \|f_s\|^p d\langle M\rangle_s)\langle M\rangle_a^{\frac{p}{2}-1}] < \infty.$$

Then there is a sequence of step processes f_t^n such that

$$E[(\int_0^a \|f_s - f_s^n\|^p d\langle M\rangle_s)\langle M\rangle_a^{\frac{p}{2}-1}] \longrightarrow 0$$

Proof is similar to that of Lemma 6.7. Define a measure $\hat{P}$ on $([0,a]\times\Omega, \underline{\underline{P}})$ by

$$\hat{P}(A) = E[(\int_0^a I_A(s,\omega)d\langle M\rangle_s)\langle M\rangle_a^{\frac{p}{2}-1}]$$

and $f^n = \hat{E}[f|\underline{\underline{B}}_n]$, where $\underline{\underline{B}}_n$, n=1,2,... are σ-fields defined in the proof of Lemma 6.7. Then $\hat{E}[\|f - f^n\|^p] \longrightarrow 0$ by Lemma 6.4. This proves the lemma.

We can now prove the following theorem similarly as that of Theorem 6.8.

7.5. Theorem. Let $p \ge 2$. Let f_t be a predictable W_p^m-valued process satisfying

(3) $$\int_0^a \|f_s\|^p d\langle M\rangle_s < \infty \quad \text{a.s.}$$

Then the stochastic integral $\int_0^t f_s dM_s$ is well defined. It is a strongly continuous W_p^m-valued local martingale.

We shall apply the above theorem to the regularity problem of the real valued stochastic integral $\int_0^t f_s(\lambda)dM_s$ with parameter λ.

7.6. Theorem. Suppose $p \geq 2$ and $mp > d$. Let $f_s(\lambda)$, $\lambda \in \Lambda$ be a predictable C_b^m-valued process satisfying

$$(4) \qquad \int_0^a \|f_s\|_{\infty,m}^p d\langle M\rangle_s < \infty \qquad \text{a.s.}$$

Then the real valued stochastic integral $\int_0^t f_s(\lambda)dM_s$ with parameter λ has a modification $L_t(\lambda)$ which satisfies the following properties.

(i) $L_t(\lambda)$ is continuous in (t,λ) and ℓ-times continuously differentiable[1)] in λ where $\ell < m - \frac{d}{p}$.

(ii) If $|k| < m - \frac{d}{p}$, then $D^k L_t(\lambda)$ is continuous in (t,λ) and satisfies

$$(5) \qquad D^k L_t(\lambda) = \int_0^t D^k f_s(\lambda) dM_s, \quad \forall t \qquad \text{a.s.}$$

for any λ.

Proof. We will consider f_t as a W_p^m-valued process. Since it holds $\|f_s\| \leq K\|f_s\|_{\infty,m}$ with some positive constant K, f_t satisfies condition (3). Therefore stochastic integral $L_t = \int_0^t f_s dM_s$ is well defined as a W_p^m-valued strongly continuous local martingale. Then by Sobolev's embedding theorem, L_t is a C_b^ℓ-valued strongly continuous process where $\ell < m - \frac{d}{p}$. This means that $L_t = L_t(\lambda)$ is continuous in (t,λ) and ℓ-times continuous differentiable in λ.

We shall show that $L_t(\lambda)$ is a modification of the real valued stochastic

---

1) Derivatives are continuous in (t,λ).

integral $\int_0^t f_s(\lambda)dM_s$ with parameter λ. Note that L_t satisfies

$$(L_t,\Phi) = \int_0^t (f_s,\Phi)dM_s$$

for any Φ of $(W_p^m)'$. Take the Dirac measure δ_λ concentrated at λ as an element of $(W_p^m)'$. Then the above implies

$$L_t(\lambda) = \int_0^t f_s(\lambda)dM_s$$

immediately. Next take $D^k\delta_\lambda$ as an element of $(W_p^m)'$, where $|k| < m - \frac{d}{p}$. Then we have

$$(-1)^{|k|}(L_t,D^k\delta_\lambda) = \int_0^t (-1)^{|k|}(f_s,D^k\delta_\lambda)dM_s$$

which implies

$$(D^kL_t,\delta_\lambda) = \int_0^t (D^kf_s,\delta_\lambda)dM_s.$$

This proves (5). The proof is complete.

We shall next consider the convergence of sequence of W_p^m-valued stochastic integrals.

7.7. Theorem. Suppose that $mp > d$ and $p \geq 2$. Let $\{f_s^n\}$ be a sequence of predictable W_p^m-valued processes such that $\int_0^a \|f_s - f_s^n\|^p d\langle M\rangle_s$ converges to 0 in probability. Let $L_t^n = \int_0^t f_s^n dM_s$. Then $\sup_t \|L_t^n - L_t\|$ converges to 0 in probability as $n \to \infty$.

Proof is similar to that of Theorem 3.1. It is omitted.

Let f_s be a strongly continuous W_p^m-valued process. Associated with the partition $\Delta = \{0 = t_0 < \ldots < t_n = a\}$, we define the step process f_s^Δ

by f_{t_i} if $t_i \le s < t_{i+1}$. Then, $\int_0^t \|f_s - f_s^\Delta\|^p d\langle M\rangle_s$ converges to 0 a.s. as $|\Delta| \to 0$. Therefore, we have the following.

7.8. <u>Corollary.</u> Suppose that f_s is a predictable strongly continuous W_p^m-valued process. Then there is a sequence of partitions Δ_n of $[0,a]$ with $|\Delta_n| \longrightarrow 0$ such that

$$\sup_t \left\| \int_0^t f_s dM_s - \int_0^t f_s^{\Delta_n} dM_s \right\| \longrightarrow 0 \quad \text{a.s.}$$

If $\ell < m - \frac{d}{p}$, then $\int_0^t f_s^{\Delta_n}(\lambda) dM_s$ converges to $\int_0^t f_s(\lambda) dM_s$ by the norm $\| \ \|_{\infty,\ell}$ a.s.

7.9. <u>Corollary.</u> Let Y be a Λ-valued $\underline{F}_0$-measurable random variable. If $f_s(\lambda)$ is continuous in (s,λ) and continuously differentiable in λ, then it holds

$$\int_0^t f_s(\lambda) dM_s \Big|_{\lambda=Y} = \int_0^t f_s(Y) dM_s.$$

<u>Proof.</u> Let $\{\Delta_n\}$ be a sequence of partitions of Corollary 7.8. Then the assertion is valid for each $f_s^{\Delta_n}$ since it is a step process. Let n tend to infinity. Then we see that the equality is valid for the above f_s.

Finally we shall consider the regularity of the Stratonovich integrals.

7.10. <u>Theorem.</u> Let $f_t(\lambda)$ be a continuous random field satisfying the following properties.

(i) It is $m+1$-times continuously differentiable in λ a.s.

(ii) For each λ, $f_t(\lambda)$ is a continuous semimartingale represented as

$$f_t(\lambda) = f_0(\lambda) + \sum_{j=1}^n \int_0^t g_s^j(\lambda) dN_s^j,$$

where $N_t^1,\dots,N_t^n$ are continuous semimartingales, $g_s^j(\lambda)$ are continuous random fields satisfying

(a) $g_s^j(\lambda)$ is m+1-times continuously differentiable a.s.,

(b) For each λ, it is $\underline{\underline{F}}_t$-adapted.

Then the Stratonovich integral $\int_0^t f_s(\lambda)\circ dM_s$ has a modification which is continuous in (t,λ) and m-times continuously differentiable in λ. Furthermore, it holds for any k such that $|k| \le m$,

$$D^k\int_0^t f_s(\lambda)\circ dM_s = \int_0^t D^k f_s(\lambda)\circ dM_s .$$

__Proof.__ It holds

$$\int_0^t f_s(\lambda)\circ dM_s = \int_0^t f_s(\lambda)dM_s + \frac{1}{2}\sum_j\int_0^t g_s^j(\lambda)d\langle N^j,M\rangle_s \quad \text{a.s.}$$

for any λ. The right hand side has a modification which is continuous in (t,λ) and m-times continuously differentiable by Theorem 7.6. Hence the Stratonovich integral has a modification with the same property, too. Furthermore,

$$D^k\int_0^t f_s(\lambda)\circ dM_s = \int_0^t D^k f_s(\lambda)dM_s + \frac{1}{2}\sum_j\int_0^t D^k g_s^j(\lambda)d\langle N^j,M\rangle_s .$$

On the other hand, since $D^k f_t(\lambda) = D^k f_0(\lambda) + \sum_j\int_0^t D^k g_s^j(\lambda)dN_s^j$ holds for $|k| \le m$, we have

$$\int_0^t D^k f_s(\lambda)\circ dM_s = \int_0^t D^k f_s(\lambda)dM_s + \frac{1}{2}\sum_j\int_0^t D^k g_s^j(\lambda)d\langle N^j,M\rangle_s .$$

This proves (6). The proof is complete.

8. __ITÔ'S FORMULA__

One of the fundamental tool for studying stochastic differential equations is so called Itô's formula, which describes the differential rule for change of variables or compositon of functions. We present here a differential rule for the composition of two stochastic processes, which is a generalization of the well known Itô's formula.

8.1. Theorem. Let $F_t(x)$, $t\in[0,a]$, $x\in R^d$ be a random field continuous in (t,x) a.s., satisfying

(i) $F_t(x)$ is twice continuously differentiable in x.[1]

(ii) For each x, $F_t(x)$ is a continuous semimartingale and it satisfies

$$(1)\quad F_t(x) = F_0(x) + \sum_{j=1}^{m}\int_0^t f_s^j(x)dY_s^j,^{2)} \qquad \forall x\in R^d$$

a.s., where $Y_s^1,\dots,Y_s^m$ are continuous semimartingales, $f_s^j(x)$, $s\in[0,a]$, $x\in R^d$ are random fields which are continuous in (s,x) and satisfy

(a) $f_s^j(x)$ are twice continuously differentiable in x.

(b) For each x, $f_s^j(x)$ are adapted processes.

Let now $X_t = (X_t^1,\dots X_t^d)$ be continuous semimartingales. Then we have

$$(2)\quad F_t(X_t) = F_0(X_0) + \sum_{j=1}^{m}\int_0^t f_s^j(X_s)dY_s^j + \sum_{i=1}^{d}\int_0^t \frac{\partial F_s}{\partial x_i}(X_s)dX_s^i$$
$$+ \sum_{i=1}^{d}\sum_{j=1}^{m}\int_0^t \frac{\partial f_s^j}{\partial x_i}(X_s)d\langle Y^j,X^i\rangle_s$$
$$+ \frac{1}{2}\sum_{i,j=1}^{d}\int_0^t \frac{\partial^2 F_s}{\partial x_i\partial x_j}(X_s)d\langle X^i,X^j\rangle_s .$$

Observe that the above formula is not like the classical formula for the differential of composite functions, where the last two temrs do not appear. We will see later that if we replace Itô integrals by Stratonovich integrals, then we have a rule similar to the classical rule. See Theorem 8.3.

---

1) Derivatives are continuous in (t,x).

2) Modifications continuous in x. See Theorem 7.6.

If we take $F_t(x)$ as a C^2 function $F(x)$ in the thoerem, we obtain a well known Itô's formula.

8.2. Corollary. Let $F : R^d \longrightarrow R^1$ be a C^2 function and let $X_t = (X_t^1,\dots,X_t^d)$ be continuous semimartingales. Then we have

$$F(X_t) = F(X_0) + \sum_{i=1}^{d}\int_0^t \frac{\partial F}{\partial x_i}(X_s)dX_s^i$$

$$+\frac{1}{2}\sum_{i,j=1}^{d}\int_0^t \frac{\partial^2 F}{\partial x_i \partial x_j}(X_s)d\langle X^i,X^j\rangle_s.$$

For the proof of the theorem, take a partition $\Delta_n = \{0=t_0<\dots<t_n=t\}$ and divide $F_t(X_t) - F(X_0)$ into the sum of small differences $F_{t_{k+1}}(X_{t_{k+1}}) - F_{t_k}(X_{t_k})$. Then,

$$F_t(X_t) - F_0(X_0)$$

$$= \sum_{k=0}^{n-1}\{F_{t_{k+1}}(X_{t_k}) - F_{t_k}(X_{t_k})\} + \sum_{k=0}^{n-1}\{F_{t_{k+1}}(X_{t_{k+1}}) - F_{t_{k+1}}(X_{t_k})\}$$

$$= \sum_{k=0}^{n-1}\sum_{j=1}^{m}\int_{t_k}^{t_{k+1}} f_s^j(x)dY_s^j\Big|_{x=X_{t_k}}$$

$$+ \sum_{k=0}^{n-1}\sum_{i=1}^{d}\Big(\frac{\partial}{\partial x_i}F_{t_{k+1}}(X_{t_k}) - \frac{\partial}{\partial x_i}F_{t_k}(X_{t_k})\Big)(X_{t_{k+1}}^i - X_{t_k}^i)$$

$$+ \sum_{k=0}^{n-1}\sum_{i=1}^{d}\frac{\partial}{\partial x_i}F_{t_k}(X_{t_k})(X_{t_{k+1}}^i - X_{t_k}^i)$$

$$+\frac{1}{2}\sum_{i,j=1}^{d}\sum_{k=0}^{n-1}\frac{\partial^2}{\partial x_i \partial x_j}F_{t_{k+1}}(\xi_k)(X_{t_{k+1}}^i - X_{t_k}^i)(X_{t_{k+1}}^j - X_{t_k}^j)$$

$$= I_1^{(n)} + I_2^{(n)} + I_3^{(n)} + I_4^{(n)},$$

where ξ_k are random variables such that $|\xi_k - X_{t_k}| \le |X_{t_{k+1}} - X_{t_k}|$. We will prove in the sequel

(3) $\lim_{n\to\infty} I_1^{(n)} = \sum_{j=1}^{m}\int_0^t f_s^j(X_s)dY_s^j$,

(4) $\lim_{n\to\infty} I_2^{(n)} = \sum_{i=1}^{d}\sum_{j=1}^{m}\int_0^t \frac{\partial f_s^j}{\partial x_i}(X_s)d\langle Y^j, X^i\rangle_s$,

(5) $\lim_{n\to\infty} I_3^{(n)} = \sum_{i=1}^{d}\int_0^t \frac{\partial F_s}{\partial x_i}(X_s)dX_s^i$,

(6) $\lim_{n\to\infty} I_4^{(n)} = \frac{1}{2}\sum_{i,j=1}^{d}\int_0^t \frac{\partial^2}{\partial x_i \partial x_j} F_s(X_s)d\langle X^i, X^j\rangle_s$.

Then the formula of the theorem will follow.

Proof of (3) and (5). If Y_t^j is a process of bounded variation, the assertion is obvious. So we will consider the case that Y_t^j is a continuous local martingale. Set $\overset{\Delta}{X_s}{}^n = X_{t_k}$ if $t_k \le s < t_{k+1}$. Then it holds by Corollary 7.9

$$I_1^{(n)} = \sum_{j=1}^{m}\int_s^t f_s^j(\overset{\Delta}{X_s}{}^n)dY_s^j .$$

Since

$$\int_0^t |f_s^j(\overset{\Delta}{X_s}{}^n) - f_s^j(X_s)|^2 d\langle Y^j\rangle_s \xrightarrow[n\to\infty]{} 0 \quad \text{in probability,}$$

assertion (3) follows from Theorem 3.1. The proof of (5) is immediate.

Proof of (4). It holds by Theorem 7.6

$$\frac{\partial}{\partial x_i} F_{t_{k+1}}(x) - \frac{\partial}{\partial x_i} F_{t_k}(x) = \sum_{j=1}^{m}\int_{t_k}^{t_{k+1}} \frac{\partial f_s^j}{\partial x_i}(x)dY_s^j$$

a.s. for each x. Since $\frac{\partial f_s^j}{\partial x_i}(x)$ is continuous in (s,x) and continuously differentiable in x, it holds by Corollary 7.9 [1])

---

1) $\int_0^t \frac{\partial f_s^j}{\partial x_i}(x)dY_s^j$ is understood as a modification in W_p^1 with $p > d$.

$$\frac{\partial}{\partial x_i} F_{t_{k+1}}(X_{t_k}) - \frac{\partial}{\partial x_i} F_{t_k}(X_{t_k}) = \sum_{j=1}^{m} \int_{t_k}^{t_{k+1}} \frac{\partial f_s^j}{\partial x_i}(X_{t_k}) dY_s^j .$$

Now, setting for convenience

$$L_t^{ij}(\Delta_n) = \int_0^t \frac{\partial f_s^j}{\partial x_i}(X_s^{\Delta_n}) dY_s^j ,$$

$I_2^{(n)}$ is written as the sum of joint quadratic variations $\langle L^{ij}(\Delta_n), X^i\rangle_t^{\Delta_n}$ associated with the partition Δ_n. Thus it is enough to prove that each $\langle L^{ij}(\Delta_n), X^i\rangle_t^{\Delta_n}$ converges to $\langle L^{ij}, X^i\rangle_t$ where

$$L_t^{ij} = \int_0^t \frac{\partial f_s^j}{\partial x_i}(X_s) dY_s^j .$$

We will show this in case where X^i, Y^j are local martingales. It holds

$$|\langle L^{ij}(\Delta_n), X^i\rangle_t^{\Delta_n} - \langle L^{ij}, X^i\rangle_t|$$

$$\leq |\langle L^{ij}, X^i\rangle_t^{\Delta_n} - \langle L^{ij}, X^i\rangle_t| + |\langle L^{ij}(\Delta_n) - L^{ij}, X^i\rangle_t^{\Delta_n}| .$$

The first member of the right hand side converges to 0 as $n \to \infty$. The second member converges to 0 as $n \to \infty$ by Theorem 4.4. The proof is complete.

Proof of (6). Set $h_s(x) = \frac{\partial^2 F_s}{\partial x_i \partial x_j}(x)$ and

$$K_t = \int_0^t h_s(X_s) dX_s^i , \qquad K_t^n = \int_0^t h_s^{\Delta_n}(X_s^{\Delta_n}) dX_s^i .$$

Then $\langle K^n, X^j\rangle_t^{\Delta_n}$ converges to $\langle K, X^j\rangle_t$ as $n \to \infty$, which can be shown quite similarly as the proof of (4). On the other hand, it holds

$$|\langle K^n, X^j\rangle_t^{\Delta_n} - \sum_{k=0}^{n-1} h_{t_{k+1}}(\xi_k)(X_{t_{k+1}}^i - X_{t_k}^i)(X_{t_{k+1}}^j - X_{t_k}^j)|$$

$$\leq \sup_k |h_{t_k}(X_{t_k}) - h_{t_{k+1}}(\xi_k)| \{\langle X^i\rangle_t^{\Delta_n} \langle X^j\rangle_t^{\Delta_n}\}^{\frac{1}{2}} .$$

This converges to 0 as $|\Delta_n| \to 0$. Therefore we have

$$\sum_{k=0}^{n-1} h_{t_{k+1}}(\xi_k)(X^i_{t_{k+1}} - X^i_{t_k})(X^j_{t_{k+1}} - X^j_{t_k}) \xrightarrow[n\to\infty]{} \int_0^t h_s(X_s)d\langle X^i, X^j\rangle_s .$$

The proof is complete.

In applications it is sometimes useful to rewrite the above formula using Stratonovich integral. The new formula is close to the classical formula for the differential rule of composite function. We need, however, additional assumption for processes.

8.3. <u>Theorem.</u> Let $F_t(x)$, $t \in [0,a]$, $x \in R^d$ be a random field continuous in (t,x) a.s., satisfying

(i) For each t, $F_t(\cdot)$ is a C^3-map from R^d into R^1 a.s. ω

(ii) For each x, $F_t(x)$ is a continuous semimartingale and it satisfies

$$(7)\quad F_t(x) = F_0(x) + \sum_{j=1}^{m} \int_0^t f^j_s(x)\circ dY^j_s , \qquad \forall x \in R^d \quad \text{a.s.},$$

where $Y^1_s, \dots, Y^m_s$ are continuous semimartingales, $f^j_t(x)$ are random fields satisfying conditions (i) and (ii) of Theorem 8.1.

Let now $X_t = (X^1_t, \dots, X^d_t)$ be continuous semimartingales. Then we have

$$(8)\quad F_t(X_t) = F_0(X_0) + \sum_{j=1}^{m} \int_0^t f^j_s(X_s)\circ dY^j_s + \sum_{i=1}^{d} \int_0^t \frac{\partial F_s}{\partial x_i}(X_s)\circ dX^i_s .$$

<u>Proof.</u> We will write the form of $f^j_t(x)$ explicitly:

$$f^j_t(x) = f^j_0(x) + \sum_{k=1}^{\ell} \int_0^t g^{jk}_s(x)dZ^k_s ,$$

where Z^k are continuous semimartingales and $g^{jk}_s(x)$ are continuous in (s,x) satisfying conditions (a) and (b) of Theorem 8.1. Then using Itô integral, $F_t(x)$ of (7) is written as

$$(9)\quad F_t(x) = F_0(x) + \sum_{j=1}^{m} \int_0^t f^j_s(x)dY^j_s + \frac{1}{2} \sum_{j,k} \int_0^t g^{jk}_s(x)d\langle Z^k, Y^j\rangle_s .$$

Hence by Theorem 8.1,

$$(10)\quad F_t(X_t) = F_0(X_0) + \sum_j \int_0^t f_s^j(X_s)dY_s^j + \frac{1}{2}\sum_{j,k}\int_0^t g_s^{jk}(X_s)d\langle Z^k, Y^j\rangle_s$$

$$+ \sum_i \int_0^t \frac{\partial F_s}{\partial x_i}(X_s)dX_s^i + \sum_{i,j}\int_0^t \frac{\partial f_s^j}{\partial x_i}(X_s)d\langle Y^j, X^i\rangle_s$$

$$+ \frac{1}{2}\sum_{i,j}\int_0^t \frac{\partial^2 F_s}{\partial x_i \partial x_j}(X_s)d\langle X^i, X^j\rangle_s .$$

We shall apply Theorem 8.1 to $f_t^j(x)$ in the place of $F_t(x)$. Then we see that $f_t^j(X_t)$ is a continuous semimartingale satisfying

$$f_t^j(X_t) = f_0^j(X_0) + \sum_{k=1}^{\ell}\int_0^t g_s^{jk}(X_s)dZ_s^k + \sum_{i=1}^{d}\int_0^t \frac{\partial f_s^j}{\partial x_i}(X_s)dX_s^i$$

+ a process of bounded variation.

Therefore,

$$\langle f^j(X), Y^j\rangle_t = \sum_k \int_0^t g_s^{jk}(X_s)d\langle Z^k, Y^j\rangle_s + \sum_i \int_0^t \frac{\partial f_s^j}{\partial x_i}(X_s)d\langle X^i, Y^j\rangle_s .$$

Consequently, the Stratonovich integral $\sum_j \int_0^t f_s^j(X_s)\circ dY_s^j$ coincides with the sum of the second, the third and half of the fifth term of the right hand side of (10).

Next we shall apply Theorem 8.1 to $\frac{\partial F_t}{\partial x_i}(x)$ in place of $F_t(x)$. Noting the relation (9), we see that $\frac{\partial F_t}{\partial x_i}(X_t)$ is again a continuous semimartingale and satisfies

$$\langle \frac{\partial F}{\partial x_i}(X), X^i\rangle_t = \sum_j \int_0^t \frac{\partial f_s^j}{\partial x_i}(X_s)d\langle Y^j, X^i\rangle_s + \sum_j \int_0^t \frac{\partial^2 F_s}{\partial x_i \partial x_j}(X_s)d\langle X^i, X^j\rangle_s .$$

Therefore the Stratonovich integral $\sum_i \int_0^t \frac{\partial F_s}{\partial x_i}(X_s)\circ dX_s^i$ coincides with the sum of the fourth, half of the fifth and the sixth terms. The proof is complete.

8.4. <u>Corollary.</u> Let $F ; R^d \longrightarrow R^1$ be a C^3-class function and let

$X_t = (X_t^1, \ldots, X_t^d)$ be continuous semimartingales. Then we have

$$F(X_t) = F(X_0) + \sum_{i=1}^{d} \int_0^t \frac{\partial F}{\partial x_i}(X_s) \circ dX_s^i .$$

We are now able to prove Burkholder's inequality.

Proof of Theorem 7.3. We shall apply Corollary 8.2 by setting $F(x) = |x|^p$ $(p>2)$ and $X_t = M_t$. Then

$$|M_t|^p = p\int_0^t |M_s|^{p-1} \mathrm{sign}(M_s) dM_s + \frac{1}{2}p(p-1)\int_0^t |M_s|^{p-2} d\langle M\rangle_s .$$ [1]

Taking the expectation, we have

$$\begin{aligned} E[\,|M_t|^p] &\le \frac{1}{2}p(p-1)E[\int_0^t |M_s|^{p-2} d\langle M\rangle_s] \\ &\le \frac{1}{2}p(p-1)E[\sup_{0\le s\le t}|M_s|^{p-2}\langle M\rangle_t] \\ &\le \frac{1}{2}p(p-1)E[\sup_{0\le s\le t}|M_s|^p]^{\frac{p-2}{p}} E[\langle M\rangle_t^{\frac{p}{2}}]^{\frac{2}{p}} \\ &\le \frac{1}{2}p(p-1)q^{p-2}E[\,|M_t|^p]^{\frac{p-2}{p}} E[\langle M\rangle_t^{\frac{p}{2}}]^{\frac{2}{p}} . \end{aligned}$$

Here we have used Hölder's inequality and Doob's inequality of martingales. The assertion of Theorem 7.3 follows immediately.

Remark. Burkholder's inequality for L^p-martingale was used for the proof of the regularity of stochastic integral with respect to parameters. The latter property was then used in the proof of Theorem 8.1 to show equalities (3) and (4). However if we look at the special case of Itô's formula such as Corollary 8.2, Theorem 7.6 and Corollary 7.9 are not needed since the terms corresponding to $I_2^{(n)}$ is identically 0. That is to say, Corollary 8.2 is proved without using Burkholder's

---

1) $\mathrm{sign}(x) = 1$ if $x \ge 0$ and $= -1$ if $x < 0$.

inequality.

9. BROWNIAN MOTIONS AND STOCHASTIC INTEGRALS.

Let $B_t = (B_t^1, \ldots, B_t^m)$ be an m-dimensional standard Brownian motion defined on $(\Omega, \underline{\underline{F}}, P; \underline{\underline{F}}_t)$. We will call it an $(\underline{\underline{F}}_t)$-Brownian motion if it is $(\underline{\underline{F}}_t)$-adapted and the future of Brownian motion $B_u - B_t$; $u \geq t$ and the past σ-field $\underline{\underline{F}}_t$ are independent for any t. The following theorem characterizes $(\underline{\underline{F}}_t)$-Brownian motion by martingales and their joint quadratic variations.

9.1. Theorem. Let $B_t = (B_t^1, \ldots, B_t^m)$ be an m-dimensional $(\underline{\underline{F}}_t)$-adapted continuous stochastic process. It is an $(\underline{\underline{F}}_t)$-Brownian motion if and only if each $B_t^1, \ldots, B_t^m$ are square integrable martingales such that $\langle B^i, B^j \rangle_t = \delta_{ij} t$.

Proof. Suppose that $B_t = (B_t^1, \ldots, B_t^m)$ is an $(\underline{\underline{F}}_t)$-Brownian motion. Then it is square integrable. Since $B_t - B_s$ is independent of $\underline{\underline{F}}_s$,

$$E[B_t^i - B_s^i | \underline{\underline{F}}_s] = E[B_t^i - B_s^i] = 0$$

holds for any $s < t$. Hence B_t^i are square integrable martingales. It holds

$$E[(B_t^i)^2 - (B_s^i)^2 | \underline{\underline{F}}_s] = E[(B_t^i - B_s^i)^2] = t - s,$$

$$E[B_t^i B_t^j - B_s^i B_s^j | \underline{\underline{F}}_s] = E[(B_t^i - B_s^i)(B_t^j - B_s^j)] = 0 \quad \text{if } i \neq j.$$

Therefore $B_t^i B_t^j - \delta_{ij} t$ is a martingale, proving $\langle B^i, B^j \rangle_t = \delta_{ij} t$.

Conversely suppose that B_t^i are square integrable martingales satisfying $\langle B^i, B^j \rangle_t = \delta_{ij} t$. We will apply Itô's formula (Corollay 8.2) to $F(t,x) = \exp\{i(\alpha,x)+\frac{1}{2}|\alpha|^2 t\}$ and $M_t = (t,B_t)$ where $\alpha=(\alpha_1,\ldots,\alpha_m)$ and $i = \sqrt{-1}$. Then, we have

$$F(t,B_t)$$
$$= 1 + \int_0^t \frac{\partial F}{\partial s}(s,B_s)ds + \sum_{k=1}^m \int_0^t \frac{\partial F}{\partial x_k}(s,B_s)dB_s^k + \frac{1}{2}\sum_{k,\ell}\int_0^t \frac{\partial^2 F}{\partial x_k \partial x_\ell}(s,B_s)d\langle B^k,B^\ell\rangle_s$$
$$= 1 + i\sum_{k=1}^m \alpha_k \int_0^t F(s,B_s)dB_s^k,$$

since the sum of the second and the fourth terms is 0. This shows that $F(t,B_t)$ is a martingale with mean 1. Therefore,

$$E[F(t,B_t)F(s,B_s)^{-1}|\underline{\underline{F}}_s] = 1,$$

from which we have

$$E[e^{i(\alpha,B_t-B_s)}|\underline{\underline{F}}_s] = e^{-\frac{1}{2}|\alpha|^2|t-s|}.$$

Consequently, $B_t - B_s$ is Gaussian with mean 0, covariance $(t-s)I$ and further it is independent of $\underline{\underline{F}}_s$. Therefore, B_t is an $(\underline{\underline{F}}_t)$-Brownian motion. The proof is complete.

Continuous local martingale can be transformed to a Brownian motion via time-change if the quadratic variation is a strictly increasing process. Let M_t, $t\in[0,a]$ be a continuous local martingale. Assume that the quadratic variation $\langle M\rangle_t(\omega)$ is a strictly increasing function of t a.s. and $\lim_{t\uparrow a}\langle M\rangle_t = \infty$ a.s. Let $\tau_t(\omega)$ be the inverse function of $\langle M\rangle_t$:

$$\tau_s(\omega) = \sup\{r \le s\ ;\ \langle M\rangle_r \le s\}.$$

Then it is a strictly increasing function of s and moreover, for each s, it is an $(\underline{\underline{F}}_t)$-stopping time. Then by Doob's optional sampling theorem, the time-changed process $X_s = M_{\tau_s}$, $s \in [0,\infty)$ is a continuous martingale relative to $(\hat{\underline{F}}_s) = (\underline{F}_{\tau_s})$. The quadratic variation of X_s is $\langle X\rangle_s = \langle M\rangle_{\tau_s} = s$. Therefore X_s is an $(\hat{\underline{F}}_s)$-Brownian motion. A similar property is valid for multidimensional local martingales.

9.2. Theorem. Let $M_t = (M_t^1,\ldots,M_t^m)$ be a continuous local martingale. Suppose that there is a strictly increasing process A_t with $\lim_{t\uparrow a} A_t = \infty$ a.s. such that $\langle M^i,M^j\rangle_t = \delta_{ij}A_t$. Let τ_s be the inverse function of A_t. Then the time-changed process $\hat{M}_s = (M^1_{\tau_s},\ldots,M^m_{\tau_s})$ is a standard Brownian motion..

Remark. In case A_t is strictly increasing but $\lim_{t\uparrow\infty} A_t = A_\infty$ is not infinite a.s., the quadratic variation of $\hat{M}$ is

$$\langle \hat{M}^i,\hat{M}^j\rangle_s = \delta_{ij}A_\infty \wedge s.$$

Therefore $\hat{M}_s$ is a stopped Brownian motion at time A_∞.

Now let us define Itô integral by 1-dimensional Brownian motion B_t. Let $f(t)$ be a predictable process such that $\int_0^t f(r)^2 dr < \infty$ a.s., then Itô integral $\int_0^t f(r)dB_r$ is well defined and it is a continuous local martingale. If $f(r)$ is square integrable, $E[\int_0^t f(r)^2 dr] < \infty$, then $\int_0^t f(r)dB_r$ is a square integrable martingale. It holds

(1) $E[\int_0^t f(r)dB_r] = 0,$

(2) $E[|\int_0^t f(r)dB_r|^2] = E[\int_0^t f(r)^2 dr],$

(3) $E[|\int_0^t f(r)dB_r|^p] \leq (\frac{1}{2}p(p-1)q^{p-2})^{\frac{p}{2}} E[|\int_0^t f(r)^2 dr|^{\frac{p}{2}}]$

$$\leq (\tfrac{1}{2}p(p-1)q^{p-2})^{\frac{p}{2}} t^{\frac{p}{2}-1} E[\int_0^t |f(r)|^p dr].$$

The last inequality follows from Burkholder's inequality and Hölder's inequality.

The Stratonovich integral by dB_t is well defined for continuous semimartingale $f(r)$. It holds

$$\int_0^t f(r)\circ dB_r = \int_0^t f(r)dB_r + \frac{1}{2}\langle f,B\rangle_t .$$

We denote by $\underline{\underline{F}}_{s,t}$ the least complete σ-field for which $B_u - B_v$; $s \leq u \leq v \leq t$ are measurable. Then it is increasing in t and is decreasing in s, i.e. $\underline{\underline{F}}_{s,t} \subset \underline{\underline{F}}_{s',t'}$ is satisfied if $t < t'$ and $s' < s$.

Let t be a fixed time in $[0,a]$ and let $f(r)$, $r \in [0,t]$ be a continuous stochastic process which is $\underline{\underline{F}}_{r,t}$-measurable for each r. The Itô backward integral is defined as

$$\int_s^t f(r)\hat{d}B_r \equiv \lim_{|\Delta|\to 0} \sum_{k=0}^{n-1} f(t_{k+1})(B_{t_{k+1}} - B_{t_k}).$$

If $f(r)$ is square integrable $E[\int_s^t f(r)^2 dr] < \infty$, then the integral is a square integrable backward martingale, i.e. $Y_s = \int_s^t f(r)\hat{d}B_r$ is $\underline{\underline{F}}_{s,t}$-measurable and satisfies $E[Y_u | \underline{\underline{F}}_{s,t}] = Y_s$ if $u < s$. The following properties are obvious.

$$E[\int_s^t f(r)\hat{d}B_r] = 0,$$

$$E[|\int_s^t f(r)\hat{d}B_r|^2] = E[\int_s^t f(r)^2 dr].$$

Itô's backward integral can be defined for backward predictable process $f(r)$ such that $\int_s^t f(r)^2 dr < \infty$ a.s.

The Stratonovich backward integral is defined similarly. Let $f(r)$, $r \in [0,t]$ be a continuous backward semimartingale relative to $\underline{\underline{F}}_{r,t}$, where t is fixed. Then the Stratonovich backward integral is defined by

$$\int_s^t f(r)\circ\hat{d}B_r = \lim_{|\Delta|\to 0} \sum_{k=0}^{n-1} \frac{1}{2}(f(t_{k+1}) + f(t_k))(B_{t_{k+1}} - B_{t_k}).$$

10. APPENDIX. KOLMOGOROV'S THEOREM

We shall introduce a criterion for the Hölder continuity of random fields, which is a generalization of the well known Kolmogorov's criterion for the continuity of stochstic processes. It will provide us another method of deriving the regularity of stochastic integrals with respect to the parameter.

10.1. Theorem. Let $X_\lambda(\omega)$ be a real valued random field with parameter $\lambda = (\lambda_1,\dots,\lambda_d) \in \Lambda = [0,1]^d$. Suppose that there are constants $\gamma > 0$, $\alpha_i > d$, $i=1,\dots,d$ and $C > 0$ such that

(1) $$E[|X_\lambda - X_\mu|^\gamma] \le C \sum_{i=1}^{d} |\lambda_i - \mu_i|^{\alpha_i}, \qquad \forall \lambda, \mu \in \Lambda .$$

Then X_λ has a continuous modification $\widetilde{X}_\lambda$.

Let β_i, $i=1,\dots,d$ be arbitrary positive numbers less than $\alpha_i(\alpha_0 - d)\times \alpha_0^{-1}\gamma^{-1}$, $i=1,\dots,d$ respectively, where $\alpha_0 = \min_i \alpha_i$. Then for almost all ω, there is a positive integer $m_0(\omega)$ and positive constant K such that

(2) $$|\widetilde{X}_\lambda(\omega) - \widetilde{X}_\mu(\omega)| \le K \sum_{i=1}^{d} |\lambda_i - \mu_i|^{\beta_i}$$

holds for any λ, μ of Λ such that $\sum_i |\lambda_i - \mu_i|^{\beta_i} \le 2^{-m_0(\omega)}$.

Before the proof, we introduce some terminologies. Let q be a positive number less than 1. A positive number represented by $\Sigma_{i\ge 1}\, a_i q^i$ (finite sum) where a_i are nonnegative integers less than q^{-1} is called a q-adic number. The set of all q-adic numbers less than 1 is dense in $[0,1]$.

Now let β_i, $i=1,\dots,d$ be positive numbers stated in the theorem. Set $\beta_0 = \min_i \beta_i$ and $q_i = 2^{-\beta_0\beta_i^{-1}}$, $i=1,\dots,d$. Then $\frac{1}{2} \le q_i < 1$. Define $\underline{q} = (q_1,\dots,q_d)$. A vector $\lambda = (\lambda_1,\dots,\lambda_d)$ of Λ is called q-adic if each λ_i is a q_i-adic number. The set of all $\underline{q}$-adic numbers in Λ

is denoted by Δ. Δ_m is defined by the subset of Δ such that each component λ_i is written as $\Sigma_{k\le m} a_k q_i^k$. It holds $\bigcup_m \Delta_m = \Delta$. Note that the number of elements in Δ_m is at most 2^{md}, since a_k of the above takes values 0 or 1 only. Two points λ, μ of Δ_m are called <u>neighbors in</u> Δ_m if $|\lambda_i - \mu_i| = 0$ or q_i^m for any i=1,...,d.

<u>Proof of Theorem.</u> Suppose that λ, μ are neighbors in Δ_m. Then by Chebischev's inequality and inequality (1),

$$\begin{aligned} P(|X_\lambda - X_\mu| > 2^{-m\beta_0}) &\le 2^{m\beta_0\gamma} E|X_\lambda - X_\mu|^\gamma \\ &\le C2^{m\beta_0\gamma} \sum_i |\lambda_i - \mu_i|^{\alpha_i} \\ &\le C2^{m\beta_0\gamma} \sum_i 2^{-\beta_0\beta_i^{-1} m\alpha_i} \\ &\le C\sum_i 2^{m\beta_0(\gamma - \beta_i^{-1}\alpha_i)}. \end{aligned}$$

Set

$$A_m = \{\omega \mid |X_\lambda(\omega) - X_\mu(\omega)| > 2^{-m\beta_0} \text{ for some neighbors } \lambda, \mu \text{ in } \Delta_m\}.$$

Since the number of the pair (λ,μ) which is neighbor in Δ_m is at most $3^d 2^{md}$, we have

$$P(A_m) \le 3^d \sum_i 2^{m\{\beta_0(\gamma-\beta_i^{-1}\alpha_i)+d\}}.$$

It suffices to consider the case that β_i, i=1,...,d satisfy $(\alpha_0 - d)\gamma^{-1}(1+\varepsilon)^{-1} < \beta_i < \alpha_i(\alpha_0 - d)\{\alpha_0\gamma(1 + d\alpha_0^{-1}\varepsilon)\}^{-1}$ for some $\varepsilon > 0$. Then it holds $\beta_i > (\alpha_0 - d)\gamma^{-1}(1+\varepsilon)^{-1}$, i=1,...,d and we have $\beta_0(\gamma - \beta_i^{-1}\alpha_i) + d < 0$. Then by Borel-Cantelli's lemma, we have $P(\underline{\lim}\, A_m^c) = 1$.

Let $\omega \in \underline{\lim}\, A_m^c$. Then there is a positive integer $m_0(\omega)$ such that for any $m \ge m_0(\omega)$,

(3) $|X_\lambda(\omega) - X_\mu(\omega)| \le 2^{-\beta_0 m}$ for any neighbors λ, μ of Δ_m.

We will fix the above ω and show that

(4) $|X_\lambda(\omega) - X_\mu(\omega)| \le K \sum_i |\lambda_i - \mu_i|^{\beta_i}$

for any λ, μ of Δ such that $\sum_i |\lambda_i - \mu_i|^{\beta_i} \le 2^{-m_0\beta_0}$. Given λ, μ of the above property, choose a positive integer k greater than or equal to m_0 such that $2^{-(k+1)\beta_0} < \Sigma_i |\lambda_i - \mu_i|^{\beta_i} \le 2^{-k\beta_0}$. Then it holds $|\lambda_i - \mu_i| \le 2^{-k\beta_0\beta_i^{-1}} = q_i^k$ for any i. Now let $\lambda = \lambda^{(k)} + A_{k+1}\underline{q}^{(k+1)} + A_{k+2}\underline{q}^{(k+2)} + \ldots$ be the $\underline{q}$-adic expansion of λ, where $\lambda^{(k)} \in \Delta_k$ and $A_{k+1}\underline{q}^{(k+1)} = (a_{k+1}^1 q_1^{k+1}, \ldots, a_{k+1}^d q_d^{k+1})$. Similarly let $\mu = \mu^{(k)} + B_{k+1}\underline{q}^{(k+1)} + B_{k+2}\underline{q}^{(k+2)} + \ldots$ be the $\underline{q}$-adic expansion of μ. Then $\lambda^{(k)}$ and $\mu^{(k)}$ are neighbors in Δ^k. Therefore we have $|X_{\lambda^{(k)}} - X_{\mu^{(k)}}| \le 2^{-k\beta_0}$ by (3). We define $\lambda^{(k+1)} = \lambda^{(k)} + A_{k+1}\underline{q}^{(k+1)}, \ldots, \lambda^{(k+n)} = \lambda^{(k+n-1)} + A_{k+n}\underline{q}^{(k+n)}$ etc. Then for each n, $\lambda^{(k+n)}$ and $\lambda^{(k+n-1)}$ are neighbors in Δ_{k+n}. Therefore,

$$
\begin{aligned}
|X_\lambda - X_{\lambda^{(k)}}| &\le \sum_{k=1}^{\infty} |X_{\lambda^{(k+n)}} - X_{\lambda^{(k+n-1)}}| \le \sum_{n=1}^{\infty} 2^{-\beta_0(k+n)} \\
&\le \frac{1}{1-2^{-\beta_0}} 2^{-\beta_0(k+1)} .
\end{aligned}
$$

By the same reasoning, we see that $|X_\mu - X_{\mu^{(k)}}|$ is dominated by the same quantity. Therefore,

$$
\begin{aligned}
|X_\lambda - X_\mu| &\le (2^{\beta_0} + \frac{2}{1-2^{-\beta_0}}) 2^{-\beta_0(k+1)} \\
&\le (2^{\beta_0} + \frac{2}{1-2^{-\beta_0}}) \sum_i |\lambda_i - \mu_i|^{\beta_i} .
\end{aligned}
$$

This proves (4).

Now inequality (4) shows that $X_\lambda(\omega)$ is uniformly continuous on $\underline{q}$-adic numbers Δ. Hence it has a unique continuous extension $\widetilde{X}_\lambda(\omega)$.

It is a modification of X_λ for any λ since X_λ is continuous in probability. It is clear that $\tilde{X}_\lambda$ satisfies (2). The proof is complete.

10.2. <u>Definition.</u> The random field satisfying (2) is called $(\beta_1,\ldots,\beta_d)$-<u>Hölder continuous.</u>

If we apply Theorem 10.1, we can improve the regularity of stochastic integral with respect to parameters discussed at Section 7. The next theorem indicates that if the integrand is Hölder continuous with respect to the parameter, the same property is valid for the integral.

10.3. <u>Theorem.</u> Let $f_s(\lambda)$, $(s,\lambda) \in [0,a] \times \Lambda$ be a measurable random field satisfying the following properties.

(i) For each λ, $f_s(\lambda)$ is predicable.

(ii) For any $p > 2$, there is a positive constant $C_1^{(p)}$ such that

$$\int_0^a E[\,|f_s(\lambda)|^p]ds \leq C_1^{(p)} \quad \text{for any } \lambda.$$

(iii) For any $p > 2$, there is a positive constant $C_2^{(p)}$ such that

$$\int_0^a E[\,|f_s(\lambda) - f_s(\mu)|^p]ds \leq C_2^{(p)}|\lambda - \mu|^{\alpha p} ,$$

where $0 < \alpha \leq 1$. Let M_t be a continuous local martingale such that $\langle M\rangle_t - \langle M\rangle_s \leq t - s$ holds for any $t > s$ a.s. Then there is a modification of the stochastic integral $\int_0^t f_s(\lambda)dM_s$ which is continuous in (t,λ). Furthermore it is (β_1,β_2)-Hölder continuous in (t,λ), where β_1 is an arbitrary positive number less than half and β_2 is the one less than α.

<u>Proof.</u> Applying Burkholder's inequality, we have if $s < t$,

$$E[\,|\int_0^t f_r(\lambda)dM_r - \int_0^s f_r(\mu)dM_r|^p]$$

$$\le 2^p\{E[|\int_s^t f_r(\lambda)dM_r|^p] + E[|\int_0^s (f_r(\lambda) - f_r(\mu))dM_r|^p]\}$$

$$\le 2^p C^{(p)}\{E[|\int_s^t |f_r(\lambda)|^2 d\langle M\rangle_r|^{\frac{p}{2}}] + E[|\int_0^s |f_r(\lambda) - f_r(\mu)|^2 d\langle M\rangle_r|^{\frac{p}{2}}]\}$$

$$\le 2^p C^{(p)}\{E[|\int_s^t |f_r(\lambda)|^2 dr|^{\frac{p}{2}}] + E[|\int_0^s |f_r(\lambda) - f_r(\mu)|^2 dr|^{\frac{p}{2}}]\}$$

$$\le 2^p C^{(p)}\{|t-s|^{\frac{p}{2}-1} E[\int_s^t |f_r(\lambda)|^p dr] + |s|^{\frac{p}{2}-1} E[\int_0^s |f_r(\lambda) - f_r(\mu)|^p dr]\}$$

$$\le 2^p C^{(p)}\{C_1^{(p)}|t-s|^{\frac{p}{2}-1} + C_2^{(p)}|s|^{\frac{p}{2}-1}|\lambda - \mu|^{\alpha p}\}.$$

Take p greater than $(d+1)(\frac{1}{2}\wedge\alpha)^{-1} + 2$. Then both of $\frac{p}{2} - 1$ and αp are greater than $d+1$. Then by Theorem 10.1, $X(t,\lambda) = \int_0^t f_r(\lambda)dM_n$ has a continuous modification. Furthermore, it is (β_1, β_2)-Hölder continuous in (t,λ), where $\beta_1 < (\frac{1}{2} - \frac{1}{p})[1 - (d+1)\{(\frac{p}{2} - 1)\wedge\alpha p\}^{-1}]$ and $\beta_2 < \alpha[1 - (d+1)\{(\frac{p}{2} - 1)\wedge\alpha p\}^{-1}]$. Making p tend to infinity, we see that the Hölder continuity is valid for any β_1 less than half and β_2 less than α. The proof is complete.

Remark. If M_t is a general continuous local martingale, assumptions required to the integrand f_t should be followings.

(ii)' For any $p > 2$, there is a positive constant $C_1^{(p)}$ such that

$$E[\int_0^a |f_s(\lambda)|^p (d\langle M\rangle_s + ds)] \le C_1^{(p)} \qquad \forall\lambda \in \Lambda$$

(iii)' For any $p > 2$, there is a positive constant $C_2^{(p)}$ such that

$$E[\int_0^a |f_s(\lambda) - f_s(\mu)|^p (d\langle M\rangle_s + ds)] \le C_2^{(p)}|\lambda - \mu|^{\alpha p}.$$

Then the integral has a modification continuous in (t,λ). For the proof, we proceed as follows. Let $\tau_t(\omega)$ be the inverse time of the strictly increasing process $A_t \equiv \langle M\rangle_t + t$. Set $\widetilde{M}_t = M_{\tau_t}$, $\widetilde{\underline{F}}_t = \underline{F}_{\tau_t}$, $\widetilde{f}_t = f_{\tau_t}$. Then, by Doob's optional sampling theorem, $\widetilde{M}_t$ is a local martingale relative to

$(\widetilde{\underline{F}}_t)$, and satisfies $\langle\widetilde{M}\rangle_t = \langle M\rangle_{\tau_t}$. Hence it holds $\langle\widetilde{M}\rangle_t - \langle\widetilde{M}\rangle_s \leq t - s$. Further, the process $\widetilde{f}_t$ satisfies

$$E\Big[\int_0^{A_a} |\widetilde{f}_s(\lambda)|^p ds\Big] \leq C_1^{(p)}, \qquad E\Big[\int_0^{A_a} |\widetilde{f}_s(\lambda) - \widetilde{f}_s(\mu)|^p ds\Big] \leq C_2^{(p)} |\lambda - \mu|^{\alpha p}.$$

Therefore $\int_0^t \widetilde{f}_s d\widetilde{M}_s$ has a modification continuous in (t, λ). This implies the continuity (modification) of the integral $\int_0^t f_s(\lambda) dM_s$ because of the relation

$$\int_0^{A_t} \widetilde{f}_s(\lambda) d\widetilde{M}_s = \int_0^t f_s(\lambda) dM_s .$$

We may replace assumptions (ii) and (iii) of Theorem 10.3 by a "local" assumption.

10.4. <u>Theorem</u>. Let $f_s(\lambda)$, $(s,\lambda) \in [0,a]\times\Lambda$ be a measurable random field satisfying (i) of Theorem 10.3 and followings.

(ii") For any $p > 1$, $\sup_\lambda \int_0^a |f_s(\lambda)|^p ds < \infty$,

(iii") There are $0 < \alpha \leq 1$ and $C_s(\lambda, \mu) > 0$ such that

$$|f_s(\lambda) - f_s(\mu)| \leq C_s(\lambda, \mu) |\lambda - \mu|^\alpha, \qquad \forall \lambda, \mu \in \Lambda$$

where $C_s(\lambda, \mu)$ satisfies

$$\sup_{\lambda,\mu} \int_0^a |C_s(\lambda, \mu)|^p ds < \infty, \qquad \forall p > 1.$$

Let M_t be a continuous local martingale such that $\langle M\rangle_t - \langle M\rangle_s \leq t - s$ holds for any $t > s$ a.s. Then the stochastic integral $\int_0^t f_s(\lambda) dM_s$ has a continuous modification, which satisfies the same Hölder continuity as that of Theorem 10.3.

<u>Proof.</u> Define a sequence of stopping times T_n;

$$T_n = \inf\{t>0 : \sup_\lambda \int_0^t |f_s(\lambda)|^p ds + \sup_{\lambda,\mu} \int_0^t |C_s(\lambda,\mu)|^p ds \geq n\}$$

($= \infty$ if $\{\dots\} = \phi$). For each n, it holds

$$\sup_\lambda E[\int_0^{a\wedge T_n} |f_s(\lambda)|^p ds] \leq n, \quad E[\int_0^{a\wedge T_n} |f_s(\lambda) - f_s(\mu)|^p ds] \leq n|\lambda - \mu|^{\alpha p}.$$

Therefore $\int_0^{t\wedge T_n} f_s(\lambda) dM_s$ has a continuous modification which is Hölder continuous by Theorem 10.3. Therefore $\int_0^t f_s(\lambda) dM_s$ has also a Hölder continuous modification.

10.5. Corollary. If $f_s(\lambda)$ is continuous in (s,λ) and continuously differentiable in λ, then the stochastic integral has a continuous modification which is (β_1,β_2)-Hölder continuous in (t,λ), where β_1 is arbitrary positive number less than half and β_2 the one less than 1.

Proof. The condition (ii") is obviously satisfied. By the mean value theorem, we have $|f_s(\lambda) - f_s(\mu)| \leq C_s(\lambda,\mu)|\lambda - \mu|$, where

$$C_s(\lambda,\mu) = (\sum_i |\frac{\partial}{\partial \lambda_i} f_s(\mu + \theta(\lambda - \mu))|^2)^{\frac{1}{2}}.$$

Therefore the condition (iii") is satisfied with $\alpha = 1$. Then the assertion follows from the theorem.

We will next consider the differentiability of the integral with respect to the parameter.

10.6. Theorem. Let $f_s(\lambda)$ be a measurable random field satisfying (1) of Theorem 10.3 and the following.

(iv) $f_s(\lambda)$ is m-times continuously differentiable in λ for all s, a.s. and derivatives $D^k f_s(\lambda)$, $|k| \leq m$ satisfy conditions (ii) and (iii) of Theorem 10.3 or conditions (ii") and (iii") of Theorem 10.4.

Let M_t be a continuous local martingale such that $\langle M\rangle_t - \langle M\rangle_s \leq t-s$ for any $t > s$ a.s. Then there is a modification of the integral which is continuous in (t,λ) and m-times continuously differentiable. Furthermore, it holds

$$D^k\int_0^t f_s(\lambda)dM_s = \int_0^t D^k f_s(\lambda)dM_s$$

for any D^k such that $|k| \leq m$.

Proof. We prove the case $m = 1$, assuming that $\partial_\ell f_s$, $\ell=1,\dots,d$ satisfy conditions (ii) and (iii) of Theorem 10.3. Let $e_\ell = (0,\dots,0,1,0,\dots,0)$ (1 is the ℓ-th component) be a unit vector in R^d. For $y \in R^1 - \{0\}$ such that $\lambda + ye_\ell \in \Lambda$, set

$$N_t(\lambda,y) = \frac{1}{y}\{\int_0^t f_s(\lambda+ye_\ell)dM_s - \int_0^t f_s(\lambda)dM_s\}.$$

In order to prove the continuous differentiability of $\int_0^t f_s(\lambda)dM_s$ with respect to λ, it is enough to show that $N_t(\lambda,y)$ has a continuous extension at $y = 0$. We will prove this by applying Kolmogorov theorem.

For this purpose, we have to make an L^p-estimate of $N_t(\lambda,y) - N_{t'}(\lambda',y')$. Suppose $t < t'$. Since $N_t(\lambda,y)$ is written as

$$N_t(\lambda,y) = \int_0^t \{\int_0^1 \partial_\ell f_s(\lambda + vye_\ell)dv\}dM_s, \tag{5}$$

we have

$$E[\,|N_t(\lambda,y) - N_{t'}(\lambda',y')|^p] \tag{6}$$

$$\leq 2^p E[\,|\int_0^t \{\int_0^1 (\partial_\ell f_s(\lambda + vye_\ell) - \partial_\ell f_s(\lambda' + vy'e_\ell))dv\}dM_s|^p]$$

$$+ 2^p E[\,|\int_t^{t'} \{\int_0^1 \partial_\ell f_s(\lambda' + vy'e_\ell)dv\}dM_s|^p].$$

Using Burkholder's inequality and Hölder's inequality, the first member of the right hand side is dominated by

$$2^p C(p) E[\,|\int_0^t |\int_0^1 (\ \dots\ ) dv|^2 d\langle M\rangle_s|^{\frac{p}{2}}]$$

$$\leq 2^p C(p) E[\int_0^t |\int_0^1 (\ \dots\ ) dv|^p d\langle M\rangle_s \langle M\rangle_t^{\frac{p}{2}-1}\ ]$$

$$\leq 2^p C(p) t^{\frac{p}{2}-1} \int_0^1 \int_0^t E[\,|(\ \dots\ )|^p] ds dv.$$

Here $(\ \dots\ ) = (\partial_\ell f_s(\lambda + vye_\ell) - \partial_\ell f_s(\lambda' + vy'e_\ell))$. From the assumption (iii), $\int_0^t E[\,|(\ \dots\ )|^p] ds$ is dominated by a constant times $(|\lambda - \lambda'| + v|y - y'|)^{\alpha p}$. Therefore, the first member of the right hand side of (6) is dominated by const.$\times(|y - y'|^{\alpha p} + |\lambda - \lambda'|^{\alpha p})$.

On the other hand, the second member of the right hand side of (6) is dominated by

$$2^p C(p) E[\,|\int_t^{t'} |\{\ \dots\ \}|^2 d\langle M\rangle_s|^{\frac{p}{2}}]$$

$$\leq 2^p C(p) E[\int_t^{t'} |\{\ \dots\ \}|^p d\langle M\rangle_s (\langle M\rangle_{t'} - \langle M\rangle_t)^{\frac{p}{2}-1}\ ]$$

$$\leq 2^p C(p)\, |t' - t|^{\frac{p}{2}-1} \int_t^{t'} E[\,|\{\ \dots\ \}|^p] ds$$

where $\{\ \dots\ \} = \int_0^1 \partial_\ell f_s(\lambda' + vy'e_\ell) dv$. By the assumption (ii) for $\partial_\ell f_s(\lambda)$, we see that $\int_t^{t'} E[\,|\{\ \dots\ \}|^p] ds$ is bounded.

Summing up the above two estimates, we arrive at

$$E[\,|N_t(\lambda, y) - N_{t'}(\lambda', y')|^p] \leq \text{const.}\times(|y - y'|^{\alpha p} + |\lambda - \lambda'|^{\alpha p} + |t' - t|^{\frac{p}{2}-1}).$$

Choose p greater than $(d+2)(\frac{1}{2}+\alpha)^{-1} + 2$. Then both of αp and $\frac{p}{2} - 1$ are greater than $d + 2$, so that $N_t(\lambda, y)$ has a continuous extension at any $(t, \lambda, 0)$. This means that $\int_0^t f_s(\lambda) dM_s$ is continuously differentiable in λ and the derivative $\partial_\ell \int_0^t f_s(\lambda) dM_s$ is continuous in (t, λ). Further, letting y tend ot 0 in (5), we have $N_t(\lambda, 0) = \int_0^t \partial_\ell f_s(\lambda) dM_s$. This proves

$$\partial_\ell \int_0^t f_s(\lambda)dM_s = \int_0^t \partial_\ell f_s(\lambda)dM_s \qquad \forall(t,\lambda) \quad \text{a.s.}$$

The proof is complete in case $m = 1$.

10.7. Corollary. Suppose $f_s(\lambda)$ is continuous in (s,λ) and $m+1$-times continuously differentiable in λ. Then $\int_0^t f_s(\lambda)dM_s$ is continuous in (t,λ) and m-times continuously differentiable in λ. Furthermore, the m-th derivatives $D^k\int_0^t f_s(\lambda)dM_s$, $|k| = m$ are β-Hölder continuous in λ for any β less than 1.

In the next chapter, the Kolmogorov's criterion will be used extensively for deriving the smoothness of the solutions of stochastic differential equations.

BIBLIOGRAPHICAL NOTES

Doob's inequality and the optional sampling theorem for martingales (Theorem 1.2) are in any text book on martingales. See the book by Doob, Meyer or Neveu. The optional sampling theorem stated in Theorem 1.2 is a slight generalization of Doob's theorem.

A topology is introduced to the space $\underline{\underline{M}}_c^{loc}$ of continuous local martingale so that the quadratic variation is continuous in $\underline{\underline{M}}_c^{loc}$. Emery [8] introduced a topology to the space of semimartingales, which is equivalent to our topology.

Stochastic integrals by continuous semimartingales are adapted from Kunita-Watanabe [28], Meyer [34], and Ikeda-Watanabe [13]. Theorem 5.2, Corollary 8.2, Theorem 9.1 and 9.2 can be found in the above literatures.

Itô's formula (Theorem 8.1 and 8.3) are in Kunita [24]. A special case of Theorem 8.1 is studied in Ventcel [42] and Rozovsky [38]. Bismut [2] contains a similar formula. A recent work by A-S, Sznitman, "Martingales dépendant d'un paramètre: une formula d'Itô", Z.W. 60 (1982) discusses a similar formula. The author is indebted to T. Salisbury for Theorem 10.1.

CHAPTER II

STOCHASTIC DIFFERENTIAL EQUATIONS AND STOCHASTIC FLOWS OF HOMEOMORPHISMS

1. STOCHASTIC DIFFERENTIAL EQUATION WITH LIPSCHITZ CONTINUOUS COEFFICIENTS

A primitive and intuitive way of expressing a stochastic differential equation could be

$$\frac{d\xi_t}{dt} = X_0(t,\xi_t) + \sum_{k=1}^{m} X_k(t,\xi_t)\dot{B}_t^k ,$$

where $\dot{B}_t^k$, $k=1,\dots,m$ are independent white noises. It is intended to describe the motion of a particle driven by random forces or the motion perturbed by random noises. However, the equation fails to have a rigorous meaning, since $X_k(t,\xi_t)\dot{B}_t^k$ are not well defined. For the rigorous argument, we will introduce Itô's stochastic differential equation.

Let $B_t = (B_t^1,\dots,B_t^m)$, $t\in[0,a]$ be an m-dimensional Brownian motion defined on a probability space $(\Omega,\underline{F},P)$. For a pair s, t of $[0,a]$ such that $s < t$, we denote by $\underline{F}_{s,t}$ the least complete σ-field for which all $B_u - B_v$; $s \le v \le u \le t$ are measurable. Then the family of σ-fields $\{\underline{F}_{s,t}\}$ is increasing in t, decreasing in s; $\underline{F}_{s,t} \subset \underline{F}_{s',t'}$ if $s' \le s$ and $t \le t'$. Then $B_t - B_s$, $t \ge s$ is an $\underline{F}_{s,t}$-martingale for any s.

Given continuous mappings $X_k(t,x)$, $k=0,\dots,m$; $[0,a]\times R^d \longrightarrow R^d$, we shall consider an Itô's stochastic differential equation (SDE)

$$(1)\qquad d\xi_t = \sum_{k=1}^{m} X_k(t,\xi_t)dB_t^k + X_0(t,\xi_t)dt.$$

1.1. Definition. Given a time $s\in[0,a]$ and a state $x\in R^d$, a continuous stochastic process ξ_t, $t\in[s,a]$ with values in R^d is called a solution of (1) with the initial condition $\xi_s = x$, if it is $(\underline{F}_{s,t})$-adapted for each $t\geq s$ and satisfies

$$(2)\qquad \xi_t = x + \sum_{k=1}^{m}\int_s^t X_k(r,\xi_r)dB_r^k + \int_s^t X_0(r,\xi_r)dr.$$

For the convenience of notations, we will often write dt as dB_t^0 and write SDE (2) as

$$(3)\qquad \xi_t = x + \sum_{k=0}^{m}\int_s^t X_k(r,\xi_r)dB_r^k.$$

In this section we will show following Itô [16] that equation (3) has a unique solution for any initial condition if coefficients $X_0,\dots,X_m$ are globally Lipschitz continuous, i.e., there is a positive constant L such that

$$(4)\qquad |X_k(t,x)-X_k(t,y)| \leq L|x-y|$$

holds for all $t\in[0,a]$ and $x,y\in R^d$.

1.2. Theorem. Suppose that coefficients $X_0,\dots,X_m$ of equation (3) are globally Lipschitz continuous. Then the equation has a unique solution for any given initial condition. Further it is in L^p for any $p\geq 1$.

Proof. We shall construct a solution starting from x at time s, by the method of successive approximation. Define a sequence of $(\underline{F}_{s,t})$-adapted continuous stochastic processes by induction:

$$\xi_t^0 = x$$

$$\xi_t^n = x + \sum_{k=0}^{m} \int_s^t X_k(r, \xi_r^{n-1}) dB_r^k, \qquad n \geq 1.$$

Then it holds

$$\xi_t^{n+1} - \xi_t^n = \sum_{k=0}^{m} \int_s^t \{X_k(r, \xi_r^{n-1}) - X_k(r, \xi_r^{n-1})\} dB_r^k .$$

Therefore we have for $p \geq 2$,

$$E[\sup_{s \leq u \leq t} |\xi_u^{n+1} - \xi_u^n|^p]$$

$$\leq (m+1)^p \sum_{k=0}^{m} E[\sup_{s \leq u \leq t} |\int_s^u \{X_k(r, \xi_r^{n-1}) - X_k(r, \xi_r^n)\} dB_r^k|^p].$$

By Doob's inequality and Burkholder's inequality, each term corresponding to $k \geq 1$ is dominated by

$$q^p E[|\int_s^t \{\dots\} dB_r^k|^p] \leq q^p C^{(p)} |t-s|^{\frac{p}{2}-1} E[\int_s^t |\{\dots\}|^p dr]$$

$$\leq q^p C^{(p)} |t-s|^{\frac{p}{2}-1} L^p E[\int_s^t |\xi_r^n - \xi_r^{n-1}|^p dr].$$

The term corresponding to $k = 0$ is dominated by

$$|t-s|^{\frac{p}{q}} L^p E[\int_s^t |\xi_r^n - \xi_r^{n-1}|^p dr].$$

Therefore we get

$$(5) \quad E[\sup_{s \leq u \leq t} |\xi_u^{n+1} - \xi_u^n|^p] \leq c_1 E[\int_s^t |\xi_r^n - \xi_r^{n-1}|^p dr].$$

Denote the left hand side by $\rho_t^{(n)}$. Then the above implies $\rho_t^{(n)} \leq c_1 \int_s^t \rho_r^{(n-1)} dr$. By iteration, we get $\rho_t^{(n)} \leq \frac{c_1^n}{n!} a^n \rho_t^{(0)}$. Then

$$\sum_{n=0}^{\infty} E[\sup_{s \leq u \leq t} |\xi_u^{n+1} - \xi_u^n|^p]^{\frac{1}{p}} \leq \sum_{n=0}^{\infty} \{\frac{c_1^n}{n!} a^n \rho_t^{(0)}\}^{\frac{1}{p}} < +\infty,$$

since $\rho_t^{(0)} < \infty$. Therefore, $\{\xi_t^n\}$ converges uniformly in $[s,t]$ a.s. and in L^p-norm. Denote the limit as ξ_t. It is a continuous $(\underline{\underline{F}}_{s,t})$-adapted process. Furthermore, $\int_s^t X_k(r,\xi_r^n)dB_r^k$ converges to $\int_s^t X_k(r,\xi_r)dB_r^k$ in L^p-norm, since the quadratic variation of $\int_s^t \{X_k(r,\xi_r^n) - X_k(r,\xi_r)\}dB_r^k$ converges to 0 in L^p-norm. The convergence is valid for $k = 0$, obviously. Consequenctly ξ_t is a solution of equation (3).

We will next prove the uniqueness of the solution. Let ξ_t and $\tilde{\xi}_t$ be solutions of equation (3). Define $T_n = \inf\{t > 0 \,;\, |\xi_t| \geq n$ or $|\tilde{\xi}_t| \geq n\}$ $(= \infty$ if $\{\dots\} = \phi)$. Then it holds

$$\xi_t^{T_n} - \tilde{\xi}_t^{T_n} = \sum_{k=0}^{n} \int_s^{t\wedge T_n} \{X_k(r,\xi_r^{T_n}) - X_k(r,\tilde{\xi}_r^{T_n})\}dB_r^k .$$

Then by a similar calculation as the above, we obtain

$$E[\sup_{s\leq u\leq t} |\xi_u^{T_n} - \tilde{\xi}_u^{T_n}|^p] \leq c_1 E[\int_s^{t\wedge T_n} |\xi_r^{T_n} - \tilde{\xi}_r^{T_n}|^p dr].$$

Set $\rho_t = E[\sup_{s\leq u\leq t} |\xi_u^{T_n} - \tilde{\xi}_u^{T_n}|^p]$, where n is fixed. Then we get $\rho_t \leq c_1\int_s^t \rho_r dr$. By Gronwall's lemma, we get $\rho_t \equiv 0$. This proves $\xi_t^{T_n} = \tilde{\xi}_t^{T_n}$. Since $T_n \uparrow \infty$, we have $\xi_t = \tilde{\xi}_t$. The proof is complete.

1.3. Definition. The unique solution is denoted by $\xi_{s,t}(x)$.

The solution $\xi_{s,t}(x)$ has many properties analogous to those of ordinary differential equation. Instead of (3), consider a control system of ordinary differential equation on R^d;

$$\frac{d\phi_t}{dt} = X_0(t,\phi_t) + \sum_{k=0}^{m} X_k(t,\phi_t)u_t^k , \tag{6}$$

where $u_t = (u_t^1,\dots,u_t^m)$ is a piecewise smooth function. We denote the solution starting from (s,x) as $\phi_{s,t}(x)$. It is a well known fact that if coefficients $X_0,\dots,X_m$ are globally Lipschitz continuous, $\phi_{s,t}$ defines a flow of hemoemorphisms:

(E.1) $\phi_{s,t}(x)$ is Lipschitz continuous in (s,t,x),

(E.2) For $r < s < t$, $\phi_{r,t}(x) = \phi_{s,t} \circ \phi_{r,s}(x)$,

(E.3) For each $s < t$, $\phi_{s,t}$; $R^d \longrightarrow R^d$ is a homeomorphism.

In the subsequent sections we will prove the similar property for the solution $\xi_{s,t}(x)$ of equation (3). In Section 2, we will prove the Hölder continuity of $\xi_{s,t}(x)$ in (s,t,x). In Section 3, more smoothness of the solution with respect to x will be shown under additional smoothness assumptions for coefficients $X_0, \ldots, X_m$. The homeomorphic property of the map $\xi_{s,t}$; $R^d \longrightarrow R^d$ will be shown at Section 4.

We will introduce some notations for a class of smooth functions.

1.4. Definition. Let k be a nonnegative integer and let α be a number such that $0 < \alpha \leq 1$. A real function f on R^d is called a $C^{k,\alpha}$ function if it is k-th continuously differentiable and the k-th derivatives are locally Hölder continuous of order α. If the k-th derivatives are globally Hölder continuous we will call it a $C_g^{k,\alpha}$ function. In particular if $k=0$, $C^{0,\alpha}$ (or $C_g^{0,\alpha}$) function is a locally (or globally) Hölder continuous function.

The $C^{k,\alpha}$ map from a C^∞ manifold M into a C^∞ manifold N is defined similarly.

2. CONTINUITY OF THE SOLUTION WITH RESPECT TO THE INITIAL DATA

Let $\xi_{s,t}(x)$ be the solution of Itô's stochastic differential equation with globally Lipschitz continuous coefficients starting from (s,x);

$$(1) \quad \xi_{s,t}(x) = x + \sum_{k=0}^{m} \int_s^t X_k(r, \xi_{s,r}(x))dB_r^k.$$

The purpose of this section is to prove that there is a continuous modification of the solution $\xi_{s,t}(x)$ and Itô integrals $\int_s^t X_k(r,\xi_{s,r}(x))dB_r^k$ with respect to three variables (s,t,x) so that the equation (1) is satisfied for all (s,t,x) a.s. Our argument is based on the following L^p-estimate of the solution.

2.1. Theorem. For any p greater than 2, there is a positive constant $C_1^{(p)}$ such that

(2) $$E|\xi_{s,t}(x) - \xi_{s',t'}(x')|^p$$
$$\leq C_1^{(p)}\{|x-x'|^p+(1+|x|^p+|x'|^p)(|t-t'|^{\frac{p}{2}}+|s-s'|^{\frac{p}{2}})\}$$

holds for all (s,t,x) and (s',t',x') such that $s<t$ and $s'<t'$.

Remark. If coefficients $X_0,\ldots,X_m$ of equation (1) are bounded functions, we have an estimate

$$E|\xi_{s,t}(x)-\xi_{s',t'}(x')|^p \leq C_2^{(p)}\{|x-x'|^p+|t-t'|^{\frac{p}{2}}+|s-s'|^{\frac{p}{2}}\}.$$

C.f. Blagovescensky-Fleidlin [4], Funaki [10].

The following will be immediate from the above, applying Kolmogorov's theorem. (See Appendix in Chap. I).

2.2. Theorem. There are modifications of the solution and the stochastic integrals in (1) with following properties. $\xi_{s,t}(x)$ and $\int_s^t X_k(r,\xi_{s,r}(x))dB_r^k$, $k=0,\ldots,m$ are continuous in (s,t,x) and the equality (1) holds for any s,t,x a.s.

Furthermore, the solution $\xi_{s,t}(x)$ is (β,β,α)-Hölder continuous in (s,t,x), where β is an arbitrary number less than $\frac{1}{2}$ and α is an arbitrary number less than 1.

The rest of this section is devoted to the proof of Theorem 2.1. We will consider the case $s < s' < t < t'$ only. Other cases will be treated quite similarly. Since

$$\xi_{s',t'}(x') = x' + \sum_{k=0}^{m} \int_{s'}^{t} X_k(r,\xi_{s',r}(x'))dB_r^k + \sum_{k=0}^{m} \int_{t}^{t'} X_k(r,\xi_{s',r}(x'))dB_r^k,$$

$$\xi_{s,t}(x) = \xi_{s,s'}(x) + \sum_{k=0}^{m} \int_{s'}^{t} X_k(r,\xi_{s,r}(x))dB_r^k,$$

we have

$$|\xi_{s,t}(x) - \xi_{s',t'}(x')|^p \le (2m+3)^p \{ \sum_{k=0}^{m} |\int_t^{t'} X_k(r,\xi_{s',r}(x'))dB_r^k|^p$$

$$+ |\xi_{s,s'}(x) - x'|^p + \sum_{k=0}^{m} |\int_{s'}^{t} \{X_k(r,\xi_{s,r}(x)) - X_k(r,\xi_{s',r}(x'))\}dB_r^k|^p \}.$$

Consequently it is sufficient to prove the following three estimates:

(3) $\quad E[|\int_t^{t'} X_k(r,\xi_{s',r}(x'))dB_r^k|^p] \le C_3 |t'-t|^{\frac{p}{2}}(1+|x'|^p),$

(4) $\quad E|\xi_{s,s'}(x) - x'|^p \le C_4\{|x-x'|^p + |s-s'|^{\frac{p}{2}}(1+|x|^p)\},$

(5) $\quad E[|\int_{s'}^{t} \{X_k(r,\xi_{s,r}(x)) - X_k(r,\xi_{s',r}(x'))\}dB_r^k|^p]$

$\quad \le C_5\{|x-x'|^p + |s-s'|^{\frac{p}{2}}(1+|x|^p)\}.$

For the proofs of (3) and (4), we claim a lemma.

2.3. <u>Lemma.</u> Let p be any real number and $\varepsilon > 0$. Then there is a positive constant $C_6^{(p,\varepsilon)}$ such that

$$E[(\varepsilon + |\xi_{s,t}(x)|^2)^p] \le C_6^{(p,\varepsilon)}(\varepsilon + |x|^2)^p$$

holds for all $s,t \in [0,a]$ and $x \in R^d$.

<u>Proof.</u> Set $f(x) = (\varepsilon + |x|^2)$ and apply Itô's formula (Corollary 8.2

in Chap. I) to $F(x) = f(x)^p$ and $M_t = \xi_t = \xi_{s,t}(x)$, where (s,x) is fixed. Set $x = (x_1,\dots,x_d)$ and observe

$$\frac{\partial F}{\partial x_i}(x) = 2pf(x)^{p-1}x_i,$$

$$\frac{\partial^2 F}{\partial x_i \partial x_j}(x) = 2pf(x)^{p-2}\{f(x)\delta_{ij} + 2(p-1)x_i x_j\}.$$

By setting $X_k(r,x) = (X_k^1(r,x),\dots,X_k^d(r,x))$ and $\xi_t = (\xi_t^1,\dots,\xi_t^d)$,

$$(6)\quad F(\xi_t) - F(x)$$

$$= 2p \sum_{i,k\geq 1}\int_s^t f(\xi_r)^{p-1}\xi_r^i X_k^i(r,\xi_r)dB_r^k$$

$$+ 2p \sum_{i\geq 1}\int_s^t f(\xi_r)^{p-1}\xi_r^i X_0^i(r,\xi_r)dr$$

$$+ p \sum_{i,j\geq 1}\int_s^t f(\xi_r)^{p-2}\{f(\xi_r)\delta_{ij} + 2(p-1)\xi_r^i\xi_r^j\}(\sum_{k\geq 1} X_k^i(r,\xi_r)X_k^j(r,\xi_r))dr.$$

Here we have used the relation

$$d\langle \xi^i,\xi^j\rangle_t = \sum_{k,\ell\geq 1} X_k^i(t,\xi_t)X_\ell^j(t,\xi_t)d\langle B^k,B^\ell\rangle_t$$

$$= \sum_{k\geq 1} X_k^i(t,\xi_t)X_k^j(t,\xi_t)dt.$$

The first member of the right hand side of (6) is of mean 0. Observe the inequalities $|X_k^i(r,x)| \leq Cf(x)^{1/2}$, $|x^i| \leq f(x)^{1/2}$ etc. Then we see that the second and the third members are dominated by a constant times $\int_0^t F(\xi_r)dr$. Therefore, taking expectations in (6), we have

$$E[F(\xi_t)] - F(x) \leq C_7^{(p,\varepsilon)}\int_s^t E[F(\xi_r)]dr,$$

where $C_7^{(p,\varepsilon)}$ is a positive constant. By Gronwall's lemma, we get $E[F(\xi_t)] \leq F(x)\exp C_7^{(p,\varepsilon)}(t-s)$. The proof is complete.

<u>Proof</u> of (3). Let $k \geq 1$. By Burkholder's inequality (see (2) of

Section 7, Chap. I), we have

$$E[\,|\int_t^{t'} X_k(r,\xi_{s',r}(x'))dB_r^k|^p] \le C_0^{(p)}|t'-t|^{\frac{p}{2}-1}\int_t^{t'} E[|X_k(r,\xi_{s',r}(x'))|^p]dr.$$

Since it holds $|X_k(r,x)| \le C(1+|x|)$ with some positive constant C, Lemma 2.3 implies inequality (3) immediately. The case $k=0$ can be proved similarly.

Proof of (4). Since

$$\xi_{s,s'}(x) - x' = x - x' + \sum_{k=0}^{m}\int_s^{s'} X_k(r,\xi_{s,r}(x))dB_r^k,$$

we have, using (3),

$$\begin{aligned} E[|\xi_{s,s'}(x)-x'|^p] &\le (m+2)^p\{|x-x'|^p + \sum_{k=0}^{m} E[|\int_s^{s'} X_k(r,\xi_{s,r}(x))dB_r^k|^p]\} \\ &\le (m+2)^p\{|x-x'|^p+(m+1)C_3|s'-s|^{\frac{p}{2}}(1+|x|^p)\}. \end{aligned}$$

This proves (4).

For the proof of estimate (5), we require a lemma.

2.4. Lemma. For any real number p, there is a positive constant $C_8^{(p)}$ not depending on $\varepsilon > 0$ such that

(7) $$E[(\varepsilon+|\xi_{s,t}(x)-\xi_{s,t}(y)|^2)^p] \le C_8^{(p)}(\varepsilon+|x-y|^2)^p$$

holds for all $s < t$ and x,y.

Proof. Apply Itô's formula to $F(x) = f(x)^p$, $f(x) = \varepsilon+|x|^2$ and $M_t = \eta_t = \xi_{s,t}(x)-\xi_{s,t}(y)$, where s,x,y are fixed. Since

$$\eta_t = x-y+\sum_{k=0}^{m}\int_s^t \{X_k(r,\xi_{s,r}(x))-X_k(r,\xi_{s,r}(y))\}dB_r^k,$$

we have

$$(8) \quad F(\eta_t) - F(\eta_s)$$

$$= 2p \sum_{i,k} \int_s^t f(\eta_r)^{p-1} \eta_r^i \{X_k^i(r, \xi_{s,r}(x)) - X_k^i(r, \xi_{s,r}(y))\} dB_r^k$$

$$+ p \sum_{i,j} \int_s^t f(\eta_r)^{p-2} (f(\eta_r)\delta_{ij} + 2(p-1)\eta_r^i \eta_r^j)$$

$$\times \{\sum_{k \geq 1} (X_k^i(r, \xi_{s,r}(x)) - X_k^i(r, \xi_{s,r}(y)))(X_k^j(r, \xi_{s,r}(x)) - X_k^j(r, \xi_{s,r}(y)))\} dr.$$

The expectation of the first of the right hand side is 0 except for the term corresponding to k=0. Observe $|\eta_r^i| \leq f(\eta_r)^{1/2}$ and

$$|X_k^i(r, \xi_{s,r}(x)) - X_k^j(r, \xi_{s,r}(y)| \leq L|\eta_r| \leq L f(\eta_r)^{\frac{1}{2}}$$

by the Lipschitz condition. Then the expectation of the term $\int \ldots dB_r^0$ plus that of the last member in (8) is dominated by $C_9 \int_s^t E[F(\eta_r)] dr$. Then we get

$$E[F(\eta_t)] - F(x-y) \leq C_9 \int_s^t E[F(\eta_r)] dr.$$

The assertion follows from Gronwall's lemma.

Remark. Let ε tend to 0 in Lemma 2.4. Then we have

$$(9) \quad E[|\xi_{s,t}(x) - \xi_{s,t}(y)|^{2p}] \leq C_8^{(p)} |x-y|^{2p}$$

for any $s < t$ and $x, y \in R^d$. Observe the inequality in case $p < 0$. If $x \neq y$, then $\xi_{s,t}(x) \neq \xi_{s,t}(y)$ a.s. for any $s < t$. We will see at Section 4 a stronger property: The map $\xi_{s,t}(\cdot, \omega)$; $R^d \longrightarrow R^d$ is one to one for any $s < t$ a.s.

Inequality (9) is a special case of inequality (2). By Kolmogorov's theorem, there is a modification of the solution which is continuous in x a.s. for any $s < t$. Further, stochastic integral $\int_s^t X_k(r,\xi_{s,r}(x))dB_r^k$ has also a modification which is continuous in x. In fact, by Burkholder's inequality and (9),

$$E[|\int_s^t X_k(r,\xi_{s,r}(x))dB_r^k - \int_s^t X_k(r,\xi_{s,r}(y))dB_r^k|^p]$$

$$\leq C_0^{(p)}|t-s|^{\frac{p}{2}-1}\int_s^t E[|X_k(r,\xi_{s,r}(x))-X_k(r,\xi_{s,r}(y))|^p]dr$$

$$\leq C_0^{(p)}L^p|t-s|^{\frac{p}{2}-1}\int_s^t E[|\xi_{s,r}(x)-\xi_{s,r}(y)|^p]dr$$

$$\leq C_0^{(p)}C_8^{(\frac{p}{2})}|t-s|^{\frac{p}{2}}|x-y|^p.$$

Taking this continuous modification, the equality

$$\xi_{s,t}(x) = x + \sum_{k=0}^{m}\int_s^t X_k(r,\xi_{s,r}(x))dB_r^k$$

holds for all x a.s. for any $s < t$.

Now let $s_0 < s$ and substitute $\xi_{s_0,s}(x)$ in the place of x. Then we have

$$\xi_{s,t}(\xi_{s_0,s}(x)) = \xi_{s_0,s}(x) + \sum_{k=0}^{m}\int_s^t X_k(r,\xi_{s,r}(\xi_{s_0,s}(x)))dB_r^k.$$

Define $\hat{\xi}_{s_0,t}$ by

$$\hat{\xi}_{s_0,t}(x) = \begin{cases} \xi_{s_0,t}(x) & \text{if } t \leq s \\ \xi_{s,t}(\xi_{s_0,s}(x)) & \text{if } t \geq s. \end{cases}$$

Then it satisfies

$$\hat{\xi}_{s_0,t}(x) = x + \sum_{k=0}^{m}\int_{s_0}^t X_k(r,\hat{\xi}_{s_0,r}(x))dB_r^k.$$

By the uniqueness of the solution, we obtain $\xi_{s_0,t}(x) = \xi_{s,t}(\xi_{s_0,s}(x))$ for all x a.s. for any $s_0 < s < t$.

<u>Proof</u> of (5). By Burkholder's inequality, we have

$$(10)\quad E[\,|\int_{s'}^{t}\{X_k(r,\xi_{s,r}(x)) - X_k(r,\xi_{s',r}(x'))\}dB_r^k|^p]$$

$$\leq C_0^{(p)}|t-s'|^{\frac{p}{2}-1}\int_{s'}^{t} E[\,|X_k(r,\xi_{s,r}(x)) - X_k(r,\xi_{s',r}(x'))|^p]dr$$

$$\leq C_0^{(p)}L^p\,|t-s'|^{\frac{p}{2}-1}\int_{s'}^{t} E[\,|\xi_{s,r}(x) - \xi_{s',r}(x')|^p]dr.$$

Note that $\xi_{s,r}(x) = \xi_{s',r}\circ\xi_{s,s'}(x)$ and that $\xi_{s',r}(y)$ and $\xi_{s,s'}(x)$ are independent. Apply Lemma 2.4 and estimate (4). Then we have

$$\begin{aligned} E[\,|\xi_{s,r}(x) - \xi_{s',r}(x')|^p] &= \int E[\,|\xi_{s',r}(y) - \xi_{s',r}(x')|^p]P(\xi_{s,s'}(x)\in dy) \\ &\leq C_8^{(p)}\int |y-x'|^p P(\xi_{s,s'}(x)\in dy) \\ &\leq C_8^{(p)} E[\,|\xi_{s,s'}(x) - x'|^p] \\ &\leq C_8^{(p)} C_4\{|x-x'|^p + |s'-s|^{\frac{p}{2}}(1+|x|^p)\}. \end{aligned}$$

Substitute the above inequality to (10), we get the estimate (5).

<u>Proof</u> of Theorem 2.2. If (2) is satisfied, then by Kolmogorov's theorem, $\xi_{s,t}(x)$ has a modification which is locally (β,β,α)-Hölder continuous with respect to (s,t,x), where $\beta < p^{-1}(\frac{p}{2} - d)$ and $\alpha < 2p^{-1}(\frac{p}{2} - d)$. Since p is arbitrary, β can take any value less than half and α can take any value less than 1.

We will next prove the continuity of the integral $\int_s^t X(r,\xi_{s,r}(x))dB_r^k$. Since the case $k = 0$ is obvious, we will consider the case $k \geq 1$. Suppose $s < s' < t < t'$ as before. Then

$$\int_s^t X_k(r,\xi_{s,r}(x))dB_r^k - \int_{s'}^{t'} X_k(r,\xi_{s',r}(x'))dB_r^k$$

$$= \int_s^{s'} X_k(r,\xi_{s,r}(x))dB_r^k + \int_{s'}^t \{X_k(r,\xi_{s,r}(x)) - X_k(r,\xi_{s',r}(x'))\}dB_r^k$$

$$- \int_t^{t'} X_k(r,\xi_{s',r}(x'))dB_r^k .$$

L_p-estimates of the first and the third terms of the right hand side have been given in (3). L_p-estimate of the second term is given by (5). Therefore, L^p-norm of the left hand side is again dominated by a quantity like the right hand side of (2). Therefore the stochastic integrals $\int_s^t X_k(r,\xi_{s,r}(x))dB_r^k$, $k=1,\ldots,m$ have the same kind of continuity as that of $\xi_{s,t}(x)$.

Other properties of the theorem will be obvious from the above.

3. SMOOTHNESS OF THE SOLUTION WITH RESPECT TO THE INITIAL DATA

We have seen in the previous section that the solution $\xi_{s,t}(x)$ of a SDE is locally Hölder continuous of order $\alpha < 1$, provided that coefficients of the SDE are Lipschitz continuous. In this section we will see more smoothness of the solution under additional smoothness assumption for coefficients.

3.1. Theorem. Suppose that coefficients $X_0,\ldots,X_m$ of an Itô SDE are $C_g^{1,\alpha}$ functions for some $\alpha > 0$ and their first derivatives are bounded. Then the solution $\xi_{s,t}(x)$ is a $C^{1,\beta}$ function of x for any β less than α for each $s < t$ a.s. Furthermore, the derivative $\partial_\ell \xi_{s,t}(x) = (\frac{\partial \xi_{s,t}(x)}{\partial x_\ell})$ satisfies the following SDE

(1) $$\partial_\ell \xi_{s,t}(x) = e_\ell + \sum_{k=0}^m \int_s^t X'_k(r,\xi_{s,r}(x))\partial_\ell \xi_{s,r}(x)dB_r^k$$

for all (s,t,x) a.s., where $X'_k(r,x)$ is a matrix valued function $(\frac{\partial X_k^i(r,x)}{\partial x_j})_{i,j=1,\ldots,d}$ and e_ℓ is the unit vector $(0,\ldots,0,1,0,\ldots,0)$ (1 is the ℓ-th component).

For $y \in R - \{0\}$, define

(2) $$\eta_{s,t}(x,y) = \frac{1}{y}\{\xi_{s,t}(x+ye_\ell) - \xi_{s,t}(x)\}.$$

Then the existence of the partial derivative $\frac{\partial \xi_{s,t}(x)}{\partial x_\ell}$ for any s,t,x, a.s. can be assured if $\eta_{s,t}(x,y)$ has a continuous extension at $y = 0$ for any s,t,x a.s. This follows from the following lemma and Kolmogorov's theorem.

3.2. Lemma. For any $p > 2$, there is a positive constant $C_{10}^{(p)}$ such that

(3) $$E|\eta_{s,t}(x,y) - \eta_{s',t'}(x',y')|^p$$
$$\le C_{10}^{(p)}\{|x-x'|^{\alpha p} + |y-y'|^{\alpha p} + (1+|x|+|x'|)^{\alpha p}(|s-s'|^{\frac{\alpha p}{2}} + |t-t'|^{\frac{\alpha p}{2}})\}.$$

Proof. We first show the boundedness of $E|\eta_{s,t}(x)|^p$. By the mean value theorem, it holds

(4) $$\eta_{s,t}(x,y) = e_\ell + \sum_{k=0}^{m}\int_s^t\{\int_0^1 X_k'(r,\xi_{s,r}(x) + v(\xi_{s,r}(x+ye_\ell) - \xi_{s,r}(x)))dv\}$$
$$\times \eta_{s,r}(x,y)dB_r^k.$$

Therefore we have

(5) $$E|\eta_{s,t}(x,y)|^p \le (m+2)^p\{1 + \sum_{k=0}^{m} E[|\int_s^t (\int_0^1 X_k'(\ldots)dv)\eta_{s,r}(x,y)dB_r^k|^p]\}.$$

Using Burkholder's inequality, we have for $k \ge 1$,

$$E[|\int_s^t (\int_0^1 X_k'(\ldots)dv)\eta_{s,r}(x,y)dB_r^k|^p]$$
$$\le C_{11}^{(p)}|t-s|^{\frac{p}{2}-1} E[\int_s^t |\int_0^1 X_k'(\ldots)dv\eta_{s,r}(x,y)|^p dr]$$
$$\le C_{11}^{(p)}|t-s|^{\frac{p}{2}-1} ||X_k'|| \int_s^t E|\eta_{s,r}(x,y)|^p dr.$$

Here $\|X_k'\| = \sup_{(r,x)} |X_k'(r,x)|$ and $|A|$ denotes the norm of the matrix $A = (a_{ij})$ defined by $|A| = \sqrt{\Sigma_{i,j} a_{ij}^2}$. Similar estimate is valid for $k = 0$. Then from (5), we obtain

$$E|\eta_{s,t}(x,y)|^p \le C_{12}^{(p)} + C_{13}^{(p)} \int_s^t E|\eta_{s,r}(x,y)|^p dr,$$

where constants $C_{12}^{(p)}$ and $C_{13}^{(p)}$ do not depend on s,t,x,y. Therefore by Gronwall's inequality, we see that $E|\eta_{s,t}(x,y)|^p$ is bounded.

We next show (3) in case $t=t'$. We assume $s<s'\le t$. Other cases will be treated similarly. Note that $\eta_{s,t}(x,y) - \eta_{s',t}(x',y')$ is a sum of the following terms:

$$(6) \qquad \int_s^{s'} \left(\int_0^1 X_k'(r, \xi_{s,r}(x) + v(\xi_{s,r}(x+ye_\ell) - \xi_{s,r}(x))dv\right) \eta_{s,r}(x,y) dB_r^k$$

$$(7) \qquad \int_{s'}^t \Big[\left(\int_0^1 X_k'(r, \xi_{s,r}(x) + v(\xi_{s,r}(x+ye_\ell) - \xi_{s,r}(x))dv\right) \eta_{s,r}(x,y)$$

$$- \left(\int_0^1 X_k'(r, \xi_{s',r}(x') + v(\xi_{s',r}(x'+y'e_\ell) - \xi_{s',r}(x'))dv\right) \eta_{s',r}(x',y')\Big] dB_r^k.$$

Using Burkholder's inequality, the expectation of the p-th power of (6) is estimated in case $k \ge 1$ as

$$E\left[\left|\int_s^{s'} \left(\int_0^1 X_k'(\ldots)dv\right) \eta_{s,r}(x,y) dB_r^k\right|^p\right]$$

$$\le C_{14}^{(p)} |s'-s|^{\frac{p}{2}-1} E\left[\int_s^{s'} \left|\left(\int_0^1 X_k'(\ldots)dv\right) \eta_{s,r}(x,y)\right|^p dr\right]$$

$$\le C_{14}^{(p)} \|X_k'\|^p |s'-s|^{\frac{p}{2}-1} \int_s^{s'} E|\eta_{s,r}(x,y)|^p dr,$$

which is dominated by $C_{15}^{(p)} |s-s'|^{p/2}$ by the argument of the previous paragraph.

We will calculate the expectation of the p-th power of (7). Note that the integrant [...] in (7) is estimated as

|integrant [...] |

$$\leq \int_0^1 |X'_k(r,\xi_{s,r}(x)+vy\eta_{s,r}(x,y))|dv\cdot|\eta_{s,r}(x,y)-\eta_{s',r}(x',y')|$$

$$+\int_0^1 |X'_k(r,\xi_{s,r}(x)+vy\eta_{s,r}(x,y))-X'_k(r,\xi_{s',r}(x')+vy'\eta_{s',r}(x',y'))|dv\cdot|\eta_{s',r}(x',y')|$$

$$\leq \|X'_k\|\,|\eta_{s,r}(x,y)-\eta_{s',r}(x',y')|$$

$$+L\int_0^1 \{(1-v)^\alpha|\xi_{s,r}(x)-\xi_{s',r}(x')|^\alpha+v^\alpha|\xi_{s,r}(x+ye_\ell)-\xi_{s',r}(x'+y'e_\ell)|^\alpha\}dv\cdot|\eta_{s',r}(x',y')|$$

$$\leq \|X'_k\|\,|\eta_{s,r}(x,y)-\eta_{s',r}(x',y')|+L|\xi_{s,r}(x)-\xi_{s',r}(x')|^\alpha\cdot|\eta_{s',r}(x',y')|$$

$$+L|\xi_{s,r}(x+ye_\ell)-\xi_{s',r}(x'+y'e_\ell)|^\alpha\cdot|\eta_{s',r}(x',y')|.$$

Here L is a Hölder constant; $|X'_k(r,x)-X'_k(r,x')|\leq L|x-x'|^\alpha$. Therefore, by Burkholder's inequality,

$$C^{(p)^{-1}}E[|\int_{s'}^t [\ldots]dB_r^k|^p]$$

$$\leq |t-s'|^{\frac{p}{2}-1}\int_{s'}^t E[|[\ldots]|^p]dr$$

$$\leq |t-s'|^{\frac{p}{2}-1}\,3^p\{\|X'_k\|^p\int_{s'}^t E[|\eta_{s,r}(x,y)-\eta_{s',r}(x',y')|^p]dr$$

$$+L^p(\int_{s'}^t E[|\xi_{s,r}(x)-\xi_{s',r}(x')|^{2\alpha p}]^{\frac{1}{2}}E[|\eta_{s',r}(x',y')|^{2p}]^{\frac{1}{2}}dr$$

$$+\int_{s'}^t E[|\xi_{s,r}(x+ye_\ell)-\xi_{s',r}(x'+y'e_\ell)|^{2\alpha p}]^{\frac{1}{2}}E[|\eta_{s',r}(x',y')|^{2p}]^{\frac{1}{2}}dr\}.$$

Apply Theorem 2.1 to $E|\xi_{s,r}(x)-\xi_{s',r}(x')|^{\alpha p}$ etc. Then the above is dominated by

$$C_{15}\{(1+|x|+|x'|)^{\alpha p}|s-s'|^{\frac{\alpha p}{2}}+|x-x'|^{\alpha p}+|y-y'|^{\alpha p}\}$$

$$+C_{16}\int_{s'}^t E[|\eta_{s,r}(x,y)-\eta_{s',r}(x',y')|^p]dr.$$

Summing up these calculations for (6) and (7), we arrive at

$$E[\,|\eta_{s,t}(x,y)-\eta_{s',t}(x',y')|^p]$$

$$\le C_{17}\{|s-s'|^{\frac{p}{2}}+(1+|x|+|x'|)^{\alpha p}|s-s'|^{\frac{\alpha p}{2}}+|x-x'|^{\alpha p}+|y-y'|^{\alpha p})\}$$

$$+C_{18}\int_{s'}^{t}E[\,|\eta_{s,r}(x,y)-\eta_{s',r}(x',y')|^p]dr.$$

By Gronwall's inequality, we have

$$E[\,|\eta_{s,t}(x,y)-\eta_{s',t}(x',y')|^p]$$

$$\le C_{17}\{(1+|x|+|x'|)^{\alpha p}|s-s'|^{\frac{\alpha p}{2}}+|x-x'|^{\alpha p}+|y-y'|^{\alpha p}\}\exp C_{18}(t-t').$$

This proves (3) in case $t=t'$.

It remains to prove (3) in case $t\neq t'$. Assuming $t<t'$, we have

$$\eta_{s,t}(x,y)-\eta_{s',t'}(x',y')$$

$$=\eta_{s,t}(x,y)-\eta_{s',t}(x',y')-\sum_{k=0}^{m}\int_{t}^{t'}\Big(\int_{0}^{1}X_k'(\ldots)dv\Big)\eta_{s',r}(x',y')dB_r^k .$$

It holds

$$C^{(p)^{-1}}E[\,|\int_{t}^{t'}\Big(\int_{0}^{1}X_k'(\ldots)dv\Big)\eta_{s',r}(x',y')dB_r^k|^p]$$

$$\le|t'-t|^{\frac{p}{2}-1}E[\int_{t}^{t'}|\Big(\int_{0}^{1}X_k'(\ldots)dv\Big)\eta_{s',r}(x',y')|^p dr]$$

$$\le|t'-t|^{\frac{p}{2}-1}\|X_k'\|^p\int_{t}^{t'}E|\eta_{s',r}(x',y')|^p dr$$

$$\le C_{19}|t'-t|^{\frac{p}{2}} .$$

Therefore we get the desired estimation (3). The proof is complete.

<u>Proof of Theorem 3.1.</u> By Kolmogorov's theorem, $\eta_{s,t}(x,y)$ has a

continuous extension at $y=0$ for all $s<t$ and $x \in R^d$ a.s. This means that $\xi_{s,t}(x)$ is continuously differentiable in the domain $\{(s,t,x) | s<t, x \in R^d\}$ and the derivative $\partial_\ell \xi_{s,t}(x)$ is β-Hölder continuous for any $\beta < \alpha$. Let y tend to 0 in (4). Then we obtain (1). The proof is complete.

3.3. Theorem. Let k be a positive integer and α be $0 < \alpha \leq 1$. Suppose that coefficients $X_0, \dots, X_m$ are $C_g^{k,\alpha}$ functions of x for some α and their derivatives up to k-th order are bounded. Then the solution $\xi_{s,t}(x)$ is a $C^{k,\beta}$ function of x for any β less than α.

Proof. We will consider the case $k=2$. Let $y \in R - \{0\}$ and set

$$\zeta_{s,t}(x,y) = \frac{1}{y}\{\partial_i \xi_{s,t}(x+ye_\ell) - \partial_i \xi_{s,t}(x)\}.$$

Then similarly as the proof of Lemma 3.2, we obtain an estimate

$$E[\,|\zeta_{s,t}(x,y) - \zeta_{s',t'}(x',y')|^p]$$

$$\leq C_{20}\{|x-x'|^{\alpha p} + |y-y'|^{\alpha p} + (1+|x|+|x'|)^{\alpha p}(|s-s'|^{\frac{\alpha p}{2}} + |t-t'|^{\frac{\alpha p}{2}})\}$$

for all $s<t$, $s'<t'$, $x,x' \in R^d$, $y,y' \in R-\{0\}$. This implies the existence of the partial derivative $\partial_\ell \partial_i \xi_{s,t}(x)$ for all $s<t$ and x a.s. and the partial derivative is β-Hölder continuous for any $\beta < \alpha$.

4. STOCHASTIC FLOW OF HOMEOMORPHISMS (I). CASE OF GLOBALLY LIPSCHITZ CONTINUOUS COEFFICIENTS

In section 2, we saw that if coefficients of an Itô SDE are globally Lipschitz continuous, then there is a modification of the solution $\xi_{s,t}(x)$ which is continuous in three variables (s,t,x) a.s. Then for any $s < t$, $\xi_{s,t}(\cdot,\omega)$ defines a continuous map $R^d \longrightarrow R^d$ for almost all ω. We will prove in this section that the map is actually a homeomorphism of R^d onto itself a.s.

We will first consider the "one to one" property of the map $\xi_{s,t}(\cdot,\omega)$. Lemma 2.4 implies the inequality

$$(1) \quad E[\,|\xi_{s,t}(x)-\xi_{s,t}(y)|^{2p}] \le C_8^{(p)}|x-y|^{2p}$$

for negative p. This shows that if $x \neq y$, then $\xi_{s,t}(x) \neq \xi_{s,t}(y)$ a.s. for any $s < t$. But this does not imply immediately that the map $\xi_{s,t}(\cdot,\omega)$ is one to one a.s. To prove the latter assertion, we require a lemma.

4.1. Lemma. Set

$$\eta_{s,t}(x,y) = \frac{1}{|\xi_{s,t}(x)-\xi_{s,t}(y)|}\,.$$

Then for any $p > 2$, there is a constant $C_{21}^{(p)}$ such that for any $\delta > 0$

$$E[\,|\eta_{s,t}(x,y)-\eta_{s',t'}(x',y')|^p]$$

$$\le C_{21}^{(p)}\delta^{-2p}\{|x-x'|^p+|y-y'|^p+(1+|x|^p+|x'|^p+|y|^p+|y'|^p)(|t-t'|^{\frac{p}{2}}+|s-s'|^{\frac{p}{2}})\}$$

holds for all $s < t$ and x,y,x',y' such that $|x-y| \ge \delta$ and $|x'-y'| \ge \delta$.

Proof. A simple computation yields

$$|\eta_{s,t}(x,y)-\eta_{s',t'}(x',y')|^p$$

$$\le 2^p\eta_{s,t}(x,y)^p\eta_{s',t'}(x',y')^p\{|\xi_{s,t}(x)-\xi_{s',t'}(x')|^p+|\xi_{s,t}(y)-\xi_{s',t'}(y')|^p\}.$$

Take expectations for both sides and use Hölder's inequality. Then,

$$E[\,|\eta_{s,t}(x,y)-\eta_{s',t'}(x',y')|^p]$$

$$\leq 2^p E[\,|\eta_{s,t}(x,y)|^{4p}]^{\frac{1}{4}} E[\,|\eta_{s',t'}(x',y')|^{4p}]^{\frac{1}{4}}$$

$$\times \{E[\,|\xi_{s,t}(x)-\xi_{s',t'}(x')|^{2p}]^{\frac{1}{2}} + E[\,|\xi_{s,t}(y)-\xi_{s',t'}(y')|^{2p}]^{\frac{1}{2}}\}.$$

It holds by (1)

$$E[\,|\eta_{s,t}(x,y)|^{4p}]^{\frac{1}{4}} \leq C_{22}|x-y|^{-p} \leq C_{22}\delta^{-p},$$

where $|x-y| \geq \delta$. Also by Theorem 2.1,

$$E[\,|\xi_{s,t}(x)-\xi_{s',t'}(x')|^{2p}]^{\frac{1}{2}} \leq C_{23}\{|x-x'|^p + (1+|x|^p+|x'|^p)(|t-t'|^{\frac{p}{2}} + |s-s'|^{\frac{p}{2}})\}.$$

Therefore we get the lemma.

We can prove the "one to one" property of the map $\xi_{s,t}$. Take p as large as $\frac{p}{2} > 2(d+1)$ in Lemma 4.1. Kolmogorov's theorem states that $\eta_{s,t}(x,y)$ is continuous in (s,t,x,y) in the domain $\{(s,t,x,y) \mid s < t, |x-y| \geq \delta\}$. Since δ is arbitrary, it is also continuous in the domain $\{(s,t,x,y) \mid s < t, x \neq y\}$. This proves that the map $\xi_{s,t} : R^d \longrightarrow R^d$ is one to one for any $0 < s < t < a$ a.s.

We will next consider the onto property of the map $\xi_{s,t}$. We claim a lemma.

4.2. <u>Lemma.</u> Let $\hat{R}^d = R^d \cup \{\infty\}$ be the one point campactification of R^d. Set $\hat{x} = |x|^{-2}x$ and define

$$\eta_{s,t}(\hat{x}) = \frac{1}{1+|\xi_{s,t}(x)|} \quad \text{if } \hat{x} \in R^d, \quad = 0 \text{ if } \hat{x} = 0.$$

Then for any positive p, there is a constant $C_{24}^{(p)}$ such that

$$E[\,|\eta_{s,t}(\hat{x}) - \eta_{s',t'}(\hat{x}')|^p] \leq C_{24}^{(p)}\{|\hat{x}-\hat{x}'|^p + |t-t'|^{\frac{p}{2}} + |s-s'|^{\frac{p}{2}}\}.$$

<u>Proof.</u> Since

$$|\eta_{s,t}(\hat{x}) - \eta_{s',t'}(\hat{x}')|^p \leq \eta_{s,t}(\hat{x})^p \eta_{s',t'}(\hat{x}')^p |\xi_{s,t}(x) - \xi_{s',t'}(x')|^p,$$

we have by Hölder's inequality

$$E[|\eta_{s,t}(\hat{x}) - \eta_{s',t'}(\hat{x}')|^p] \leq E[|\eta_{s,t}(\hat{x})|^{4p}]^{\frac{1}{4}} E[|\eta_{s',t'}(\hat{x}')|^{4p}]^{\frac{1}{4}}$$
$$\times E[|\xi_{s,t}(x) - \xi_{s',t'}(x')|^{2p}]^{\frac{1}{2}}.$$

Apply Lemma 2.3 and Theorem 2.1. Then the right hand side is dominated by

$$C_{25}(1+|x|)^{-p}(1+|x'|)^{-p}\{|x-x'|^p + (1+|x|+|x'|)^p(|t-t'|^{\frac{p}{2}} + |s-s'|^{\frac{p}{2}})\}$$
$$\leq C_{25}\{|\hat{x}-\hat{x}'|^p + |t-t'|^{\frac{p}{2}} + |s-s'|^{\frac{p}{2}}\},$$

if x and x' are finite. Here we have used the inequality $(1+|x|)^{-1} \times (1+|x'|)^{-1}|x-x'| \leq |\hat{x}-\hat{x}'|$. In case $x = \infty$, we have

$$E[|\eta_{s',t'}(\hat{x}')|^p] \leq C_{26}(1+|x'|)^{-p} \leq C_{26}|\hat{x}'|^p.$$

Therefore the inequality of the lemma follows.

The "onto" property of the map $\xi_{s,t}$ follows from Lemma 4.2. Take p greater than $2(d+3)$. Then by Kolmogorov's theorem, $\eta_{s,t}(\hat{x})$ is continuous at $\hat{x} = 0$. Therefore, $\xi_{s,t}(\cdot,\omega)$ can be extended to a continuous map from $\hat{R}^d$ into itself for any $s < t$ a.s. The extension $\tilde{\xi}_{s,t}(x,\omega)$ is continuous in (s,t,x) a.s. We will fix such ω. The map $\tilde{\xi}_{s,t}(\cdot,\omega)$; $\hat{R}^d \longrightarrow \hat{R}^d$ is then homotopic to the identity map $\tilde{\xi}_{s,s}(\cdot,\omega)$, so that it is an onto map by a well known theorem of homotopic theory. The restriction of $\tilde{\xi}_{s,t}(\cdot,\omega)$ to R^d is again an "onto" map since $\tilde{\xi}_{s,t}(\infty,\omega) = \infty$.

The map $\xi_{s,t}(\cdot,\omega)$; $R^d \longrightarrow R^d$ is one to one and onto. Hence the inverse map $\xi_{s,t}^{-1}(\cdot,\omega)$; $R^d \longrightarrow R^d$ is also one to one and onto. It is continuous. Indeed, the inverse map $\tilde{\xi}_{s,t}^{-1}(\cdot,\omega)$; $\hat{R}^d \longrightarrow \hat{R}^d$ is continuous since $\tilde{\xi}_{s,t}(\cdot,\omega)$ is a one to one, continuous map from the compact space $\hat{R}^d$ into itself.

We will summarize the result.

4.3. Theorem. Suppose that coefficients of an Itô SDE are globally Lipschitz continuous. Then there is a modification of the solution, denoted by $\xi_{s,t}(x,\omega)$ which satisfies the following properties.

(F.1) For each $s<t$ and x, $\xi_{s,t}(x,\cdot)$ is $(\underline{\underline{F}}_{s,t})$-measurable.

(F.2) For almost all ω, $\xi_{s,t}(x,\omega)$ is continuous in (s,t,x) and satisfies $\lim_{t\downarrow s} \xi_{s,t}(x,\omega) = x$.

(F.3) For almost all ω, $\xi_{s,t+u}(x,\omega) = \xi_{t,t+u}(\xi_{s,t}(x,\omega),\omega)$ is satisfied for all $s<t$ and $u>0$.

(F.4) For almost all ω, the map $\xi_{s,t}(\cdot,\omega)$; $R^d \longrightarrow R^d$ is an onto homeomorphism for all $s<t$.

4.4. Theorem. Let k be a positive integer. Suppose that coefficients of an Itô equation are $C_g^{k,\alpha}$ functions for some $\alpha>0$ and their derivatives up to k-th order are bounded. Then the map $\xi_{s,t}(\cdot,\omega)$; $R^d \longrightarrow R^d$ is a C^k-diffeomorphism for all $s<t$ a.s.

Proof. Smoothness of the map $\xi_{s,t}$; $R^d \longrightarrow R^d$ was shown in Theorem 3.3. It is enough to show that the Jacobian matrix $\partial\xi_{s,t}(x) = (\frac{\partial \xi_{s,t}(x)}{\partial x})$ is nonsingular for any x a.s. If it were shown then the implicit function theorem states that the inverse map is again of C^k-class. Now by Theorem 3.1, the Jacobian matrix satisfies following linear SDE;

$$\partial \xi_{s,t} = I + \sum_{k=0}^{m} \int_s^t X_k'(r,\xi_{s,r}(x))\partial \xi_{s,r} dB_r^k .$$

Consider an adjoint equation of the above:

$$K_{s,t}(x) = I - \sum_{k=0}^{m} \int_s^t K_{s,r}(x)X'_k(r,\xi_{s,r}(x))dB_r^k$$
$$- \sum_{k=1}^{m} \int_s^t K_{s,r}(x)X'_k(r,\xi_{s,r}(x))^2 dr.$$

Obviously it has unique matrix solution $K_{s,t}(x)$. We can prove similarly as before

$$E[\,|K_{s,t}(x) - K_{s',t'}(x')|^p] \le C_{27}^{(p)}\{|x-x'|^{\alpha p} + (1+|x|+|x'|)^{\alpha p}(|s'-s|^{\frac{\alpha p}{2}} + |t'-t|^{\frac{\alpha p}{2}})\}.$$

Hence $K_{s,t}(x)$ is continuous in (s,t,x) a.s. By Itô's formula, it holds

$$K_{s,t}(x)\partial\xi_{s,t}(x) = I + \int_s^t (dK_{s,r}(x))\partial\xi_{s,r}(x) + \int_s^t K_{s,r}(x)d\partial\xi_{s,r}(x) + \langle K(x), \xi(x)\rangle_t$$
$$= I.$$

Therefore $\partial\xi_{s,t}(x)$ has the inverse matrix $K_{s,t}(x)$ for any (s,t,x), proving $\partial\xi_{s,t}(x,\omega)$ is nonsingular for any (s,t,x) a.s.

5. SDE WITH LOCALLY LIPSCHITZ CONTINUOUS COEFFICIENTS

In Sections 1-4 we have considered SDE's whose coefficients are globally Lipschitz continuous. In the sequel, we will not assume the global Lipschitz condition for the equation. Then, as in the case of ordinary differential equation, the equation may not have a global solution; the explosion may occur at a finite time.

In this section we shall construct the solution up to the explosion time. Let $X_0(t,x),\dots,X_m(t,x)$, $t\in[0,a]$, $x\in R^d$ be R^d valued continuous functions. We assume that these are locally Lipschitz continuous, i.e., Lipschitz conditions are satisfied on any bounded domain of R^d. Consider an Itô's SDE on R^d:

(1) $\quad d\xi_t = \sum_{k=0}^{m} X_k(t,\xi_t)dB_t^k,$

where $(B_t^1,\dots,B_t^m)$ is an m-dimensional Brownian motion and $B_t^0 \equiv t$ as before.

5.1. Definition. Let $\xi_{s,t}(x)$, $x \in R^d$, $0 < s < t < T(s,x,\omega) \wedge a$ be a random field with values in R^d. It is called a local solution of (1) with the initial condition $\xi_s = x$ if the following four conditions are satisfied.

(i) $T(s,x,\omega)$ is a measurable accessible $(\underline{\underline{F}}_{s,t})$-stopping time[1], strictly greater than s, where $(\underline{\underline{F}}_{s,t}) = \sigma(B_u - B_v \,;\, s \le u \le v \le t)$.

(ii) $\xi_{s,t}(x)$ is continuous in (s,t,x)[2].

(iii) $\xi_{s,t}(x)$ is an $(\underline{\underline{F}}_{s,t})$-semimartingale for each (s,x).

(iv) For $s < t < T(s,x)$,

$$\xi_{s,t}(x) = x + \sum_{k=0}^{m} \int_s^t X_k(r,\xi_{s,r}(x))dB_r^k . \tag{2}$$

Furthermore, if $\lim_{t \uparrow T(s,x)} \xi_{s,t}(x) = \infty$ is satisfied if $T(s,x) < \infty$ a.s., then $\xi_{s,t}$ is called a maximal solution and $T(s,x)$ is called the explosion time.

5.2. Theorem. Suppose that coefficients $X_0,\dots,X_m$ of equation (1) are locally Lipschitzian. Then the maximal solution exists uniquely.

Proof. For each natural number N, we may choose globally Lipschitz continuous R^d-valued functions $X_k^N(t,x)$, $k=0,\dots,m$ such that $X_k(t,x) = X_k^N(t,x)$, $t \in [0,a]$ and $|x| \le N$. Then equation (2) corresponding to coefficients $\{X_k^N\}$ has a unique solution, which we will denote as $\xi_{s,t}^N(x)$. Let $T_N(s,x)$ be the hitting time for the set $\{x \,;|x| \ge N\}$ of the process $\xi_{s,t}^N(x)$; $T_N(s,x) = \inf\{t < a; |\xi_{s,t}^N(x)| \ge N\}$ $(= \infty$ if $\{\dots\} = \phi)$.

---

1) There is an increasing sequence of stopping times $T_N(s,x)$, $N=1,2,\dots$ which are measurable in (s,x) such that $T_N(s,x) < T(s,x)$ and $T_N(s,x) \uparrow T(s,x)$ hold for any s,x a.s.

2) Each stopped process $\xi_{s,t}^{T_N(s,x)}(x)$ is a semimartingale.

Then it is an $(\underline{\underline{F}}_{s,t})$-stopping time for each (s,x). For $t \leq T_N(s,x)$ it holds

$$\xi^N_{s,t}(x) = x + \sum_{k=0}^{m} \int_s^t X_k(r, \xi^N_{s,r}(x))dB^k_r .$$

Then by the uniqueness of the solution, we have $\xi^N_{s,t}(x) = \xi^M_{s,t}(x)$ for $t < T_N(s,x)$ if $N < M$. Set $T(s,x) = \lim_{N\to\infty} T_N(s,x)$ and define $\xi_{s,t}(x)$ for $s \leq t < T(s,x)$ by $\xi_{s,t}(x) = \xi^N_{s,t}(x)$ if $s \leq t < T_N(s,x)$. Then the process $\xi_{s,t}$ with the explosion time $T(s,x)$ is a desired solution. The uniqueness will be obvious.

Stopping times $T_N(s,x)$ are lower semicontinuous in x a.s., i.e., $\{x\ ;\ T_N(s,x)>c\}$ is open for any $c > 0$. In fact if x_0 belongs to the set, it holds $|\xi^N_{s,t}(x_0)| < N$ for all $t \leq c$. Then the same inequality holds for x belonging to a suitable neighborhood of x_0. Now $T(s,x)$ is also lower semicontinuous since $T_N(s,x) \uparrow T(s,x)$. Consequently, the set

$$D_{s,t} = D_{s,t}(\omega) = \{x \mid T(s,x) > t\}$$

is an open set for any $s < t$ a.s. We may regard that $\xi_{s,t}(\cdot,\omega)$ is a continuous map from $D_{s,t}(\omega)$ into R^d for each $s < t$ a.s. We denote by $R_{s,t} = R_{s,t}(\omega)$ the range of the map $\xi_{s,t}$;

$$R_{s,t} = R_{s,t}(\omega) = \{\xi_{s,t}(x,\omega) \mid x \in D_{s,t}(\omega)\}.$$

5.3. <u>Theorem.</u> (i) Both of $D_{s,t}(\omega)$ and $R_{s,t}(\omega)$ are open for any $s < t$ a.s. The map $\xi_{s,t}(\cdot,\omega)\ ;\ D_{s,t}(\omega) \longrightarrow R_{s,t}(\omega)$ is a homeomorphism for any $s < t$ a.s.

(ii) It holds $D_{s,t} \subset D_{s,r}$ and $\{\xi_{s,r}(x)\ ;\ x \in D_{s,t}\} \subset D_{r,t}$ for any $s < r < t$.

The map $\xi_{s,t}$ satisfies $\xi_{s,t} = \xi_{r,t} \circ \xi_{s,r}$ on $D_{s,t}$ for any $s < r < t$ a.s.

Proof. Set $D^N_{s,t} = \{x \mid T^N(s,x) > t\}$. It is an open set and $\bigcup_N D^N_{s,t} = D_{s,t}$. Since $\xi_{s,t}(\cdot,\omega) = \xi^N_{s,t}(\cdot,\omega)$ holds on $D^N_{s,t}(\omega)$, the map $\xi_{s,t}(\cdot,\omega)$; $D^N_{s,t}(\omega) \longrightarrow R^d$ is an into homeomorphism by Theorem 4.3. Therefore the map is a homeomorphism from $D_{s,t}$ into R^d. The range $R_{s,t}$ then becomes an open set by the theorem of the invariance of the domain.

It holds $\xi_{s,t} = \xi_{r,t} \circ \xi_{s,r}$ on $D^N_{s,t}$ for any $s < r < t$ a.s., since $\xi_{s,t} = \xi^N_{s,t}$, $\xi_{r,t} = \xi^N_{r,t}$ and $\xi_{s,r} = \xi^N_{s,r}$ are satisfied. See (F.3) of Theorem 4.3. The assertion (ii) follows immediately.

5.4. Theorem. Suppose that coefficients $X_0, \ldots, X_m$ are $C^{k,\alpha}$ functions of x for $k \geq 1$ and $0 < \alpha < 1$. Then the solution $\xi_{s,t}(x)$ is a $C^{k,\beta}$ function for any β less than α for any $s < t$ a.s. Furthermore, the map $\xi_{s,t}$; $D_{s,t} \longrightarrow R_{s,t}$ is a C^k diffeomorphism for any $s < t$ a.s.

The above smoothness property is valid for each $\xi^N_{s,t}$ because of Theorem 3.4. Hence the property is also valid for $\xi_{s,t}$.

5.5. Definition. The maximal solution $\xi_{s,t}(x)$ is called conservative if $P(T(s,x) = \infty) = 1$ holds for all (s,x). It is called strictly conservative if $P(T(s,x) = \infty$ for all $(s,x)) = 1$ is satisfied.

If $\xi_{s,t}$ is conservative, then $D_{s,t}(\omega)$ is an open dense subset of R^d a.s. for any $s < t$. If $\xi_{s,t}$ is strictly conservative, then $D_{s,t}(\omega) = R^d$ holds for any $s < t$ a.s.

The conservativeness does not imply the strict conservativeness in general. A counter example was given by Elworthy [7]. Although the example is stated in terms of manifold, we will refer it here for better understanding.

5.6. Example. Let $M = R^d - \{0\}$ be the punctured space. Let B_t be a d-dimensional Brownian motion and let $\xi_{s,t}(x) = B_t - B_s + x$. It is conservative for any s,x since it does not hit the origin with probability 1. However, it is not strictly conservative. In fact, for any ω, $T(s,x,\omega) < a$ holds for some x, since for any $t \in [0,a]$, there is x such that $B_t(\omega) - B_s(\omega) + x = 0$.

On the other hand, in case of one dimensional SDE, the conservativeness implies the strict conservativeness to be shown below. Let $\xi_{s,t}(x)$, $x \in R^1$, $s < t < T(s,x)$ be a solution of one dimensional SDE. Then $D_{s,t}(\omega)$ is an open interval for any $s < t$ a.s. In fact if $y_1, y_2 \in D_{s,t}(\omega)$, $\xi_{s,t}(y_1) < \xi_{s,t}(y) < \xi_{s,t}(y_2)$ is satisfied for any y of (y_1, y_2) a.s., since $\xi_{s,t}$; $D_{s,t}(\omega) \longrightarrow R^1$ is a homeomorphism for any $s < t$ a.s. Therefore $(y_1, y_2) \subset D_{s,t}(\omega)$ if $y_1, y_2 \in D_{s,t}(\omega)$. Now suppose that the process is conservative for all (s,x). Then $D_{s,t}(\omega) = R^1$ holds a.s. for any $s < t$ since $D_{s,t}(\omega)$ is an open dense interval. Noting the continuity of $\xi_{s,t}(x)$ in (s,t,x), we see that $D_{s,t}(\omega) = R^1$ holds for any $s < t$ a.s.

A necessary and sufficient condition for conservativeness of one dimensional SDE was given by W. Feller, in case that coefficients do not depend on time t. Consider a one dimensional SDE

(3) $$d\xi_t = b(\xi_t)dt + \sigma(\xi_t)dB_t,$$

where $\sigma(x) \neq 0$ is assumed. Set $c(x) = \exp 2\int_0^x \frac{b(s)}{\sigma(s)^2} ds$ and define

(4) $$K(x) = \int_0^x \frac{2}{c(r)} \int_0^r \frac{c(s)}{\sigma(s)^2} ds dr.$$

5.7. Theorem. The solution $\xi_{s,t}$ of (3) is strictly conservative if and only if $K(+\infty) = K(-\infty) = \infty$.

Proof. We follow Ikeda-Watanabe [13], p. 365. Set

$$Lu = \frac{1}{2}\sigma^2 u'' + bu'.$$

Then there is a solution of equation $(L-1)u = 0$ satisfying $u(0) = 1$ $1+K(x) \le u(x) \le \exp K(x)$. By Itô's formula,

$$e^{-(t-s)}u(\xi_{s,t}(x)) = u(x) + \int_s^t e^{-(r-s)}u'(\xi_{s,r}(x))\sigma(\xi_{s,r}(x))dB_r.$$

Hence $e^{-(t-s)}u(\xi_{s,t}(x))$ is a positive martingale. Therefore $e^{-(T(s,x)-s)}\lim_{t\uparrow T(s,x)} u(\xi_{s,t}(x))$ exists and is finite a.s. If $K(+\infty) = K(-\infty) = \infty$, then $\lim_{t\uparrow T(s,x)} u(\xi_{s,t}(x)) = \infty$, so that we have $e^{-(T(s,x)-s)} = 0$ a.s. This proves $P(T(s,x) = \infty) = 1$ for any s,x.

Next suppose $K(+\infty) < \infty$. Let $x > 0$ and $\tau = \tau(s,x)$ be the hitting time of $\xi_{s,t}(x)$ to the interval $(-\infty,0]$. Since $u(+\infty) \le \exp K(\infty) < \infty$, $e^{-(t\wedge\tau-s)}u(\xi_{s,t\wedge\tau}(x))$ is a bounded martingale by Doob's optional sampling theorem. Therefore

$$E[e^{-(T(s,x)\wedge\tau-s)}\lim_{t\uparrow T(s,x)} u(\xi_{s,t\wedge\tau}(x))] = u(x) > 0.$$

If $P(T(s,x) = \infty) = 1$ is satisfied, then we have $E[e^{-(\tau-s)}u(0)] = u(x)$ and $u(0) > u(x)$, which is a contradiction. We have thus proved $P(T(s,x) < \infty) > 0$ if $K(+\infty) < +\infty$. The proof is complete.

5.8. Definition. The point $+\infty$ or $-\infty$ is called a non-exit boundary point if $K(+\infty) = \infty$ or $K(-\infty) = \infty$ is satisfied respectively.

6. STOCHASTIC FLOW OF HOMEOMORPHISMS (II). A NECESSARY AND SUFFICIENT CONDITION

In the previous section we have seen that the map $\xi_{s,t}$; $R^d \longrightarrow R^d$ is an into homeomorphism provided that the solution is strictly conservative. This section concerns the "onto" property of the map $\xi_{s,t}$. A global Lipschitz condition for the coefficients is sufficient for this as we have seen in Section 4. However, it is not a necessary condition. We will obtain a necessary and sufficient condition, assuming some additional smoothness assumption to coefficients.

6.1. Theorem. Consider an Itô SDE such that coefficients $X_1, \ldots, X_m$ are $C^{2,\alpha}$ functions of x and X_0 is a $C^{1,\alpha}$ function for some $\alpha > 0$. Suppose that the solution is strictly conservative.

Then for almost all ω the solution $\xi_{s,t}(\cdot,\omega)$ defines a flow of $C^{1,\beta}$-diffeomorphisms of R^d for any β less than α if and only if the following SDE is also strictly conservative:

$$(1) \quad d\eta_t = -\sum_{k=1}^{m} X_k(t,\eta_t)dB_t^k - \hat{X}_0(t,\eta_t)dt,$$

where

$$(2) \quad \hat{X}_0(t,x) = X_0(t,x) - \sum_{k=1}^{m}\sum_{j=1}^{d} X_k^j(t,x)\frac{\partial}{\partial x_j}X_k(t,x).$$

Furthermore, if $\xi_{s,t}(\cdot,\omega)$ defines a flow of diffeomorphisms, then the inverse map $\xi_{s,t}^{-1}(x,\cdot)$ satisfies the following backward SDE

$$(3) \quad \xi_{s,t}^{-1}(x) = x - \sum_{k=1}^{m}\int_s^t X_k(r,\xi_{r,t}^{-1}(x))\hat{d}B_r^k - \int_s^t \hat{X}_0(r,\xi_{r,t}^{-1}(x))dr.$$

The backward stochastic integral with respect to Brownian motion was defined in Chapter I. Before we proceed to the proof of the theorem, we give two formulas concerning the inverse map $\xi_{s,t}^{-1}$ and the backward integrals. Let $R_{s,t}(\omega)$ be the image of R^d of the map $\xi_{s,t}(\cdot,\omega)$. It holds $R_{s,t}(\omega) \subset R_{r,t}(\omega)$ if $s < r < t$ a.s., because $\xi_{s,t}(x,\omega) = \xi_{r,t}(\xi_{s,r}(x,\omega),\omega)$.

6.2. Lemma. (1) Let $g(r,x)$ be a continuous function of (r,x). Then

$$(4) \qquad \int_s^t g(r,\xi_{s,r}(y))dr\Big|_{y=\xi_{s,t}^{-1}(x)} = \int_s^t g(r,\xi_{r,t}^{-1}(x))dr$$

holds on $\{\omega \mid x \in R_{s,t}(\omega)\}$.

(2) Suppose that $g(r,x)$ is a C^1 function of x. If $k \geq 1$, then it holds on $\{\omega \mid x \in R_{s,t}(\omega)\}$

$$(5) \qquad \int_s^t g(r,\xi_{s,r}(y))dB_r^k\Big|_{y=\xi_{s,t}^{-1}(x)}$$

$$= \int_s^t g(r,\xi_{r,t}^{-1}(x))\hat{d}B_r^k - \int_s^t X_k(r)g(r,\xi_{r,t}^{-1}(x))dr,$$

where $X_k(r)$ is the first order differential operator defined by

$$(6) \qquad X_k(r) = \sum_{i=1}^{d} X_k^i(r,x)\frac{\partial}{\partial x_i}.$$

Proof. Let $\Delta = \{s = t_0 < \dots < t_n = t\}$ be partitions of $[s,t]$. Then

$$\int_s^t g(r,\xi_{s,r}(y))dr\Big|_{y=\xi_{s,t}^{-1}(x)} = \lim_{|\Delta|\to 0}\sum_{i=0}^{n-1} g(t_i,\xi_{s,t_i}(y))(t_{i+1}-t_i)\Big|_{y=\xi_{s,t}^{-1}(x)}$$

$$= \lim_{|\Delta|\to 0}\sum_{i=0}^{n-1} g(t_i,\xi_{t_i,t}^{-1}(x))(t_{i+1}-t_i)$$

$$= \int_s^t g(r,\xi_{r,t}^{-1}(x))dr.$$

We will next prove (5) assuming that $g(r,x)$ is a C^1-function of t and a C^2 function of x. By Corollary 7.8, Chapter I, there is a sequence of partitions $\{\Delta_n\}$ with $|\Delta_n|\to 0$ such that

$$\int_s^t g(r,\xi_{s,r}(y))dB_r^k = \lim_{n\to\infty}\int_s^t g(r,\xi_{s,r}(y))^{\Delta_n} dB_r^k$$

holds for all y a.s. Here, $g(r,\xi_{s,r}(y))^{\Delta_n}$ is a step process defined by $g(t_i,\xi_{s,t_i}(y))$ if $t_i \le r < t_{i+1}$, where $\Delta_n = \{s = t_0 < \dots < t_n = t\}$. (We omit the index n from $t_i^{(n)}$ etc.). Then it holds

$$(7)\quad \int_s^t g(r,\xi_{s,r}(y))dB_r^k\Big|_{y=\xi_{s,t}^{-1}(x)}$$
$$= \lim_{n\to\infty}\sum_{i=0}^{n-1} g(t_i,\xi_{t_{i+1},t}^{-1}(x))(B_{t_{i+1}}^k - B_{t_i}^k)$$
$$- \lim_{n\to\infty}\sum_{i=0}^{n-1}\{g(t_i,\xi_{t_{i+1},t}^{-1}(x)) - g(t_i,\xi_{t_i,t}^{-1}(x))\}(B_{t_{i+1}}^k - B_{t_i}^k).$$

The first member of the right hand side exists and equals the backward Itô integral $\int_0^t g(r,\xi_{r,t}^{-1}(x))\hat{d}B_r^k$. The second member is written, using Stratonovich integral, as

$$-2\Big(\int_s^t g(r,\xi_{s,r}(y))\circ dB_r^k - \int_s^t g(r,\xi_{s,r}(y))dB_r^k\Big)\Big|_{y=\xi_{s,t}^{-1}(x)}$$
$$= \langle g(t,\xi_{s,t}(y)),B_t^k - B_s^k\rangle\Big|_{y=\xi_{s,t}^{-1}(x)}.$$

By Itô's formula, we have

$$g(t,\xi_{s,t}(y)) = g(s,y) + \sum_{k=1}^m \int_s^t X_k(r)g(r,\xi_{s,r}(y))dB_r^k + \text{process of bounded variation.}$$

Therefore, we have

$$\langle g(t,\xi_{s,t}(y)),B_t^k - B_s^k\rangle = \int_s^t X_k(r)g(r,\xi_{s,r}(y))dr.$$

Substitute $y = \xi_{s,t}^{-1}(x)$ to the above and apply the formula (4). Then we find that the second member of the right hand side of (7) equals

$$- \int_s^t X_k(r) g(r, \xi_{r,t}^{-1}(x)) dr.$$

Therefore the formula (5) is valid if $g(r,x)$ is smooth enough.

It remains to prove (5) for general g. Choose a sequence of smooth functions $g_n(r,x)$ such that $g_n(r,x) \longrightarrow g(r,x)$ and $\frac{\partial}{\partial x_i} g_n(r,x) \longrightarrow \frac{\partial}{\partial x_i} g(r,x)$ locally uniformly. Then formulas (5) are valid to all g_n. Making n tend to infinity, we get the formula (5) for this g by Theorem 7.7 of Chapter I. The proof is complete.

Proof of Theorem. Let us first prove that the inverse $\xi_{s,t}^{-1}(x)$ satisfies (3) on the set $\{\omega | x \in R_{s,t}(\omega)\}$. Substitute $y = \xi_{s,t}^{-1}(x)$ to the equation

$$\xi_{s,t}(y) = y + \sum_{k=1}^m \int_s^t X_k(r, \xi_{s,r}(y)) dB_r^k + \int_s^t X_0(r, \xi_{s,r}(y)) dr$$

and apply Lemma 6.2. Then,

$$x = \xi_{s,t}^{-1}(x) + \sum_{k=1}^m \int_s^t X_k(r, \xi_{r,t}^{-1}(x)) \hat{d}B_r^k - \sum_{k=1}^m \int_s^t X_k(r) X_k(r, \xi_{r,t}^{-1}(x)) dr$$
$$+ \int_s^t X_0(r, \xi_{r,t}^{-1}(x)) dr,$$

where

$$X_k(r) X_k^i(r,x) = \sum_{j=1}^d X_k^j(r,x) \frac{\partial}{\partial x_j} X_k^i(r,x).$$

Therefore we have (3).

Now if $\xi_{s,t}$ are onto maps a.s. then equation (3) is satisfied for all x a.s. This means that equation (1) is strictly conservative.

Suppose conversely that (1) is strictly conservative. Then the solution of the following backward equation is also strictly conservative:

$$\hat{\xi}_{s,t}(x) = x - \sum_{k=1}^m \int_s^t X_k(r, \hat{\xi}_{r,t}(x)) \hat{d}B_r^k - \int_s^t \hat{X}_0(r, \hat{\xi}_{r,t}(x)) dr.$$

Clearly it is an extension of $\xi_{s,t}^{-1}$, i.e. $\hat{\xi}_{s,t}(x) = \xi_{s,t}^{-1}(x)$ holds on $x \in R_{s,t}(\omega)$. Now in order to prove the onto property of the map $\xi_{s,t}$, it is sufficient to prove that $R_{s,t}(\omega)$ is closed, since $R_{s,t}(\omega)$ is a non-void open set from Theorem 5.3. Let $\overline{R_{s,t}(\omega)}$ be the closure of $R_{s,t}(\omega)$ and let $y \in \overline{R_{s,t}(\omega)}$. Choose a sequence $\{y_n\}$ from $R_{s,t}(\omega)$ converging to y. Then it holds

$$y_n = \xi_{s,t} \circ \xi_{s,t}^{-1}(y_n) = \xi_{s,t} \circ \hat{\xi}_{s,t}(y_n)$$

for all n. Making n tend to infinity, we see $y = \xi_{s,t} \circ \hat{\xi}_{s,t}(y)$, since $\hat{\xi}_{s,t}(\cdot,\omega)$; $R^d \longrightarrow R^d$ is a continuous map. Therefore $y \in R_{s,t}(\omega)$. The map $\xi_{s,t}^{-1}$ is also $C^{1,\beta}$ because of (3). The proof is complete.

7. STRATONOVICH STOCHASTIC DIFFERENTIAL EQUATIONS

In this section we shall consider SDE's described in terms of Stratonovich integrals. As we will see soon, the Stratonovich SDE can be rewritten as an Itô SDE. Hence most properties of the Stratonovich SDE can be derived from those of Itô equation. A reason that we consider Stratonovich SDE's is that formulas involving Stratonovich integrals take forms similar to those of ordinary differential equations. Actually, we will see in Theorem 7.3 that the backward Stratonovich SDE governing the inverse $\xi_{s,t}^{-1}$ is simpler than the backward Itô SDE stated in Theorem 6.1 and it is close to the case of ordinary differential equation.

Let $X_0(t,x),\dots,X_m(t,x)$ be continuous d-vector functions on $[0,a] \times R^d$. We assume that $X_1,\dots,X_m$ are C^1 functions of t and C^2 functions of x, and X_0 is a C^1 function of x. A Stratonovich SDE takes the form

$$(1) \qquad d\xi_t = \sum_{k=0}^{m} X_k(t,\xi_t) \circ dB_t^k ,$$

where $\circ\, dB_t^k$ denotes the Stratonovich integral and $dB_t^0 = dt$ as before.

7.1. Definition. A random field $\xi_{s,t}(x)$, $x \in R^d$, $s \le t < T(s,x,\omega)$ is called a local solution of (1) with the initial condition $\xi_s = x$ if it satisfies (i)-(iii) of Definition 5.1 and the following (iv)': It holds

$$(2) \qquad \xi_{s,t}(x) = x + \sum_{k=0}^{m} \int_s^t X_k(r, \xi_{s,r}(x)) \circ dB_r^k$$

for any $s < t < T(s,x,\cdot)$. A maximal solution is defined similarly.

The existence and uniqueness of the solution (1) is reduced to those of an Itô equation. In fact, consider an Itô equation

$$(3) \qquad d\xi_t = \sum_{k=1}^{m} X_k(t,\xi_t)dB_t^k + X_0^*(t,\xi_t)dt,$$

where

$$(4) \qquad X_0^*(t,x) = X_0(t,x) + \frac{1}{2}\sum_{k=1}^{m}\sum_{i=1}^{d} X_k^i(t,x)\frac{\partial}{\partial x_i}X_k(t,x).$$

Then $X_0^*, X_1, \ldots, X_m$ are all C^1 functions of x. Hence it has a unique maximal solution $\xi_{s,t}(x)$ by Theorem 5.2. It is a continuous $(\underline{\underline{F}}_{s,t})$-adapted semimartingale. Then $X_k(r,\xi_{s,r}(x))$ is also a continuous semimartingale by Itô's formula. The local martingale part is given by Itô's formula as

$$\sum_{j=1}^{m}\int_s^t \sum_{i=1}^{d} X_j^i(r,\xi_{s,r}(x))\frac{\partial X_k}{\partial x_i}(r,\xi_{s,r}(x))dB_r^j .$$

Therefore, the Stratonovich integral $\int_s^t X_k(r,\xi_{s,r}(x)) \circ dB_r^k$ is well defined and equals

$$\int_s^t X_k(r,\xi_{s,r}(x))dB_r^k + \frac{1}{2}\sum_{i,j}\int_s^t X_j^i(r,\xi_{s,r}(x))\frac{\partial X_k}{\partial x_i}(r,\xi_{s,r}(x))d\langle B^j,B^k\rangle_r$$

$$= \int_s^t X_k(r,\xi_{s,r}(x))dB_r^k + \frac{1}{2}\int_s^t \sum_{i=1}^{d} X_k^i(r,\xi_{s,r}(x))\frac{\partial}{\partial x_i}X_k(r,\xi_{s,r}(x))dr,$$

because $\langle B^j,B^k\rangle_t = \delta_{jk}t$. Therefore the solution of Itô equation (3) satisfies Stratonovich equation (2).

The uniqueness of the solution of the Stratonovich equation is also reduced to that of the Itô equation. Hence the Stratonovich equation (2) has a unique maximal solution $\xi_{s,t}(x)$, $x \in R^d$, $s < t < T(s,x)$.

A Stratonovich version of Itô's formula is as follows.

7.2. Theorem. Let $\xi_{s,t}$ be the solution of a Stratonovich equation (1). Let $F ; R^d \longrightarrow R^1$ be a C^3 function. Then it holds

$$(5) \quad F(\xi_{s,t}(x)) - F(x) = \sum_{k=0}^{m} \int_s^t X_k(r)F(\xi_{s,r}(x)) \circ dB_r^k$$

$$(6) \quad F(\xi_{s,t}(x)) - F(x) = \sum_{k=1}^{m} \int_s^t X_k(r)F(\xi_{s,r}(x))dB_r^k$$

$$+ \int_s^t (\frac{1}{2} \sum_{k=0}^{m} X_k(r)^2 + X_0(r))F(\xi_{s,r}(x))dr.$$

Proof. The first formula (5) is immediate from Corollary 8.4 in Chapter I. We shall derive (6) from (5). The process $X_k(r)F(\xi_{s,r}(x))$ is a continuous semimartingale with respect to $\underline{\underline{F}}_{s,r}$ for each s. The martingale part is obtained from Itô's formula: It is

$$\sum_{j=1}^{m} \int_s^r X_j(u)X_k(u)F(\xi_{s,u}(x))dB_u^j.$$

Therefore,

$$\int_s^t X_k(r)F(\xi_{s,r}(x)) \circ dB_r^k = \int_s^t X_k(r)F(\xi_{s,r}(x))dB_r^k$$

$$+ \frac{1}{2} \sum_{j=1}^{m} \int_s^t X_j(r)X_k(r)F(\xi_{s,r}(x))d\langle B^j,B^k\rangle_r.$$

Since $d\langle B^k,B^j\rangle_r = \delta_{kj}dr$, the last member of the above is $\frac{1}{2}\int_s^t X_k(r)^2 \times F(\xi_{s,r}(x))dr$. Thus the formula (6) follows from (5).

We will next obtain a necessary and sufficient condition that the Stratonovich equation induces a stochastic flow of homeomorphisms. The condition is stated simpler than the case of Itô equation. Indeed, we have the following.

7.3. Theorem. Suppose that coefficients $X_1, \dots, X_m$ (or X_0) of the Stratonovich equation (2) are $C^{k+1,\alpha}$ (or $C^{k,\alpha}$) functions for some $k \geq 1$ and $\alpha > 0$. Then the maximal solution defines a $C^{k,\beta}$ diffeomorphism from $D_{s,t} = \{x ; T(s,x) > t\}$ into R^d for any β less than α.

Suppose further that the solution is strictly conservative. Then the solution defines a stochastic flow of homeomorphisms if and only if the following adjoint equation is strictly conservative

$$(7) \quad d\eta_t = - \sum_{k=0}^{m} X_k(t, \eta_t) \circ dB_t^k .$$

Furthermore, if the solution $\xi_{s,t}$ defines a flow of homeomorphisms, the inverse map $\xi_{s,t}^{-1}$ satisfies the following Stratonovich backward equation

$$(8) \quad \xi_{s,t}^{-1}(x) = x - \sum_{k=0}^{m} \int_s^t X_k(r, \xi_{r,t}^{-1}(x)) \circ \hat{d}B_r^k .$$

Proof. Theorem 6.1 tells us that the solution $\xi_{s,t}$ of Itô's equation (3) defines a flow of homeomorphisms if and only if the following adjoint equation is strictly conservative:

$$(9) \quad d\eta_t = - \sum_{k=1}^{m} X_k(t, \eta_t) dB_t^k - \hat{X}_0(t, \eta_t) dt,$$

where

$$(10) \quad \hat{X}_0(t,x) = X_0^*(t,x) - \sum_{i,j} X_j^i(t,x) \frac{\partial}{\partial x_i} X_j(t,x)$$

$$= X_0(t,x) - \frac{1}{2} \sum_{i,j} X_j^i(t,x) \frac{\partial}{\partial x_i} X_j(t,x).$$

Obviously Itô equation (9) with coefficient (10) is equal to Stratonovich equation (7). Therefore the assertion follows from Theorem 6.1.

Remark. It is worth mentioning that

$$\int_s^t g(r,\xi_{s,r}(y))\circ dB_r^k \Big|_{y=\xi_{s,t}^{-1}(x)} = \int_s^t g(r,\xi_{r,t}^{-1}(x))\circ \hat{d}B_r^k$$

is satisfied if $g(r,x)$ is a C^2 function of x. In fact, by Lemma 6.2, the left hand side is written as

$$\int_s^t g(r,\xi_{s,r}(y))dB_r^k + \frac{1}{2}\int_s^t X_k(r)g(r,\xi_{s,r}(y))dr\Big|_{y=\xi_{s,t}^{-1}(x)}$$

$$= \int_s^t g(r,\xi_{r,t}^{-1}(x))\hat{d}B_r^k - \frac{1}{2}\int_s^t X_k(r)g(r,\xi_{r,t}^{-1}(x))dr.$$

This coincides with the right hand side.

7.4. Example. One dimensional SDE. Consider

$$\text{(9)} \qquad d\xi_t = b(\xi_t)dt + \sigma(\xi_t)\circ dB_t$$

$$= \{b(\xi_t) + \frac{1}{2}\sigma(\xi_t)\sigma'(\xi_t)\}dt + \sigma(\xi_t)dB_t.$$

The solution is strictly conservative if $K(+\infty) = K(-\infty) = \infty$, where K is defined by (4) in Section 5 using

$$c(x) \equiv \exp\left(\int_0^x \frac{(2b(s)+\sigma(s)\sigma'(s))}{\sigma(s)^2}\, ds\right) = \frac{\sigma(x)}{\sigma(0)} \exp\left(\int_0^x \frac{2b(s)}{\sigma(s)^2} ds\right).$$

Consider next, the adjoint equation

$$d\hat{\xi}_t = -\, b(\hat{\xi}_t)dt - \sigma(\hat{\xi}_t)\circ dB_t$$

$$= (-\, b(\hat{\xi}_t) + \frac{1}{2}\sigma(\hat{\xi}_t)\sigma'(\hat{\xi}_t))dt - \sigma(\hat{\xi}_t)dB_t.$$

The solution is strictly conservative if $\hat{K}(+\infty) = \hat{K}(-\infty) = \infty$, where $\hat{K}$ is is defined by (4) in Section 5 using

$$\hat{c}(x) = \exp\left(\int_0^x \frac{-2b(s)+\sigma(s)\sigma'(s)}{\sigma(s)^2} ds\right) = \frac{\sigma(x)}{\sigma(0)} \exp\left(-\int_0^x \frac{2b(s)}{\sigma(s)^2} ds\right)$$

$$= \frac{\sigma(x)^2}{\sigma(0)^2} c(x)^{-1} .$$

Hence $\hat{\xi}_t$ is strictly conservative if and only if

$$\int_0^\infty \frac{c(r)}{\sigma(r)^2} \int_0^r \frac{1}{c(s)} ds\,dr = \infty, \quad \int_{-\infty}^0 \frac{c(r)}{\sigma(r)^2} \int_0^r \frac{1}{c(r)} ds\,dr = \infty .$$

The above condition states that $+\infty$ and $-\infty$ are non-entrance boundary points of ξ_t, according to W. Feller. Therefore, the solution of (9) defines a flow of homeomorphisms if and only if $+\infty$ and $-\infty$ are natural (non-exit and non-entrance) boundary points.

8. STOCHASTIC DIFFERENTIAL EQUATIONS ON MANIFOLD

We shall define stochastic differential equations (SDE) on manifolds. Let M be a connected, paracompact C^∞-manifold of dimension d. Suppose we are given $m+1$-vector fields on M with parameter $t \in [0,a]$; $X_0(t)$, $\dots, X_m(t)$. We assume that $X_1, \dots, X_m$ are C^2-vector fields continuously differentiable in t, i.e., with a local coordinate $(x_1, \dots, x_d)$, these vector fields are expressed as

$$(1) \qquad X_k(t) = \sum_{i=1}^d X_k^i(t,x) \frac{\partial}{\partial x_i} ,$$

where $X_k^i(t,x)$, $i=1,\dots,d$, $k=1,\dots,m$ are C^1 functions of t and C^2 functions of x. As to vector field X_0, we assume that $X_0^i(t,x)$ are continuous in t continuously differentiable in x.

Let $B_t = (B_t^1, \dots, B_t^m)$ be an m-dimensional standard Brownian motion defined on the probability space $(\Omega, \underline{\underline{F}}, P)$. We consider a Stratonovich SDE on the manifold M:

(2) $d\xi_t = \sum_{k=1}^{m} X_k(t,\xi_t)\circ dB_t^k + X_0(t,\xi_t)dt.$

For the convenience we will write dt as dB_t^0. Then the above is written as

(2') $d\xi_t = \sum_{k=0}^{m} X_k(t,\xi_t)\circ dB_t^k.$

8.1. <u>Definition.</u> A random field $\xi_{s,t}(x)$, $x\in M$, $s \le t < T(s,x,\omega)\wedge a$ with values in M is called a <u>local solution</u> of equation (2) with the initial condition $\xi_s = x$, if it satisfies following conditions.

(i) $T(s,x,\omega)$ is (s,x,ω)-measurable and for any fixed (s,x), it is an $\underline{\underline{F}}_{s,t}$-stopping time, where $\underline{\underline{F}}_{s,t} = \sigma(B_u - B_v\ ;\ s \le u \le v \le t)$.

(ii) $\xi_{s,t}(x)$ is continuous in (s,t,x).

(iii) For any C^3 function F, $F(\xi_{s,t}(x))$ is a continuous $(\underline{\underline{F}}_{s,t})$-semimartingale and satisfies for all $t < T(s,x)\wedge a$

(3) $F(\xi_{s,t}(x)) = F(x) + \sum_{k=0}^{m}\int_s^t X_k(r)F(\xi_{s,r}(x))\circ dB_r^k.$

The solution is called <u>maximal</u> if $T(s,x) = \infty$ for all x a.s. for any s in case M is compact, or if $\lim_{t\uparrow T(s,x)} \xi_{s,t}(x) = \infty$ holds for $T(s,x) < \infty$ a.s. in case M is non-compact. Here ∞ is the infinity of M adjoined as one point compactification.

Let us express equation (2) or (3) using a local coordinate. Let $(x_1, \dots, x_d)$ be a local coordinate in a coordinate neighborhood U. Set $F(x) = x_i$. Then it holds $X_k(r)F(x) = X_k^i(r,x)$. Therefore if $\xi_{s,t}$ satisfies (3), then $\xi_{s,t}^i(x) \equiv x_i(\xi_{s,t}(x))$ satisfies

(4) $\xi_{s,t}^i(x) = x_i + \sum_{k=0}^{m}\int_s^t X_k^i(r,\xi_{s,r}(x))\circ dB_r^k$

for $t < T_U\wedge a$, where $T_U = \inf\{t > s \mid \xi_{s,t}(x)\notin U\}$ ($= \infty$ if $\{\dots\} = \phi$).

Conversely, by solving equation (4) we can construct a solution of equation (3) in each coordinate neighborhood. Consider the SDE (4), where $\xi_{s,r}(x) = (\xi^1_{s,r}(x),\dots,\xi^d_{s,r}(x))$. It has a unique solution up to the time $T_U(s,x)$. It is continuous in (s,t,x). See Section 7. By Itô's formula (5) in Section 7, the solution satisfies (3) for any C^3 function F.

8.2. <u>Lemma.</u> The solution of equation (4) does not depend on the choice of local coordinates.

<u>Proof.</u> Let $(\bar{x}_1,\dots,\bar{x}_d)$ be another local coordinate in the same neighborhood U. Then vector fields $X_k(t)$ are written as

$$X_k(t) = \sum_{i=1}^{d} \bar{X}^i_k(t,x)\frac{\partial}{\partial \bar{x}_i} .$$

Coefficients $X^i_k(t,x)$ and $\bar{X}^i_k(t,x)$ are related by

$$(5) \qquad X^i_k(t,x) = \sum_{j=1}^{d} \bar{X}^j_k(t,x)\frac{\partial x_i}{\partial \bar{x}_j} , \qquad i=1,\dots,d$$

for each $k=0,\dots,m$. Let $\bar{\xi}_{s,t}(x) = (\bar{\xi}^1_{s,t}(x),\dots,\bar{\xi}^d_{s,t}(x))$ be the solution of equation (4) with coefficients $\bar{X}^i_k(t,x)$. Apply Itô's formula to $F(\bar{x}_1, \dots,\bar{x}_d) = x_i(\bar{x}_1,\dots,\bar{x}_d)$. Then we have

$$x_i(\bar{\xi}_{s,t}) = x_i + \sum_{k=0}^{m}\int_s^t \sum_j \frac{\partial x_i}{\partial \bar{x}_j}(\bar{\xi}_{s,r})\bar{X}^j_k(r,\bar{\xi}_{s,r})\circ dB^k_r$$

$$= x_i + \sum_{k=0}^{m}\int_s^t X^i_k(r,\bar{\xi}_{s,r})\circ dB^k_r .$$

Therefore $x_i(\bar{\xi}_{s,t})$ coincides with $\xi^i_{s,t}$. The proof is complete.

8.3. <u>Theorem.</u> There is a unique maximal solution.

<u>Proof</u>. Let $\{U_i\}$, $\{V_i\}$ and $\{W_i\}$ be countable families of coordinate neighborhoods of M satisfying the following properties. For each i, U_i, V_i and W_i are balls with a same center with radius ε, 2ε, 3ε, respectively, where ε is a sufficiently small positive number; $\{U_i\}$ is a covering of M, i.e., $\bigcup_i U_i = M$. Then for each i, there is a unique solution $\xi_{s,t}^{(i)}(x)$ of equation (3) starting at $x \in W_i$ up to time $t < T_{W_i}(s,x) \wedge a$, where $T_{W_i}(s,x)$ is the first leaving time of $\xi_{s,t}(x)$ from W_i. If $W_i \cap W_j \neq \phi$, then $\xi_{s,t}^{(i)}(x) = \xi_{s,t}^{(j)}(x)$ holds for $x \in W_i \cap W_j$ and $t < T_{W_i}(s,x) \wedge T_{W_j}(s,x) \wedge a$.

Define $\xi_{s,t}(x)$, $x \in M$, $s \leq t \leq T_1(s,x)$ by

(6) $\quad T_1(s,x) = T_{V_j}(s,x) \qquad$ if $\quad x \in U_j - \bigcup_{i<j} U_i$,

(7) $\quad \xi_{s,t}(x) = \xi_{s,t}^{(j)}(x) \qquad$ if $\quad x \in U_j - \bigcup_{i<j} U_i$ and $t \leq T_1(s,x)$.

Clearly it is a local solution of equation (2).

We will prolong the solution so as to get a maximal solution. Define a process $\xi_{s,t}(x)$, $x \in M$, $s \leq t < T_\infty(s,x)$ by induction as follows.

(8) $\quad \xi_{s,t}(x) = \xi_{T_1(s,x),t} \circ \xi_{s,T_1(s,x)}(x) \qquad$ if $\; T_1(s,x) \wedge a \leq t \leq T_2(s,x) \wedge a$

where $T_2(s,x) = T_1(T_1(s,x), \xi_{s,T_1(s,x)}(x))$

$\quad = \xi_{T_{n-1}(s,x),t} \circ \xi_{s,T_{n-1}(s,x)}(x)$ if $T_{n-1}(s,x) \wedge a \leq t \leq T_n(s,x) \wedge a$

where $T_n(s,x) = T_1(T_{n-1}(s,x), \xi_{s,T_{n-1}(s,x)}(x))$

and $T_\infty(s,x) = \lim_{n\to\infty} T_n(s,x)$.

We will show that it is continuous in (s,t,x) a.s. Let $\tilde{\Omega}$ be the set of all ω which satisfies the followings. (a) $\xi_{s,t}(x)$ is continuous in $\{(s,t,x) \,;\, x \in M, s \leq t \leq T_1(s,x)\}$ and (b) $\xi_{r,t}(\xi_{s,r}(x,\omega),\omega) = \xi_{s,t}(x,\omega)$ holds for all (s,r,t,x) such that $s < r < t < T_1(s,x)$. Then it holds $P(\tilde{\Omega}) = 1$. From the definition (8), the property (b) is valid for all (s,r,t,x)

such that $s < r < t < T_\infty(s,x)$. Now take $\omega \in \tilde{\Omega}$ and consider the trajectory $\{\xi_{s_0,r}(x_0,\omega)\ ;\ r \in [s_0,t_0]\}$ where $t_0 < T_\infty(s_0,x_0,\omega)$. We may choose a chain of coordinate neighborhoods $V_{i_1},\dots,V_{i_n}$ from $\{V_i\}$ and a partition $s_0 = t_1 < \dots < t_{n+1} = t_0$ such that the trajectory $\{\xi_{s_0,r}(x_0,\omega) \mid r \in [t_k,t_{k+1}]\}$ is included in V_{i_k} for each $k=1,\dots,n$. Obviously it holds $\xi_{s_0,t_0}(x_0,\omega) = \xi_{t_n,t_{n+1}} \circ \xi_{t_{n-1},t_n} \circ \dots \circ \xi_{t_1,t_2}(x_0,\omega)$ and each $\xi_{t_{i_k},t_{i_{k+1}}}(x,\omega)$ is a continuous map of $x \in V_{i_k}$ since it is defined on a coordinate neighborhood. Therefore $\xi_{s,t}(x,\omega)$ is continuous in (s,t,x) at a suitable neighborhood of (s_0,t_0,x_0).

We will next prove that $\xi_{s,t}(x)$ of (8) is a maximal solution in case where M is non compact. Let F be a C^3 function with compact support. Then it holds for $s < t < T_\infty(s,x)$

$$F(\xi_{s,t}(x)) = F(x) + \sum_{j=1}^{m}\int_s^t X_j(r)F(\xi_{s,r}(x))dB_r^j + \int_s^t \Big(\frac{1}{2}\sum_{j=1}^{m} X_j(r)^2 + X_0(r)\Big)F(\xi_{s,r}(x))dr.$$

Let t tend to $T_\infty(s,x)\wedge a$. Then each term of the right hand side has a limit a.s. by the martingale convergence theorem. Therefore the limit of $F(\xi_{s,t}(x))$ as $t \uparrow T_\infty(s,x)\wedge a$ exists a.s. for each s,x. Now take a countable family $\{F_n\}$ of such functions separating any two points of M. Then we see that the following alternative holds. (a) $\exists \lim \xi_{s,t}(x) = \infty$ as $t \uparrow T_\infty(s,x)\wedge a$, or (b) $\exists \lim \xi_{s,t}(x) \in M$ as $t \uparrow T_\infty(s,x)\wedge a$. But (b) implies $T_\infty(s,x) = \infty$. Indeed, suppose on the contrary that $T_\infty(s,x) < \infty$ and $\xi_{s,T_\infty(s,x)}(x) \equiv \lim \xi_{s,t}(x)$ as $t \uparrow T_\infty(s,x)$ exists in M. Then there are positive integers j and n such that $\xi_{s,t}(x) \in U_j - \bigcup_{i<j} U_i$ for all $t \geq T_n(s,x)$. This implies $T_{n+1}(s,x) = \infty$ which is a contradiction. We have thus seen that only (a) can occur if $T_\infty(s,x) < \infty$. This proves that the solution is maximal.

If the manifold M is compact, the above argument shows actually that $T_\infty(s,x) = \infty$ a.s. for any s,x, i.e., the solution is conservative. We will prove that it is strictly conservative. For this we will define the solution in a different manner.

Suppose now that M is a compact manifold. We may assume that the number of coordinate neighborhoods $\{U_j\}$ which covers M is finite, say $\{U_j\}_{j=1}^m$. We define vector fields $X_j(t)$ for $t \geq a$ by $X_j(t) = X_j(a)$.

Set

$$T_1(s) = \inf_{x \in \bar{U}_1} T_{V_1}(s,x) \wedge \inf_{x \in \overline{U_2 - U_1}} T_{V_2}(s,x) \wedge \cdots \wedge \inf_{x \in \overline{U_m - \bigcup_{j<m} U_j}} T_{V_m}(s,x),$$

$$T_n(s) = T_1(T_{n-1}(s)).$$

Then $\{T_n\}$ is an increasing sequence of stopping times. Denote the limit by $T_\infty(s)$.

We shall prove $P(T_\infty(s) = \infty) = 1$ for any s. Observe first that $T_1(s) - s$ is strictly positive and lower semicontinuous in s, a.s. Indeed, for $\omega \in \tilde{\Omega}$ it holds

$$\{s\ ;\ T_1(s) > b\} = \bigcap_{j=1}^{m} \{s \mid \{\xi_{s,t}(x)\ ;\ s \leq t \leq b,\ x \in \overline{U_j - \bigcup_{i<j} U_i}\} \subset V_j\},$$

so that the set $\{s\ ;\ T_1(s) > b\}$ is open for any b. Then there is s_0 of $[0,a]$ such that $\inf_{s \in [0,a]} T_1(s) - s \geq T_1(s_0) - s_0$. This implies $\inf_{s \in [0,a]} E[T_1(s) - s] = c > 0$. If $s > a$, the value $E[T_1(s) - s]$ does not depend on s, so that we have $\inf_{0 \leq s < \infty} E[T_1(s) - s] = c$. Next observe that $T_1(s) - s,\ T_2(s) - T_1(s), \ldots, T_n(s) - T_{n-1}(s), \ldots$ are independent random variables and

$$E[T_n(s) - T_{n-1}(s)] = \int E[T_1(t) - t] P(T_{n-1}(s) \in dt) \geq c > 0.$$

Therefore, by the law of large numbers, we have

$$T_\infty(s) = \sum_{n=1}^{\infty} (T_n(s) - T_{n-1}(s)) = \infty \qquad \text{a.s.}$$

We will define $\hat{\xi}_{s,t}(x)$, $t\in[s,\infty)$ by

$$\hat{\xi}_{s,t}(x) = \xi_{s,t}(x) \quad \text{if} \quad s \le t \le T_1(s)$$

$$= \xi_{T_{n-1}(s),t} \circ \xi_{s,T_{n-1}(s)} \quad \text{if} \quad T_{n-1}(s) \le s \le T_n(s).$$

Then $\hat{\xi}_{s,t}$ is a solution of (2), which is strongly conservative. The proof is complete.

Remark. If we want to consider an Itô equation on a manifold, we should not regard coefficients of the equation as vector fields. Let us consider an Itô SDE with local coordinate $(x_1,\dots,x_d)$.

$$(7) \qquad \xi^i_{s,t}(x) = x_i + \sum_{k=0}^{m} \int_s^t X^i_k(r,\xi_{s,r}(x))dB^k_r .$$

Then by Itô's formula it satisfies for any C^2-function F,

$$(8) \qquad F(\xi_{s,t}(x)) = F(x) + \sum_{k=1}^{m} \int_s^t X_k(r)F(\xi_{s,r}(x))dB^k_r + \int_s^t L(r)F(\xi_{s,r}(x))dr,$$

where

$$X_k(r)F = \sum_{i=1}^{d} X^i_k(r,x)\frac{\partial F}{\partial x_i}$$

$$L(r)F = \sum_i X^i_0(r,x)\frac{\partial F}{\partial x_i} + \frac{1}{2}\sum_{i,j}(\sum_k X^i_k(r,x)X^j_k(r,x))\frac{\partial^2 F}{\partial x_i \partial x_j} .$$

Now let $(\bar{x}_1,\dots,\bar{x}_d)$ be another coordinate and consider another Itô SDE:

$$(9) \qquad \bar{\xi}^i_{s,t}(x) = \bar{x}_i + \sum_{k=0}^{m} \int_s^t \bar{X}^i_k(r,\bar{\xi}_{s,r}(x))dB^k_r .$$

In order that (7) and (9) define the same equation, coefficients X^i_k and $\bar{X}^i_k$ should be related by

$$(10) \qquad \bar{X}^i_k(t,x) = \sum_{j=1}^{d} X^j_k(t,x)\frac{\partial \bar{x}_i}{\partial x_j} \quad \text{for } k=1,\dots,m,$$

$$(11) \qquad \bar{X}^i_0(t,x) = \sum_{j=1}^{d} X^j_0(t,x)\frac{\partial \bar{x}_i}{\partial x_j} + \frac{1}{2}\sum_{j,\ell=1}^{d}(\sum_{k=1}^{m} X^j_k(t,x)X^\ell_k(t,x))\frac{\partial^2 \bar{x}_i}{\partial x_j \partial x_\ell} .$$

Indeed, apply Itô's formula (8) to $F(x_1,\dots,x_d) = \bar{x}_i(x_1,\dots,x_d)$ and

$\xi_{s,t}$. Then

$$\bar{x}_i(\xi_{s,t}) = \bar{x}_i + \sum_{k=1}^{m} \int_s^t \sum_j X_k^j(r,\xi_{s,r}(x)) \frac{\partial \bar{x}_i}{\partial x_j}(r,\xi_{s,r}(x))dB_r^j$$

$$+ \int_s^t (\sum_j X_0^j \frac{\partial \bar{x}_i}{\partial x_j} + \frac{1}{2} \sum_{j,\ell} (\sum_k X_k^j X_k^\ell) \frac{\partial^2 \bar{x}_i}{\partial x_j \partial x_\ell}(\xi_{s,r}(x))dr.$$

Therefore, $\bar{\xi}_{s,t}$ and $\bar{x}(\xi_{s,t})$ satisfy the same equation if (10) and (11) are satisfied.

9. STOCHASTIC FLOW OF HOMEOMORPHISMS (III). CASE OF MANIFOLD

Let $\xi_{s,t}(x)$, $x \in M$, $s \leq t < T(s,x) \wedge a$ be the maximal solution of the SDE (2) of the preceding section defined on the manifold M. For each $s < t$, $\xi_{s,t}(\cdot,\omega)$ may be considered as a continuous map from M into itself. We shall denote the domain and the range of $\xi_{s,t}$ by $D_{s,t}$ and $R_{s,t}$, respectively:

9.1. Theorem. (i) Both of $D_{s,t}$ and $R_{s,t}$ are open subsets of M for any $s < t$ a.s. The map $\xi_{s,t}(\cdot)$; $D_{s,t} \longrightarrow R_{s,t}$ is a homeomorphism for any $s < t$ a.s.

(ii) The maps $\xi_{s,t}$, $\xi_{r,t}$ and $\xi_{s,r}$ satisfy $\xi_{s,t} = \xi_{r,t} \circ \xi_{s,r}$ on $D_{s,t}$ for any $s < r < t$ a.s.

Proof. We shall first prove that the explosion time $T(s,x,\omega)$ is lower semicontinuous in (s,x). Let $\{G_n\}$ be a sequence of open subsets of M with compact closure such that $\bar{G}_n \subset G_{n+1}$ and $\bigcup_n G_n = M$. Let $T_n(s,x)$ be the first hitting time for the set G_n^c. Then $T_n(s,x)$ is lower semicontinuous for each n and the sequence $\{T_n(s,x)\}$ is increasing. Thus $T(s,x) = \lim_{n\to\infty} T_n(s,x)$ is lower semicontinuous. A consequence is that

$D_{s,t}(\omega)$ is open.

The "one to one" of the map $\xi_{s,t}(\cdot,\omega)$ is a rather local property as we will see below. Let U_n, n=1,2,... be coordinate neighborhoods of M such that $\bigcup_n U_n = M$. Let S_m, m=1,2,... be a set of open time intervals generating all open sets in (s,a). We denote by $N_{n,m}$ the set of all ω such that there are x, y (x≠y) of M and $\sigma(\omega) \in S_m$ such that $\xi_{s,t}(x) = \xi_{s,t}(y)$ for $t \geq \sigma(\omega)$, $\xi_{s,t}(x) \neq \xi_{s,t}(y)$ for $t < \sigma(\omega)$ and $\xi_{s,\sigma}(x) = \xi_{s,\sigma}(y)$ is in U_n. In the coordinate neighborhood, we see by Theorem 5.3 that $N_{n,m}$ is a null set. Therefore, $\bigcup_{n,m} N_{n,m}$ is a null set. Note that if $\xi_{s,t}(\cdot,\omega)$ is not a one to one map for some t, then ω belongs to some $N_{n,m}$.

We will next prove the local homeomorphism of the map. Consider a trajectory $\{\xi_{s,r}(x_0,\omega),\ r \in [s,t]\}$, where (s,t,x_0) is fixed. We may choose a chain of coordinate neighborhoods $V_0,\dots,V_n$ such that $x_0 \in V_0$ and for any i=1,...,n

$$\bigcup_{x \in V_0} \{\xi_{s,r}(x) \mid r \in [t_i, t_{i+1}]\} \subset V_i,$$

where $t_i = \frac{i}{n}(t-s)$. Then since

$$(1) \qquad \xi_{s,t} = \xi_{t_{n-1},t} \circ \xi_{t_{n-2},t_{n-1}} \circ \cdots \circ \xi_{s,t_1}$$

and each ξ_{t_{i-1},t_i} defines a local homeomorphism in V_i, we see that $\xi_{s,t}(\cdot,\omega)$ is a local homeomorphism.

We have thus seen that $\xi_{s,t}\ ;\ D_{s,t} \longrightarrow R_{s,t}$ is a homeomorphism for any s < t a.s. This implies $R_{s,t}$ is open. The assertion (ii) can be proved easily. The proof is complete.

9.2. <u>Theorem.</u> Suppose that coefficients $X_1,\dots,X_m$ of the Stratonovich equation are $C^{k+1,\alpha}$ vector fields for $k \geq 1$ and $\alpha > 0$. Suppose further that coefficients X_0 is a $C^{k,\alpha}$ vector field. Then the map $\xi_{s,t} ; D_{s,t} \longrightarrow R_{s,t}$ is a $C^{k,\beta}$ diffeomorphism for any β less than α for any $s < t$.

Further suppose that the solution is strictly conservative. Then the solution defines a flow of $C^{k,\beta}$ diffeomorphisms of M if and only if the solution of the following adjoint equation is strictly conservative

$$(2) \qquad d\hat{\xi}_t = - \sum_{k=0}^{m} X_k(t,\hat{\xi}_t) \circ dB_t^k .$$

Furthermore, if $\xi_{s,t}$ is a diffeomorphism, the inverse $\xi_{s,t}^{-1}(x)$ coincides with the solution of the backward equation

$$(3) \qquad \hat{d}\hat{\xi}_s = \sum_{k=0}^{m} X_k(s,\hat{\xi}_s) \circ \hat{d}B_s^k$$

with the terminal condition $\hat{\xi}_t = x$.

Proof. Consider the composition (1) of $\xi_{s,t}$. Each $\xi_{t_{i-1},t_i}(x)$ is a local $C^{k,\beta}$-diffeomorphism for any β less than α in each coordinate neighborhood V_i by Theorem 7.3. Then the composite map $\xi_{s,t}$ is a local $C^{k,\beta}$ diffeomorphism. This together with Theorem 9.1 implies the first assertion.

Suppose next that the solution is strictly conservative. Then $\xi_{s,t}(\cdot,\omega)$ is a $C^{k,\beta}$-diffeomorphism from M to $R_{s,t}(\omega)$ for any $s<t$ a.s. We shall obtain a backward equation for the inverse $\xi_{s,t}^{-1}$. Let F be a C^2 function. Then

$$F(\xi_{s,t}(y)) - F(y) = \sum_{k=1}^{m} \int_s^t (X_k(r)F)(\xi_{s,r}(y))dB_r^k + \int_s^t (L(r)F)(\xi_{s,r}(y))dr.$$

Substitute $y = \xi_{s,t}^{-1}(x)$ and apply Lemma 6.2. Then we get

$$(4) \quad F(\xi_{s,t}^{-1}(x)) - F(x) = -\sum_{k=1}^{m}\int_s^t X_k(r)F(\xi_{r,t}^{-1}(x))\hat{d}B_r^k$$

$$- \int_s^t (X_0(r) - \frac{1}{2}\sum_{k=1}^{m} X_k(r)^2)F(\xi_{r,t}^{-1}(x))dr$$

on the set $\{\omega \mid x \in R_{s,t}(\omega)\}$ a.s. The backward martingale part of $X_k(r)F(\xi_{r,t}^{-1}(x))$ is

$$-\sum_{j=1}^{m}\int_r^t X_j(u)X_k(u)F(\xi_{u,t}^{-1}(x))\hat{d}B_u^j$$

by the above formula. Therefore we have

$$\int_s^t X_k(r)F(\xi_{r,t}^{-1}(x))\circ\hat{d}B_r^k = \int_s^t X_k(r)F(\xi_{r,t}^{-1}(x))\hat{d}B_r^k - \frac{1}{2}\int_s^t X_k(r)^2F(\xi_{r,t}^{-1}(x))dr.$$

Formula (4) is then written as

$$(5) \quad F(\xi_{s,t}^{-1}(x)) - F(x) = -\sum_{k=0}^{m}\int_s^t X_k(r)F(\xi_{r,t}^{-1}(x))\circ\hat{d}B_r^k .$$

Now let $\hat{\xi}_{s,t}$ be the solution of the backward equation

$$\hat{d}\xi_s = \sum_{k=0}^{m} X_k(s,\hat{\xi}_s)\circ\hat{d}B_s^k$$

with the terminal condition $\hat{\xi}_t = x$. Then it holds

$$F(\hat{\xi}_{s,t}(x)) - F(x) = -\sum_{k=0}^{m}\int_s^t X_k(r)F(\hat{\xi}_{r,t}(x))\circ\hat{d}B_r^k .$$

Hence $\hat{\xi}_{s,t}(x)$ is an extension of $\xi_{s,t}^{-1}(x)$. We can then prove the theorem similarly as Theorem 6.1.

The following is immediate from the above theorem and Theorem 8.3.

9.3. <u>Corollary.</u> If M is a compact manifold, the solution defines a flow of homeomorphisms of M a.s.

BIBLIOGRAPHICAL NOTES

Estimates similar to Theorem 2.1 are studied in several literatures. Blagovescenskii-Freidlin [4] announce an estimate in case $s = s'$ and coefficients are bounded. Proofs are found in Stroock-Varadhan [41], Bismut [3] etc. The present estimate is close to Funaki [10].

Theorems 2.1 and 3.1 stating the continuity and the differentiability of the solution with respect to the initial data, are certain refinement of [4], where they showed that the solution $\xi_{s,t}(x)$ is a $C^{k,0}$ function if coefficients are $C^{k+1,0}$ functions. Ikeda-Watanabe [13] and Bismut [3] employ different methods for proving the similar results.

Section 4 is adapted from Kunita [23]. He is indebted to S. R. Varadhan for Theorem 4.3. Sections 6 and 9 are adapted from Kunita [26]. Consult Baxendale [1], Bismut [3], Elworthy [7], Ikeda-Watanabe [13], Malliavin [30] for different approaches to the problem of stochastic flows of diffeomorphisms.

CHAPTER III

DIFFERENTIAL GEOMETRIC ANALYSIS OF STOCHASTIC FLOWS

1. ITÔ'S FORWARD AND BACKWARD FORMULA FOR STOCHASTIC FLOWS

As in Chapter II, we shall consider an Itô SDE on R^d;

$$(1)\qquad d\xi_t = \sum_{k=0}^{m} X_k(t,\xi_t)dB_t^k$$

where $B_t = (B_t^1,\ldots,B_t^m)$, $t\in[0,a]$ is an m-dimensional Brownian motion and $B_t^0 \equiv t$. $X_0(t,x),\ldots,X_m(t,x)$ are d-vector functions, continuous in (t,x) and locally Lipschitz continuous in x. The solution starting from x at time s is denoted by $\xi_{s,t}(x)$, $s \leq t < T(s,x)\wedge a$, where $T(s,x)$ is the explosion time. It is continuous in (s,t,x). The solution is said to be strictly conservative if $P(T(s,x)=\infty$ for all $(s,x)) = 1$ is satisfied.

In Chapter I, we obtained Itô's formula for continuous semimartingales. If we apply Corollary 8.2, Chapter I to the solution $\xi_{s,t}(x)$, regarding it as a continuous semimartingale, we obtain the following: Let $F(x)$, $x\in R^d$ be a C^2 function. Then it holds

$$(2)\qquad F(\xi_{s,t}(x)) - F(x) = \sum_{k=1}^{m}\int_s^t X_k(r)F(\xi_{s,r}(x))dB_r^k + \int_s^t L(r)F(\xi_{s,r}(x))dr$$

for any (s,t,x) satisfying $s \leq t < T(s,x)\wedge a$, where

$$X_k(r) = \sum_{i=1}^{d} X_k^i(r,x)\frac{\partial}{\partial x_i}\ ,$$

$$L(r)=\frac{1}{2}\sum_{i,j}\{\sum_{k=1}^{m} X_k^i(r,x)X_k^j(r,x)\}\frac{\partial^2}{\partial x_i \partial x_j}+\sum_{i=1}^{d} X_0^i(r,x)\frac{\partial}{\partial x_i}.$$

The above formula describes the differential rule of $\xi_{s,t}$ for the forward variable t. In this section, we will give a differential rule of $\xi_{s,t}$ for the backward variable s. The backward formula will require some additional smoothness assumption to coefficients $X_0,\dots,X_m$, so that the solution $\xi_{s,t}(\cdot,\omega)$ defines a C^2-map of R^d.

1.1. Theorem. Suppose that coefficients $X_0,\dots,X_m$ are $C^{2,\alpha}$ functions for some $\alpha > 0$. Let $F ; R^d \longrightarrow R^1$ be a C^2 function. Then it holds

$$(3) \qquad F(\xi_{s,t}(x)) - F(x) = \sum_{k=1}^{m}\int_s^t X_k(r)(F\circ\xi_{r,t})(x)\hat{d}B_r^k + \int_s^t L(r)(F\circ\xi_{r,t})(x)dr$$

for any $x \in D_{s,t} = \{x\,|\,T(s,x) > t\}$.

Proof. We shall consider the case where the first and the second derivatives of coefficients $X_0,\dots,X_m$ together with those of the function F are all bounded: The general case can be easily reduced to this. The solution $\xi_{s,t}$ is then strictly conservative and it defines a stochastic flow of C^2-diffeomorphisms of R^d by Theorem 4.4 and Theorem 5.4 in Chapter II.

We will fix the forward variable t and write $\xi_{r,t}$ as ξ_r. Let $\Delta = \{0=s_0<\dots<s_n=t\}$ be partitions of $[0,t]$. Let $s\in[0,t]$. We may and do assume that s is contained in Δ, say $s = s_\ell \in \Delta$. Then

$$\begin{aligned} F(\xi_{s,t}(x)) - F(x) &= \sum_{k=\ell}^{n-1}(F\circ\xi_{s_{k+1}}\circ\xi_{s_k,s_{k+1}} - F\circ\xi_{s_{k+1}})(x) \\ &= \sum_{i=1}^{d}\sum_{k=\ell}^{n-1}\frac{\partial}{\partial x_i}(F\circ\xi_{s_{k+1}})(x)(\xi^i_{s_k,s_{k+1}}(x) - x_i) \end{aligned}$$

$$+\frac{1}{2}\sum_{i,j}\sum_{k=\ell}^{n-1}\frac{\partial^2}{\partial x_i\partial x_j}(F\circ\xi_{s_{k+1}})(x+\eta_k)(\xi^i_{s_k,s_{k+1}}(x)-x_i)(\xi^j_{s_k,s_{k+1}}(x)-x_j),$$

where η_k are random variables such that $|\eta_k|\le|\xi_{s_k,s_{k+1}}(x)-x|$. In the sequel we will prove the following convergences;

$$(4)\quad \sum_{k=\ell}^{n-1}\frac{\partial}{\partial x_i}(F\circ\xi_{s_{k+1}})(x)(\xi^i_{s_k,s_{k+1}}(x)-x_i)$$

$$\xrightarrow[|\Delta|\to 0]{}\sum_{j=0}^{m}\int_s^t X^i_j(r,x)\frac{\partial}{\partial x_i}(F\circ\xi_{r,t})(x)\hat{d}B^j_r.$$

$$(5)\quad \sum_{k=\ell}^{n-1}\frac{\partial^2}{\partial x_i\partial x_j}(F\circ\xi_{s_{k+1}})(x)(\xi^i_{s_k,s_{k+1}}(x)-x_i)(\xi^j_{s_k,s_{k+1}}(x)-x_j)$$

$$\xrightarrow[|\Delta|\to 0]{}\int_s^t\Big(\sum_{k=1}^{m}X^i_k(r,x)X^j_k(r,x)\Big)\frac{\partial^2}{\partial x_i\partial x_j}(F\circ\xi_{r,t})(x)dr.$$

$$(6)\quad \sum_{k=\ell}^{n-1}\Big\{\frac{\partial^2}{\partial x_i\partial x_j}(F\circ\xi_{s_{k+1}})(x+\eta_k)-\frac{\partial^2}{\partial x_i\partial x_j}(F\circ\xi_{s_{k+1}})(x)\Big\}$$

$$\times(\xi^i_{s_k,s_{k+1}}(x)-x_i)(\xi^j_{s_k,s_{k+1}}(x)-x_j)$$

$$\xrightarrow[|\Delta|\to 0]{} 0.$$

The formula (3) will follow immediately from the above.

Proof of (4). The left hand side is written as

$$(7)\quad \sum_{j=0}^{m}\Big\{\sum_{k=\ell}^{n-1}\frac{\partial}{\partial x_i}(F\circ\xi_{s_{k+1}})(x)\int_{s_k}^{s_{k+1}}X^i_j(r,\xi_{s_k,r}(x))dB^j_r\Big\}.$$

The 0-th term (j=0) converges to $\int_s^t X^i_0(r,x)\frac{\partial}{\partial x_i}(F\circ\xi_{r,t})(x)dr$. To see the convergence of the j-th ($j\ge 1$) term, we will extend it to a continuous backward martingale, by setting for $r\in[0,t]$

$$I^\Delta_r=\sum_{k=p+1}^{n-1}\frac{\partial}{\partial x_i}(F\circ\xi_{s_{k+1}})(x)\int_{s_k}^{s_{k+1}}X^i_j(u,\xi_{s_k,u}(x))dB^j_u$$

$$+\frac{\partial}{\partial x_i}(F\circ\xi_{s_{p+1}})(x)\int_r^{s_{p+1}}\mu^x_{s_p,r}(X^i_j(u,\xi_{r,u}))dB^j_u$$

where s_p is the number of Δ such that $s_p \le r < s_{p+1}$ and $\mu^x_{s_p,r}$ is the distribution of $\xi_{s_p,r}(x)$ and $\mu^x_{s_p,r}(X^i_j(u,\xi_{r,u}))$ is the integral of $X^i_j(u,\xi_{r,u}(y))$ by the measure $\mu^x_{s_p,r}(dy)$. If $r = s_\ell$, it coincides with the j-th term of (7). One can check directly that I^Δ_r is a continuous square integrable backward martingale.

Define another square integrable backward martingale L^Δ_r by

$$L^\Delta_r = \sum_{k=p+1}^{n-1} \frac{\partial}{\partial x_i}(F\circ\xi_{s_{k+1}})(x)X^i_j(s_{k+1},x)(B^j_{s_{k+1}} - B^j_{s_k})$$
$$+ \frac{\partial}{\partial x_i}(F\circ\xi_{s_{p+1}})(x)X^i_j(s_{p+1},x)(B^j_{s_{p+1}} - B^j_r).$$

We may compute the quadratic variation of $L^\Delta_r - I^\Delta_r$ associated with the the partition Δ. It is

$$\langle L^\Delta - I^\Delta\rangle^\Delta_t = \sum_{k=0}^{n-1} \left|\frac{\partial}{\partial x_i}(F\circ\xi_{s_{k+1}})(x)\right|^2\left(\int_{s_k}^{s_{k+1}} (X^i_j(r,\xi_{s_k,r}(x)) - X^i_j(r,x)^\Delta)dB^j_r\right)^2.$$

This converges to 0 in L^1-norm as $|\Delta|\to 0$. Then $\{L^\Delta_r - I^\Delta_r\}$ converges to 0 uniformly in L^2-norm as $|\Delta|\to 0$. On the other hand, L^Δ_r converges to

$$L_r = \int_r^t \frac{\partial}{\partial x_i}(F\circ\xi_u)(x)X^i_j(u,x)\hat{d}B^j_u.$$

Therefore I^Δ_r converges to L_r for any r. Summing up these convergences for $j=0,\ldots,m$, we obtain (4).

<u>Proof</u> of (5). We use the same notation as the proof of (4). Define continuous backward semimartingales J^Δ_r and K^Δ_r by

$$(8)\quad J^\Delta_r = \sum_{k=p+1}^{n-1} \frac{\partial^2}{\partial x_i \partial x_j}(F\circ\xi_{s_{k+1}})(x)(\xi^i_{s_k,s_{k+1}}(x) - x_i)$$
$$+ \frac{\partial^2}{\partial x_i \partial x_j}(F\circ\xi_{s_{p+1}})(x)(\mu^x_{s_p,r}(\xi^i_{r,s_{p+1}}) - x_i),$$

$$(9)\quad K^\Delta_r = \sum_{k=p+1}^{n-1} (\xi^j_{s_k,s_{k+1}}(x) - x_j) + (\mu^x_{s_p,r}(\xi^j_{r,s_{p+1}}) - x_j).$$

Then the left hand side of (5) is written as $<J^\Delta, K^\Delta>_t^\Delta - <J^\Delta, K^\Delta>_s^\Delta$.

From (4), we know that J_s^Δ and K_s^Δ converges as $|\Delta|\to 0$ to J_s and K_s, respectively, where

$$J_s = \sum_{\ell=0}^{m}\int_s^t X_\ell^i(r,x)\frac{\partial^2}{\partial x_i \partial x_j}(F\circ\xi_{r,t})(x)\hat{d}B_r^\ell,$$

$$K_s = \sum_{\ell=0}^{m}\int_s^t X_\ell^j(r,x)\hat{d}B_r^\ell .$$

Denote the martingale part of J_r^Δ etc. by $\hat{J}_r^\Delta$ etc. Then both of $<J^\Delta, K^\Delta>_t^\Delta$ and $<\hat{J}^\Delta, \hat{K}^\Delta>_t^\Delta$ converge to the same process. We will prove that it is $<\hat{J},\hat{K}>_t$. It holds

$$|<\hat{J}^\Delta,\hat{K}^\Delta>_t^\Delta - <\hat{J},\hat{K}>_t| \le |<\hat{J}^\Delta,\hat{K}^\Delta>_t^\Delta - <\hat{J},\hat{K}>_t^\Delta| + |<\hat{J},\hat{K}>_t^\Delta - <\hat{J},\hat{K}>_t|.$$

The last member converges to 0 as $|\Delta|\to 0$. The first member of the right hand side is dominated by

$$(<\hat{J}^\Delta - \hat{J}>_t^\Delta <\hat{K}^\Delta>_t^\Delta)^{\frac{1}{2}} + (<\hat{K}^\Delta - \hat{K}>_t^\Delta <\hat{J}>_t^\Delta)^{\frac{1}{2}} .$$

This converges to 0 by Theorem 3.2 of Chap. I. The proof is complete.

Proof of (6). Write K_r^Δ of (9) as $K_r^\Delta(j)$. Then the right hand side of (6) is dominated by the following, which converges to 0 as $|\Delta|\to 0$.

$$\sup_k \left|\frac{\partial^2}{\partial x_i \partial x_j}(F\circ\xi_{s_{k+1}})(x+\eta_k) - \frac{\partial^2}{\partial x_i \partial x_j}(F\circ\xi_{s_{k+1}})(x)\right| \{<K^\Delta(i)>_t^\Delta <K^\Delta(j)>_t^\Delta\}^{\frac{1}{2}} .$$

We shall next consider a Stratonovich SDE on manifold. Let M be a connected, paracompact C^∞ manifold of dimension d. Let $X_1(t),\dots,X_m(t)$, $t\in[0,a]$ be C^2 vector fields with parameter t. With a local coordinate $(x_1,\dots,x_d)$, these are represented as the first order partial differential operators

$$X_k(t) = \sum_{i=1}^{d} X_k^i(t,x)\frac{\partial}{\partial x_i}$$

at each coordinate neighborhood, where $X_k^i(t,x)$ is assumed to be C^1 function of t and C^2 funciton of x. Let $X_0(t)$ be another vector field such that $X_0^i(t,x)$ is continuous in t and a C^1 function of x. The solution of the Stratonovich equation

$$(10)\quad d\xi_t = \sum_{k=0}^{m} X_k(t,\xi_t)\circ dB_t^k$$

with the initial condition $\xi_s = x$ was defined in Chapter II, §8 via Itô's formula:

$$(11)\quad F(\xi_{s,t}(x)) - F(x) = \sum_{k=0}^{m}\int_s^t X_k(r)F(\xi_{s,r}(x))\circ dB_r^k ,$$

where $F ; M \longrightarrow R^1$ is a C^3-map. Using Itô integral, it is equivalent to

$$(12)\quad F(\xi_{s,t}(x)) - F(x) = \sum_{k=1}^{m}\int_s^t X_k(r)F(\xi_{s,r}(x))dB_r^k$$

$$+ \int_s^t L(r)F(\xi_{s,r}(x))dr,$$

where $L(r)$ is the second order operator defined by

$$(13)\quad L(r) = \frac{1}{2}\sum_{k=1}^{m} X_k(r)^2 + X_0(r).$$

The existence and uniqueness of the (maximal) solution $\xi_{s,t}(x)$, $s \leq t < T(s,x)\wedge a$ was shown in Theorem 8.3, Chapter II.

We will obtain the backward formula.

1.2. Theorem. Suppose that coefficients $X_1,\dots,X_m$ of Stratonovich SDE are $C^{3,\alpha}$ vector fields for some $\alpha > 0$ and that X_0 is a $C^{2,\alpha}$ vector field for some $\alpha > 0$. Let $F ; M \longrightarrow R^1$ be a C^2-class fucntion. Then it holds

$$(14)\quad F(\xi_{s,t}(x)) - F(x) = \sum_{k=1}^{m}\int_s^t X_k(r)(F\circ\xi_{r,t})(x)\hat{d}B_r^k$$

$$+\int_s^t (\frac{1}{2}\sum_{k=1}^m X_k(r)^2 + X_0(r))(F\circ\xi_{r,t})(x)dr$$

for $x \in D_{s,t} = \{x | T(s,x) > t\}$ a.s.

<u>Proof.</u> We shall first consider the SDE on R^d: Using Itô integral, the equation (1) is written as

$$d\xi_t = \sum_{k=1}^m X_k(t,\xi_t)dB_t^k + X_0^*(t,\xi_t)dt,$$

where

$$X_0^*(t,x) = X_0(t,x) + \frac{1}{2}\sum_{i,j} X_j^i(t,x)\frac{\partial}{\partial x_i}X_j(t,x).$$

Coefficients $X_0^*, X_1, \ldots, X_m$ satisfies conditions of Theorem 1.1. Consequently, Itô's backward formula (3) is valid, where

$$L(r) = \frac{1}{2}\sum_{i,j}\{\sum_{k=1}^m X_k^i(r,x)X_k^j(r,x)\}\frac{\partial^2}{\partial x_i \partial x_j} + \sum_{i=1}^d X_0^{*i}(r,x)\frac{\partial}{\partial x_i}$$

$$= \frac{1}{2}\sum_{k=1}^m X_k(r)^2 + X_0(r).$$

We shall next consider the case of manifold. At each coordinate neighborhood, the formula (14) is valid by the above argument. Let $\{U_n\}$ be a countable set of coordinate neighborhoods covering M. Fix the time t and define a sequence of backward stopping times $\tau_n(x)$ as

$$\tau_1(x) = \sup\{r < t\ ;\ \xi_{r,t}(x) \notin U_1\} \quad (= 0 \text{ if } \{\ldots\} = \phi),$$

$$\tau_n(x) = \sup\{r < t\ ;\ \xi_{r,t}(x) \notin U_1 \cup \ldots \cup U_n\} \quad (= 0 \text{ if } \{\ldots\} = \phi).$$

Let $s_0 < t$ be a fixed time. The formula (14) is valid for any s of $(\tau_1(x) \vee s_0, t)$. We can show by induction that (14) is valid for any s of $(\tau_n(x) \vee s_0, t)$. Since $\lim_{n\to\infty} \tau_n(x) \le s_0$ holds for $x \in D_{s_0,t} = D_{s_0,t}(\omega)$, the formula (14) is valid for any s of (s_0,t). This proves the theorem.

If coefficients $X_0,\dots,X_m$ satisfy additional smoothness assumptions, we can rewrite the formula (14) using Stratonovich integral.

1.3. Theorem. Suppose that $X_1,\dots,X_m$ are $C^{4,\alpha}$ vector fields with some $\alpha>0$ and X_0 is a $C^{3,\alpha}$ vector field with the same α. Let F ; $R^d \longrightarrow R^1$ be a C^3 function. Then it holds

$$(15)\quad F(\xi_{s,t}(x)) - F(x) = \sum_{k=0}^{m} \int_s^t X_k(r)(F\circ\xi_{r,t})(x)\circ \hat{d}B_r^k .$$

Proof. In order that the right hand side is well defined, we have to check that $X_k(r)(F\circ\xi_{r,t})(x)$ is a backward semimartingale. Note that $\xi_{u,t}(x)$ is a C^3 function of x. (See Theorem 9.2 in Chapter II). We shall operate $\partial_i = \frac{\partial}{\partial x_i}$ to (14). Then, by Theorem 7.6, Chapter I,

$$\partial_i(F\circ\xi_{r,t})(x) - \partial_i F(x) = \sum_{j=1}^{m}\int_r^t \partial_i(X_j(u)(F\circ\xi_{u,t}))(x)\hat{d}B_u^j$$
$$+ \int_r^t \partial_i(L(u)(F\circ\xi_{u,t}))(x)dr.$$

Therefore $\partial_i(F\circ\xi_{r,t})(x)$ is a backward semimartingale as we desired.

The backward martingale part of $X_k(r)(F\circ\xi_{r,t})(x)$ is

$$\sum_{j=1}^{m}\int_r^t X_k(u)X_j(u)(F\circ\xi_{u,t})(x)\hat{d}B_u^j .$$

Therefore we have

$$\int_s^t X_k(r)(F\circ\xi_{r,t})(x)\circ\hat{d}B_r^k$$
$$= \int_s^t X_k(r)(F\circ\xi_{r,t})(x)\hat{d}B_r^k + \frac{1}{2}\int_s^t X_k(r)^2(F\circ\xi_{r,t})(x)dr.$$

The formula (14) then leads to the formula (15). The proof is complete.

Remark. We have seen in Chapter II, Section 9 that the inverse map $\xi_{s,t}^{-1}$ satisfies the backward equation

$$(16) \quad F(\xi_{s,t}^{-1}(x)) - F(x) = -\sum_{k=1}^{m}\int_s^t X_k(r)F(\xi_{r,t}^{-1}(x))\hat{d}B_r^k + \int_s^t (\frac{1}{2}\sum_{k=1}^{m} X_k(r)^2 - X_0(r))F(\xi_{r,t}^{-1}(x))dr$$

$$(17) \quad = -\sum_{k=0}^{m}\int_s^t X_k(r)F(\xi_{r,t}^{-1}(x))\circ \hat{d}B_r^k .$$

Apply previous theorem to $\xi_{s,t}^{-1}$, we get the following Itô's forward formula.

$$(18) \quad F(\xi_{s,t}^{-1}(x)) - F(x) = -\sum_{k=1}^{m}\int_s^t X_k(r)(F\circ\xi_{s,r}^{-1})(x)dB_r^k + \int_s^t (\frac{1}{2}\sum_{k=1}^{m} X_k(r)^2 - X_0(r))(F\circ\xi_{s,r}^{-1})(x)\,dr$$

$$(19) \quad = -\sum_{k=0}^{m}\int_s^t X_k(r)(F\circ\xi_{s,r}^{-1})(x)\circ dB_r^k .$$

2. ITÔ'S FORMUMA FOR $\xi_{s,t}$ ACTING ON VECTOR FIELDS

The stochastic flow of diffeomorphisms determined by an SDE on a manifold acts naturally on tensor fields and defines a stochastic process with values in tensor fields. We shall be concerned with the SDE or Itô's formula governing the tensor field valued process. In this section, we will mainly deal with vector fields; the case of general tensor fields will be discussed at Section 4.

We begin with introducing the differential of the map. Let ϕ be a diffeomorphism of the manifold M. The differential ϕ_{*x} of the map ϕ is by definition the linear map from the tangent space $T_x(M)$ to the tangent space $T_{\phi(x)}(M)$ such that

$$\phi_{*x}X_x f = X_x(f\circ\phi) \qquad \forall X_x \in T_x(M).$$

Given a vector field X on M, we denote by X_x the restriction of X at the point x. We define a new vector field ϕ_*X by

$$(\phi_*X)_x = \phi_{*\phi^{-1}(x)} X_{\phi^{-1}(x)} .$$

Then it holds

$$\phi_*Xf(x) = X(f\circ\phi)(\phi^{-1}(x)), \qquad \forall x\in M$$

for any f of $C^\infty(M)$ = the space of all C^∞ functions.

We shall express ϕ_*X using a local coordinate $(x_1,\dots,x_d)$. Let $X = \Sigma\, X^i(x)\frac{\partial}{\partial x_i}$ be the coordinate expression. Then ϕ_*X is expressed as $\phi_*X = \Sigma\,(\phi_*X)^i(x)\frac{\partial}{\partial x_i}$, where

$$(\phi_*X)^i(x) = \sum_j X^j(\phi^{-1}(x))\frac{\partial\phi^i}{\partial x_j}(\phi^{-1}(x))$$

and $\phi^i(x) = x_i(\phi(x))$. This follows immediately by setting $f(x) = x^i(x)$. Thus denoting the d-vector $(X^1,\dots,X^d)$ by X and Jacobian matrix $(\frac{\partial\phi^i}{\partial x_j})$ by $D\phi(x)$, the vector $\phi_*X(x)$ with components $(\phi_*X)^i(x)$ satisfies $\phi_*X(x) = D\phi(\phi^{-1}(x))X(\phi^{-1}(x))$.

Let Y be a vector field and ϕ_t the flow of Y, i.e., the local one parameter group of local transformations of M generated by Y:

$$\frac{d}{dt}(f\circ\phi_t)(x)\Big|_{t=0} = Yf(x), \qquad \forall f\in C^\infty(M).$$

The <u>Lie derivative</u> of the vector field X with respect to Y is the vector field L_YX defined by

$$(L_YX)_x = -\lim_{t\to 0}\frac{1}{t}\{(\phi_{t*}X)_x - X_x\} = -\lim_{t\to 0}\frac{1}{t}\{\phi_{t*\phi_t^{-1}(x)} X_{\phi_t^{-1}(x)} - X_x\}.$$

The relation $L_YX = [Y,X] \equiv YX - XY$ is well known.

Now let $\xi_{s,t}(x)$ be the solution of the Stratonovich SDE (1) of Section 1. We shall assume that it defines a stochastic flow of diffeomorphisms.

Then differential $\xi_{s,t*}$ is well defined for any $s < t$ a.s.

2.1. Theorem. Assume that coefficients $X_1,\dots,X_m$ of the equation are $C^{4,\alpha}$ vector fields for some $\alpha > 0$ and X_0 is a $C^{3,\alpha}$ vector field for the same α. Then $\xi_{s,t*}$ satisfies the following formula for any C^2 vector field X.

$$(1)\qquad \xi_{s,t*}X - X = -\sum_{k=1}^{m}\int_s^t L_{X_k(r)}\xi_{s,r*}X dB_r^k + \int_s^t \Big(\frac{1}{2}\sum_{k=1}^{m} L^2_{X_k(r)} - L_{X_0(r)}\Big)\xi_{s,r*}X dr$$

$$(2)\qquad = -\sum_{k=1}^{m}\int_s^t \xi_{r,t*}L_{X_k(r)}X\hat{d}B_r^k + \int_s^t \xi_{r,t*}\Big(\frac{1}{2}\sum_{k=1}^{m} L^2_{X_k(r)} - L_{X_0(r)}\Big)X dr.$$

Further, the inverse $\xi^{-1}_{s,t*}$ satisfies

$$(3)\qquad \xi^{-1}_{s,t*}X - X = \sum_{k=1}^{m}\int_s^t \xi^{-1}_{s,r*}L_{X_k(r)}X dB_r^k + \int_s^t \xi^{-1}_{s,r*}\Big(\frac{1}{2}\sum_{k=1}^{m} L^2_{X_k(r)} + L_{X_0(r)}\Big)X dr$$

$$(4)\qquad = \sum_{k=1}^{m}\int_s^t L_{X_k(r)}\xi^{-1}_{r,t*}X\hat{d}B_r^k + \int_s^t \Big(\frac{1}{2}\sum_{k=1}^{m} L^2_{X_k(r)} + L_{X_0(r)}\Big)\xi^{-1}_{r,t*}X dr.$$

With additional assumptions on coefficients, the formulas (1)-(4) can be written using Stratonovich integrals.

2.2. Theorem. Suppose that $X_1,\dots,X_m$ are $C^{5,\alpha}$ vector fields for some $\alpha > 0$ and X_0 is a $C^{4,\alpha}$ vector field for the same α. Then $\xi_{s,t*}X$ satisfies the following formula for any C^3-vector field X.

$$(5)\quad \xi_{s,t*}X - X = -\sum_{k=0}^{m}\int_s^t L_{X_k(r)}\xi_{s,r*}X\circ dB_r^k$$

$$(6)\quad = -\sum_{k=0}^{m}\int_s^t \xi_{r,t*}L_{X_k(r)}X\circ \hat{d}B_r^k .$$

Further, $\xi_{s,t*}^{-1}$ satisfies the following formula.

$$(7)\quad \xi_{s,t*}^{-1}X - X = \sum_{k=0}^{m}\int_s^t \xi_{s,r*}^{-1}L_{X_k(r)}X\circ dB_r^k$$

$$(8)\quad = \sum_{k=0}^{m}\int_s^t L_{X_k(r)}\xi_{r,t*}^{-1}X\circ \hat{d}B_r^k .$$

Proof of Theorem 2.1 and 2.2 can be carried over by similar methods. We will give here the proof of Theorem 2.2 only, since the computation involved is simpler than that of Theorem 2.1.

Proof of Theorem 2,2. We shall first prove (5). Let $f\in C^\infty(M)$. Then it holds

$$\xi_{s,t*}Xf(x) = X(f\circ\xi_{s,t})(\xi_{s,t}^{-1}(x)).$$

Set $F_t(x) = X(f\circ\xi_{s,t})(x)$ and $M_t = \xi_{s,t}^{-1}(x) = (\xi_{s,t}^{-11}(x),\dots,\xi_{s,t}^{-1d}(x))$ (expression by local coordinate). By Itô's formula (Theorem 8.3 in Chapter I), it holds

$$d(F_t\circ M_t) = (dF_t)(M_t) + \sum_{i=1}^{d}\frac{\partial F_t}{\partial x_i}(M_t)\circ dM_t^i .$$

We have from (11) of Section 1,

$$dF_t(x) = \sum_{k=0}^{m} X(X_k(t)f\circ\xi_{s,t})(x)\circ dB_t^k .$$

Consequently

$$(dF_t)(M_t) = \sum_{k=0}^{m} X(X_k(t)f\circ\xi_{s,t})(\xi_{s,t}^{-1}(x))\circ dB_t^k$$
$$= \sum_{k=0}^{m}\xi_{s,t*}XX_k(t)f(x)\circ dB_t^k .$$

From (19) of Section 1, we have

$$dM_t^i = d\xi_{s,t}^{-1i} = -\sum_{k=0}^{m} X_k(t)(\xi_{s,t}^{-1i})(x)\circ dB_t^k$$
$$= -\sum_{k=0}^{m} \xi_{s,t*}^{-1}X_k(t)^i(\xi_{s,t}^{-1}(x))\circ dB_t^k .$$

Therefore,

$$\sum_{i=1}^{d} \frac{\partial F_t}{\partial x_i}(M_t)\circ dM_t^i = -\sum_{k=0}^{m} \xi_{s,t*}^{-1}X_k(t)\cdot X(f\circ\xi_{s,t})(\xi_{s,t}^{-1}(x))\circ dB_t^k$$
$$= -\sum_{k=0}^{m} X_k(t)\xi_{s,t*}Xf(x)\circ dB_t^k .$$

Therefore we have

$$d(F_t\circ M_t) = -\sum_{k=0}^{m} L_{X_k(t)}\xi_{s,t*}Xf(x)\circ dB_t^k .$$

This proves (5).

For the proof of (6), we will use Itô's backward formula. Set $F_s(x) = X(f\circ\xi_{s,t})(x)$ where t is fixed. It holds from Theorem 1.3,

$$\hat{d}F_s(x) = -\sum_{k=0}^{m} X\{X_k(s)(f\circ\xi_{s,t})\}(x)\circ\hat{d}B_s^k .$$

It holds from (3) of Section 9, Chapter II,

$$\hat{d}_s\xi_{s,t}^{-1i} = \sum_{k=0}^{m} X_k^i(s,\xi_{s,t}^{-1}(x))\circ\hat{d}B_s^k .$$

Therefore,

$$\sum_i \frac{\partial F_s}{\partial x_i}(\xi_{s,t}^{-1}(x))\circ\hat{d}_s\xi_{s,t}^{-1i}(x) = \sum_{k=0}^{m} X_k(s)\{X(f\circ\xi_{s,t})\}(\xi_{s,t}^{-1}(x))\circ\hat{d}B_s^k .$$

Consequently,

$$\hat{d}_s(F_s\circ\xi_{s,t}^{-1})(x) = (\hat{d}F_s)(\xi_{s,t}^{-1}(x)) + \sum_i \frac{\partial F_s}{\partial x_i}(\xi_{s,t}^{-1}(x))\circ\hat{d}_s\xi_{s,t}^{-1i}(x)$$
$$= \sum_{k=0}^{m} (X_k(s)X - XX_k(s))(f\circ\xi_{s,t})(\xi_{s,t}^{-1}(x))\circ\hat{d}B_s^k$$
$$= \sum_{k=0}^{m} \xi_{s,t*}[X_k(s),X]f(x)\circ\hat{d}B_s^k$$

$$= \sum_{k=0}^{m} \xi_{s,t*} L_{X_k(s)} Xf(x) \circ \hat{d}B_s^k .$$

This proves (6).

The proofs of (7) and (8) are similar. It is omitted.

3. COMPOSITION AND DECOMPOSITION OF THE SOLUTION

Consider two SDE's on the same manifold. Let $\xi_{s,t}(x)$, $s \leq t < T(s,x)$ and $\eta_{s,t}(x)$, $s \leq t < S(s,x)$ be solutions of equations,

(1) $$d\xi_t = \sum_{k=0}^{m} X_k(t,\xi_t) \circ dB_t^k ,$$

(2) $$d\eta_t = \sum_{k=0}^{m} Y_k(t,\eta_t) \circ dB_t^k ,$$

where $T(s,x)$ and $S(s,x)$ are explosion times of $\xi_{s,t}(x)$ and $\eta_{s,t}(x)$ respectively. We shall assume that vector fields $X_1,\dots,X_m$ (resp. $Y_1, \dots, Y_m$) are $C^{4,\alpha}$ (resp. $C^{2,\alpha}$) and X_0 (resp. Y_0) is a $C^{3,\alpha}$ (resp. $C^{1,\alpha}$) vector fields for some $\alpha > 0$.

We shall obtain an SDE governing the composite process $\zeta_{s,t}(x) = \xi_{s,t} \circ \eta_{s,t}(x)$, $s \leq t < U(s,x)$, where

$$U(s,x) = \inf\{t > s \ ; \ \eta_{s,t}(x) \notin D_{s,t}\} \wedge S(s,x)$$

and $D_{s,t} = \{x \ ; \ T(s,x) > t\}$.

3.1. Theorem. $\zeta_{s,t}(x)$, $s \leq t < U(s,x)$ is the solution of the following SDE.

(3) $$d\zeta_t = \sum_{k=0}^{m} X_k(t,\zeta_t) \circ dB_t^k + \sum_{k=0}^{m} \xi_{s,t*} Y_k(t,\zeta_t) \circ dB_t^k .$$

Proof. We shall apply Itô's formula (Theorem 8.3 in Chapter I) to

$F_t(x) = f\circ\xi_{s,t}(x)$ and $M_t = (\eta^1_{s,t},\dots,\eta^d_{s,t})$, where $f ; M \longrightarrow R^1$ is a C^3-function and $(\eta^1_{s,t},\dots,\eta^d_{s,t})$ is the coordinate expression of $\eta_{s,t}$. Then we have

$$\begin{aligned} d(F_t\circ M_t) &= (dF_t)(M_t) + \Sigma \frac{\partial F_t}{\partial x_i}(M_t)\circ dM_t^i \\ &= \sum_{k=0}^{m} X_k(t)f(\xi_{s,t}\circ\eta_{s,t})\circ dB_t^k \\ &\quad + \sum_{k=0}^{m} Y_k(t)(f\circ\xi_{s,t})(\eta_{s,t})\circ dB_t^k \\ &= \sum_{k=0}^{m} X_k(t)f(\zeta_{s,t})\circ dB_t^k + \sum_{k=0}^{m} \xi_{s,t*}Y_k(t)f(\zeta_{s,t})\circ dB_t^k \end{aligned}$$

This proves (3).

Now, Theorem 2.1 tells us that $\xi_{s,t*}Y_k(t) = Y_k(t)$ holds if and only if $L_{X_j(r)}Y_k(t) = 0$ holds for any $r\in[s,t]$ and $j=0,\dots,m$. Therefore we have the following.

3.2. Corollary. Suppose that each $Y_k(t)$ commutes with all $X_j(r)$, $j=0,\dots,m$, $r\in[s,t]$. Then the composite process $\zeta_{s,t} = \xi_{s,t}\circ\eta_{s,t}$ satisfies the following SDE

$$(4) \qquad d\zeta_t = \sum_{k=0}^{m} (X_k(t) + Y_k(t))(\zeta_t)\circ dB_t^k .$$

We shall next consider the problem of decomposing the solution of equation (1).

3.3. Theorem. Let $Y_k(t)$, $Z_k(t)$, $k=0,\dots,m$, $t\in[0,a]$ be vector fields satisfying the same smoothness condition as $X_k(t)$. Let $\eta_{s,t}(x)$ be the solution of the equation corresponding to $Y_k(t)$, $k=0,\dots,m$. Let $\kappa_{s,t}$ be the solution of

$$d\kappa_t = \sum_{k=0}^{m} \eta^{-1}_{s,t*} Z_k(t, \kappa_t) \circ dB^k_t .$$

Suppose that $X_k(t) = Y_k(t) + Z_k(t)$, $k=0,\dots,m$. Then $\xi_{s,t} = \eta_{s,t} \circ \kappa_{s,t}$ is the solution of equation (1).

<u>Proof.</u> We shall apply Itô's formula to $F_t(x) = f \circ \eta_{s,t}(x)$ and $M_t = (\kappa^1_{s,t}, \dots, \kappa^d_{s,t})$ (coordinate expression). Then we have

$$d(F_t \circ M_t) = (dF_t)(M_t) + \sum_i \frac{\partial F_t}{\partial x_i}(M_t) \circ dM^i_t$$

$$= \sum_{k=0}^{m} Y_k(t) f(\eta_{s,t} \circ \kappa_{s,t}(x)) \circ dB^k_t + \sum_{k=0}^{m} \eta^{-1}_{s,t*} Z_k(t)(f \circ \eta_{s,t})(\kappa_{s,t}(x)) \circ dB^k_t .$$

Since $\eta^{-1}_{s,t*} Z_k(t)(f \circ \eta_{s,t})(\kappa_{s,t}(x)) = Z_k(t) f(\eta_{s,t} \circ \kappa_{s,t}(x))$, we get

$$d(f \circ \eta_{s,t} \circ \kappa_{s,t}) = \sum_{k=0}^{m} X_k(t) f(\eta_{s,t} \circ \kappa_{s,t}) \circ dB^k_t .$$

This proves that $\xi_{s,t} = \eta_{s,t} \circ \kappa_{s,t}$ is the solution of equation (1).

In Chapter II, we obtained a backward SDE for the inverse map $\xi^{-1}_{s,t}$. Here we shall obtain the forward SDE for it, applying the previous theorem. Let $\kappa_{s,t}(x)$ be the solution of

$$d\kappa_t = - \sum_{k=0}^{m} \xi^{-1}_{s,t*} X_k(t, \kappa_t) \circ dB^k_t .$$

Then $\xi_{s,t} \circ \kappa_{s,t}(x) = x$ is satisfied by the previous theorem. Therefore we have $\kappa_{s,t} = \xi^{-1}_{s,t}$.

3.4. <u>Corollary.</u> The inverse $\xi^{-1}_{s,t}$ satisfies the following forward equation

$$(5) \qquad d_t \xi^{-1}_{s,t} = - \sum_{k=0}^{m} \xi^{-1}_{s,t*} X_k(t, \xi^{-1}_{s,t}) \circ dB^k_t$$

3.5. Example. Let $X_0,\dots,X_m$ be commutative complete vector fields, not depending on t. Then the solution of equation (1) is represented as

(6) $\xi_{s,t}(x) = \mathrm{Exp}\,(t-s)X_0 \circ \mathrm{Exp}\,(B_t^1 - B_s^1)X_1 \circ \dots \circ \mathrm{Exp}\,(B_t^m - B_s^m)X_m$,

where $\mathrm{Exp}\, tX_k$ is the one parameter group of transformations generated by the vector field X_k.

We shall prove first that $\xi_{s,t}^{(k)} \equiv \mathrm{Exp}\,(B_t^k - B_s^k)X_k$ is the solution of $d\xi_t^{(k)} = X_k(\xi_t^{(k)}) \circ dB_t^k$ starting from (s,x). Set $F^{(k)}(t) = f \circ \mathrm{Exp}\, tX_k$. Then by Itô's formula

$$F^{(k)}(B_t^k - B_s^k) = F^{(k)}(0) + \int_s^t \frac{\partial F^{(k)}}{\partial r}(B_r^k - B_s^k) \circ dB_r^k$$

$$= F^{(k)}(0) + \int_s^t X_k f \circ \mathrm{Exp}\,(B_r^k - B_s^k)X_k \circ dB_r^k .$$

Therefore, $\xi_{s,t}^{(k)}$ is the solution. Then the expression (6) follows from Corollary 3.2.

3.6. Example. Consider a linear SDE on R^d:

(7) $d\xi_t = A\xi_t dt + BdW_t$,

where A is a $d\times d$-matrix, B is a $d\times m$-matrix and W_t is an m-dimensional Wiener process. The equation is decomposed to

$$d\zeta_t = A\zeta_t dt, \quad d\eta_t = \zeta_{t*}^{-1}(B)dW_t,$$

i.e., if $\zeta_t(x)$ and $\eta_t(x)$ are solutions of the above equations starting at $(0,x)$, then $\xi_t \equiv \zeta_t \circ \eta_t$ is a solution of (7). It holds $\zeta_t(x) = e^{At}x$ so that $\zeta_{t*}^{-1}(B) = e^{-At}B$. Therefore $\eta_t(x) = x + \int_0^t e^{-As}BdW_s$. Hence we have the decomposition

$$\xi_t(x) = e^{At}(x + \int_0^t e^{-As} B dW_s)$$

In the system theory or the control theory, $B_t^1, \ldots, B_t^m$ in equation (1) are called inputs and the solution ξ_t is called the output. It is an important problem in applications that we can compute the output from the input explicitly. The above two examples show the way of calculating outputs. We shall consider the problem in a general framework.

The complexity of expressing the solution by the inputs depends on the structure of the Lie algebra generated by vector fields defining the equation. For simplicity we assume that vector fields $X_0, \ldots, X_m$ of equation (1) do not depend on time t. The real vector space spanned by all vector fields of the form

$$[\ldots[X_{i_1}, X_{i_2}], \ldots,], X_{i_n}], \qquad i_1, \ldots, i_n \in \{0, 1, \ldots, m\}$$

is called the Lie algebra generated by $X_0, \ldots, X_m$ and denoted by $\underline{\underline{L}}(X_0, \ldots, X_m)$ or simply by $\underline{\underline{L}}$.

We shall define a chain of subalgebras of $\underline{\underline{L}}$:

$$\underline{\underline{G}}_1 = [\underline{\underline{L}}, \underline{\underline{L}}] = \{[X, Y] \; ; \; X, Y \in \underline{\underline{L}}\}$$

$$\underline{\underline{G}}_2 = [\underline{\underline{G}}_1, \underline{\underline{G}}_1], \ldots, \underline{\underline{G}}_n = [\underline{\underline{G}}_{n-1}, \underline{\underline{G}}_{n-1}].$$

The Lie algebra $\underline{\underline{L}}$ is called <u>solvable</u> if $\underline{\underline{G}}_m = \{0\}$ for some m. If $\underline{\underline{L}}$ is a finite dimensional solvable Lie algebra, then by Lie's theorem there is a basis of $\underline{\underline{L}}$ denoted by $\{Y_1, \ldots, Y_n\}$ with following property: There is a chain of subspaces $\underline{\underline{L}}_k$, $k=0, \ldots, n$,

$$\underline{\underline{L}} = \underline{\underline{L}}_0 \supset \underline{\underline{L}}_1 \supset \ldots \supset \underline{\underline{L}}_n = \{0\}$$

such that $\{Y_{i+1}, \ldots, Y_n\}$ is a basis of $\underline{\underline{L}}_i$ and each $\underline{\underline{L}}_{i+1}$ is an ideal of $\underline{\underline{L}}_i$. On the other hand, consider another chain of subalgebras of $\underline{\underline{L}}$:

$\underline{\underline{G}}^1 = [\underline{\underline{L}},\underline{\underline{L}}],\ \underline{\underline{L}}^2 = [\underline{\underline{L}},\underline{\underline{L}}^1],\ldots \underline{\underline{L}}^n = [\underline{\underline{L}},\underline{\underline{L}}^{n-1}]$.

If $\underline{\underline{G}}^n = \{0\}$ for some n, the Lie algebra $\underline{\underline{L}}$ is called nilpotent. Nilpotent Lie algebra is solvable, obviously.

3.7. Theorem. Suppose that $X_0,\ldots,X_m$ are all complete vector fields and generate a finite dimensional solvable Lie algebra $\underline{\underline{L}}$. Let $\{Y_1,\ldots,Y_n\}$ be a basis of $\underline{\underline{L}}$ mentioned above. Then the solution $\xi_{0,t} = \xi_t$ of (1) is represented as

$$\text{(8)}\qquad \xi_t(x) = \operatorname{Exp} N_t^1 Y_1 \circ \operatorname{Exp} N_t^2 Y_2 \circ \cdots \circ \operatorname{Exp} N_t^n Y_n(x),$$

where $N_t^1,\ldots,N_t^n$ are continuous semimartingales constructed from B_t^0, $\ldots,B_t^m$ through finite repetition of the following elementary calculations.

(i) linear sums and products of $B_t^0,\ldots,B_t^m$.

(ii) Stratonovich integrals based on $B_t^0,\ldots,B_t^m$.

(iii) substitution to the exponential function e^x.

Furthermore, if $\underline{\underline{L}}$ is nilpotent $N_t^1,\ldots,N_t^n$ are constructed via (i) and (ii) only.

Remark. The algorithm of calculating $N_t^1,\ldots,N_t^n$ will be found in the proof of the theorem. It is determined by the structure constants of Lie algebra $\underline{\underline{L}}$ relative to the basis $\{Y_1,\ldots,Y_n\}$.

Before going to the proof, we remark a preliminary fact on solvable Lie algebra. Let $\{Y_1,\ldots,Y_n\}$ be the basis of $\underline{\underline{L}}$ mentioned above. Let Z be in $\underline{\underline{L}}_{k-1}$. Then $L_Z Y_i = [Z, Y_i]$ is in $\underline{\underline{L}}_k$ for any $i \geq k$ so that it is written as $\sum_{j \geq k+1} c_{ij} Y_j$ if $i \geq k$. Setting $c_{ij} = 0$ if $j = k$, we denote the $(n-k+1)\times(n-k+1)$ matrix $(c_{ij})_{i,j=k,\ldots,n}$ by ad_Z. Then the matrix $\exp ad_Z'$ is written as

$$(9)\quad \exp \mathrm{ad}'_Z = \begin{pmatrix} 1 & 0 & \dots & 0 \\ * & \dots & \dots & * \\ * & \dots & \dots & * \end{pmatrix},$$

where ad'_Z is the transpose of ad_Z.

<u>Proof.</u> Vector fields $X_0,\dots,X_r$ are written as linear sums of $Y_1, \dots, Y_n$, say $X_j = \sum_k a_{jk} Y_k$. Then SDE (1) is written as $d\xi_t = \sum_{k=1}^n Y_k(\xi_t)\circ dM_t^k$, where $M_t^k = \sum_j a_{jk} B_t^j$, $k=1,\dots,n$.

Consider two SDE's :

$$(10)\quad d\zeta_t^{(1)} = Y_1(\zeta_t^{(1)})\circ dM_t^1,$$

$$(11)\quad d\eta_t^{(1)} = \sum_{j=2}^n (\zeta_{t*}^{(1)})^{-1} Y_j(\eta_t^{(1)})\circ dM_t^j .$$

The solution of (10) is written as $\zeta_t^{(1)}(x) = \mathrm{Exp}\, M_t^1 Y_1(x)$. By Theorem 3.3 it holds $\xi_t = \zeta_t^{(1)} \circ \eta_t^{(1)}$. Furthermore, we have by Theorem 2.2,

$$(\zeta_{t*}^{(1)})^{-1} = I + \int_0^t (\zeta_{s*}^{(1)})^{-1} \mathrm{ad}_{Y_1} \circ dM_s^1 .^{1)}$$

This leads $(\zeta_{t*}^{(1)})^{-1} = \exp M_t^1 \mathrm{ad}_{Y_1}$.

With vector notations $Y = (Y_1,\dots,Y_n)$, $\hat{M}_t = (0, M_t^2, \dots, M_t^n)$ and inner product (,), equation (11) is written as

$$(12)\quad d\eta_t^{(1)} = ((\zeta_{t*}^{(1)})^{-1} Y, \circ d\hat{M}_t) = (Y, \exp M_t^1 \mathrm{ad}'_{Y_1} \circ d\hat{M}_t).$$

Define n-vector continuous semimartingale by

$$M_t^{(1)} = \int_0^t \exp M_s^1 \mathrm{ad}'_{Y_1} \circ d\hat{M}_s .$$

Since $\exp M_s^1 \mathrm{ad}'_{Y_1}$ is of the form (9), the first component of the vector $M_t^{(1)}$ is 0. Hence setting $M_t^{(1)} = (0, M_t^{(1)2}, \dots, M_t^{(1)n})$, (12) becomes

$$(13)\quad d\eta_t^{(1)} = (Y, \circ dM_t^{(1)}) = \sum_{j=2}^n Y_j(\eta_t^{(1)}) \circ dM_t^{(1)j} .$$

---

1) The linear map $(\zeta_t^{(1)})^{-1}$ on $\underline{\underline{L}}$ may be regarded as an n×n-matrix since the basis $\{Y_1,\dots,Y_n\}$ is fixed.

We shall next decompose the equation (13). Set

(14) $$d\zeta_t^{(2)} = Y_2(\zeta_t^{(2)})\circ dM_t^{(1)2},$$

(15) $$d\eta_t^{(2)} = \sum_{j=3}^{n} (\zeta_{t*}^{(2)})^{-1} Y_j(\eta_t^{(2)})\circ dM_t^{(1)j}.$$

Then it holds $\eta_t^{(1)} = \zeta_t^{(2)}\circ\eta_t^{(2)}$. We have as before that $\zeta_t^{(2)} = \mathrm{Exp}\, M_t^{(1)2}Y_2$, and equation (15) is expressed as

(16) $$d\eta_t^{(2)} = \sum_{j=3}^{n} Y_j(\eta_t^{(2)})\circ dM_t^{(2)j}.$$

Here $M_t^{(2)}$ is an n-1 vector continuous semimartingale defined by

(17) $$M_t^{(2)} = \int_0^t \exp M_s^{(1)2}\mathrm{ad}'_{Y_2}\circ d\hat{M}_s^{(1)},$$

where $\hat{M}_t^{(1)} = (0, M_s^{(1)3}, \ldots, M_t^{(1)n})$.

We can next decompose $\eta_t^{(2)}$ as $\zeta_t^{(3)}\circ\eta_t^{(3)}$, and repeating the above argument inductively we arrive at

(18) $$\xi_t = \zeta_t^{(1)}\circ\ldots\circ\zeta_t^{(n)}$$
$$= \mathrm{Exp}\, M_t^1 Y_1\circ\mathrm{Exp}\, M_t^{(1)2}Y_2\circ\ \ldots\ \circ\mathrm{Exp}\, M_t^{(n-1)n}Y_n.$$

Clearly $M_t^1, M_t^{(1)2}, \ldots, M_t^{(n-1)n}$ are constructed from $B_t^0, \ldots, B_t^m$ via (i) - (iii) of the theorem.

In case where $\underline{\underline{L}}$ is a nilpotent Lie algebra, matrices ad_{Y_i} are nilpotent. Therfore any component of matrix $\exp(M_s^{(i-1)i}\mathrm{ad}'_{Y_i})$ is a polynomial of $M_t^{(i-1)i}$, not containing exponential functions. Therefore, $M_t^{(i)i+1}$ are constructed from $B_t^0, \ldots, B_t^m$ via operations (i) and (ii) only. The proof is complete.

Remark. In case that the Lie algebra $\underline{\underline{L}}$ is nilpotent, the assertion of the theorem states that $N_t^1, \ldots, N_t^n$ in the expression (8) are linear sums of multiple Wiener integrals of the form

$$\int_0^t \cdots \int_0^{t_{m-1}} \circ dB_{t_1}^{i_1} \circ \cdots \circ dB_{t_m}^{i_m}.$$

4. ITÔ'S FORMULA FOR $\xi_{s,t}$ ACTING ON TENSOR FIELDS

In Section 2, we obtained Itô's formula for $\xi_{s,t}$ acting on vector fields. In this section, we will obtain a similar formula for $\xi_{s,t}$ acting on general tensor fields.

We begin with 1-form. Let ϕ be a diffeomorphism of M and let ϕ_{x*} be the differential of ϕ, which is a linear map from $T_x(M)$ into $T_{\phi(x)}(M)$. We denote the dual map by ϕ_x^*: It is a linear map from $T_{\phi(x)}(M)^*$ (cotangent space) to $T_x(M)^*$ such that $<\theta_{\phi(x)}, \phi_{x*}X_x> = <\phi_x^*\theta_{\phi(x)}, X_x>$ holds for any $\theta_{\phi(x)} \in T_{\phi(x)}(M)^*$ and $X_x \in T_x(M)$. Given a 1-form θ, we denote by $\phi^*\theta$ a 1-form such that $(\phi^*\theta)_x = \phi_x^*\theta_{\phi(x)}$. Then it holds $<\phi^*\theta, X>_x = <\theta, \phi_* X>_{\phi(x)}$. Let X be a complete vector field and let ϕ_t be the one parameter group of transformations generated by X. The Lie derivative of 1-form θ is defined by

$$L_X\theta = \lim_{t\to 0} \frac{1}{t}\{\phi_t^*\theta - \theta\}.$$

The following relation is well known.

$$(1) \qquad <L_X\theta, Y> + <\theta, L_X Y> = X(<\theta, Y>).$$

4.1. Theorem. Assume the same smoothness condition for vector fields $X_0(t), \ldots, X_m(t)$ as that of Theorem 2.1. Then $\xi_{s,t}^*$ satisfies the following formula for any C^2 1-form θ.

$$(2)\qquad \xi^*_{s,t}\theta-\theta=\sum_{k=1}^{m}\int_s^t \xi^*_{s,r}L_{X_k(r)}\theta dB_r^k+\int_s^t \xi^*_{s,r}(\frac{1}{2}\sum_{k=1}^{m}L^2_{X_k(r)}+L_{X_0(r)})\theta dr$$

$$(3)\qquad =\sum_{k=1}^{m}\int_s^t L_{X_k(r)}\xi^*_{r,t}\theta \hat{d}B_r^k+\int_s^t (\frac{1}{2}\sum_{k=1}^{m}L^2_{X_k(r)}+L_{X_0(r)})\xi^*_{r,t}\theta dr.$$

Further $\xi^{-1*}_{s,t}$ satisfies

$$(4)\qquad \xi^{-1*}_{s,t}\theta-\theta=-\sum_{k=1}^{m}\int_s^t L_{X_k(r)}\xi^{-1*}_{s,r}\theta dB_r^k+\int_s^t (\frac{1}{2}\sum_{k=1}^{m}L^2_{X_k(r)}-L_{X_0(r)})\xi^{-1*}_{s,r}\theta dr$$

$$(5)\qquad =-\sum_{k=1}^{m}\int_s^t \xi^{-1*}_{r,t}L_{X_k(r)}\theta \hat{d}B_r^k+\int_s^t \xi^{-1*}_{r,t}(\frac{1}{2}\sum_{k=1}^{m}L^2_{X_k(r)}-L_{X_0(r)})\theta dr.$$

4.2. Theorem. Assume the same smoothness conditions for vector fields $X_0(t),\dots,X_m(t)$ as that of Theorem 2.2. Then $\xi^*_{s,t}$ satisfies the following formula for any C^3 1-form θ.

$$(6)\qquad \xi^*_{s,t}\theta-\theta=\sum_{k=0}^{m}\int_s^t \xi^*_{s,r}L_{X_k(r)}\theta\circ dB_r^k$$

$$(7)\qquad =\sum_{k=0}^{m}\int_s^t L_{X_k(r)}\xi^*_{r,t}\theta\circ \hat{d}B_r^k$$

and

$$(8)\qquad \xi^{-1*}_{s,t}\theta-\theta=-\sum_{k=0}^{m}\int_s^t L_{X_k(r)}\xi^{-1*}_{s,r}\theta\circ dB_r^k$$

$$(9)\qquad =-\sum_{k=0}^{m}\int_s^t \xi^{-1*}_{r,t}L_{X_k(r)}\theta\circ \hat{d}B_r^k.$$

Proof. We shall prove (6) only. Other formulas will be proved by similar methods. From (5) in Section 2 it holds

$$\langle\theta,\xi_{s,t*}Y\rangle_y-\langle\theta,Y\rangle_y=-\sum_{k=0}^{m}\int_s^t \langle\theta,L_{X_k(r)}\xi_{s,r*}Y\rangle_y\circ dB_r^k.$$

Substitute $y=\xi_{s,t}(x)$ and apply Itô's formula. Then we get

$$(10)\qquad \langle\xi^*_{s,t}\theta,Y\rangle_x-\langle\theta,Y\rangle_x=-\sum_{k=0}^{m}\int_s^t \langle\theta,L_{X_k(r)}\xi_{s,r*}Y\rangle_{\xi_{s,r}(x)}\circ dB_r^k.$$

$$+ \sum_{k=0}^{m} \int_s^t X_k(r) \langle \theta, L_{X_k(r)} \xi_{s,r*} Y \rangle_{\xi_{s,r}(x)} \circ dB_r^k$$

$$= \sum_{k=0}^{m} \int_s^t \langle \xi_{s,r}^* L_{X_k(r)} \theta, Y \rangle_x \circ dB_r^k .$$

The last equality follows from (1) and the relation $\langle \phi^*\theta, X \rangle_x = \langle \theta, \phi_* X \rangle_{\phi(x)}$.

A tensor field K of type (p,q) is, by definition, an assignment of a tensor K_x of $T_q^p(x)$ to each point x of M, where

$$T_q^p(x) = T_x(M) \otimes \ldots \otimes T_x(M) \otimes T_x(M)^* \otimes \ldots \otimes T_x(M)^*$$

($T_x(M)$; p times and $T_x(M)^*$; q times). Hence for each x, K_x is a multilinear form on the product space

$$T_x(M)^* \times \ldots \times T_x(M)^* \times T_x(M) \times \ldots \times T_x(M).$$

Thus, for given 1-forms $\theta^1, \ldots, \theta^p$ and vector fileds $Y_1, \ldots, Y_q$,

$$K_x(\theta^1, \ldots, \theta^p, Y_1, \ldots, Y_q) \; (\equiv K_x(\theta_x^1, \ldots, \theta_x^p, Y_{1x}, \ldots, Y_{qx}))$$

is a scalar field. In the sequel, we assume that it is a C^2 function.

Let ϕ be a diffeomorphism of M. Given a tensor field K of type (p,q), we define a tensor field ϕ^*K by the relation

$$(11) \qquad (\phi^*K)_x(\theta^1, \ldots, \theta^p, Y_1, \ldots, Y_q)$$

$$= K_{\phi(x)}(\phi^{*-1}\theta^1, \ldots, \phi^{*-1}\theta^p, \phi_*Y_1, \ldots, \phi_*Y_q).$$

Then if K is a vector field, it holds $\phi^*K = \phi_*^{-1}K$ and if K is a 1-form, it coincides with ϕ^*K defined before.

Remark. The definition of the above ϕ^* is not equal to that of $\tilde{\phi}$ in Kobayashi-Nomizu [19], p. 28. The relation of these is $\tilde{\phi}^{-1} = \phi^*$ or $\tilde{\phi} = (\phi^{-1})^*$.

Let X be a complete vector field and ϕ_t, $t \in (-\infty,\infty)$ be the one parameter group of transformations generated by X. The Lie derivative of tensor field K with respect to X is defined by

$$L_X K = \lim_{t \downarrow 0} \frac{1}{t}\{\phi_t^* K - K\}. \tag{12}$$

If K is a tensor field of type (p,q), then it holds

$$\begin{aligned}(L_X K)_x(\theta^1,\dots,\theta^p,Y_1,\dots,Y_q) &= X(K_x(\theta^1,\dots,\theta^p,Y_1,\dots,Y_q)) \\ &- \sum_{k=1}^{p} K_x(\theta^1,\dots,L_X\theta^k,\dots,\theta^p,Y_1,\dots,Y_q) \\ &- \sum_{\ell=1}^{q} K_x(\theta^1,\dots,\theta^p,Y_1,\dots,L_X Y_\ell,\dots,Y_q).\end{aligned} \tag{13}$$

We can now state Itô's formula for $\xi_{s,t}^*$ acting on tensor field.

4.3. Theorem. Assume that coefficients $X_1,\dots,X_m$ of the equation are $C^{4,\alpha}$ vector fields for some $\alpha > 0$ and X_0 is a $C^{3,\alpha}$ vector field for the same $\alpha > 0$. Then $\xi_{s,t}^*$ satisfies the following formula for any C^2 tensor field K

$$\xi_{s,t}^* K - K = \sum_{k=1}^{m}\int_s^t \xi_{s,r}^* L_{X_k(r)} K dB_r^k + \int_s^t \xi_{s,r}^*(\frac{1}{2}\sum_{k=1}^{m} L_{X_k(r)}^2 + L_{X_0(r)})K dr \tag{14}$$

$$= \sum_{k=1}^{m}\int_s^t L_{X_k(r)}\xi_{r,t}^* K \hat{d}B_r^k + \int_s^t (\frac{1}{2}\sum_{k=1}^{m} L_{X_k(r)}^2 + L_{X_0(r)})\xi_{r,t}^* K dr. \tag{15}$$

Assume further that coefficients $X_1,\dots,X_m$ are $C^{5,\alpha}$ for some $\alpha > 0$ and X_0 is $C^{4,\alpha}$ for the same α. Let K be a C^3 tensor field of type (p,q). Then it holds

$$\xi_{s,t}^* K - K = \sum_{k=0}^{m}\int_s^t \xi_{s,r}^* L_{X_k(r)} K \circ dB_r^k \tag{16}$$

$$= \sum_{k=0}^{m}\int_s^t L_{X_k(r)}\xi_{r,t}^* K \circ \hat{d}B_r^k. \tag{17}$$

Proof. We shall first prove (16). We will write $\xi_{s,t}$ as ξ_t for convenience. Apply Itô's formula to the multilinear form K. Then, using Theorem 2.2 and 4.2,

$$K_x(\xi_t^{*-1}\theta^1,\dots,\xi_t^{*-1}\theta^p,\xi_{t*}Y_1,\dots,\xi_{t*}Y_q) - K_x(\theta^1,\dots,\theta^p,Y_1,\dots,Y_q)$$

$$= -\sum_{j=0}^{m}\{\sum_{k=1}^{p}\int_s^t K_x(\xi_r^{*-1}\theta^1,\dots,L_{X_j(r)}\xi_r^{*-1}\theta^k,\dots,\xi_{r*}Y_1,\dots,\xi_{r*}Y_q)\circ dB_r^j$$

$$-\sum_{\ell=1}^{q}\int_s^t K_x(\xi_r^{*-1}\theta^1,\dots,\xi_{r*}Y_1,\dots,L_{X_j(r)}\xi_{r*}Y_\ell,\dots,\xi_{r*}Y_q)\circ dB_r^j\}.$$

Set

$$F_t(x) = K_x(\xi_t^{*-1}\theta^1,\dots,\xi_t^{*-1}\theta^p,\xi_{t*}Y_1,\dots,\xi_{t*}Y_q)$$

and apply Itô's formula to $F_t(\xi_t(x))$. Then

$$\text{(18)}\qquad F_t(\xi_t(x)) - F_s(x)$$

$$= \sum_{j=0}^{m}\{\int_s^t X_j(r)F_r(\xi_r(x))\circ dB_r^j$$

$$-\sum_{k=1}^{p}\int_s^t K_{\xi_r(x)}(\xi_r^{*-1}\theta^1,\dots,L_{X_j(r)}\xi_r^{*-1}\theta^k,\dots,\xi_{r*}Y_1,\dots,\xi_{r*}Y_q)\circ dB_r^j$$

$$-\sum_{\ell=1}^{q}\int_s^t K_{\xi_r(x)}(\xi_r^{*-1}\theta^1,\dots,\xi_{r*}Y_1,\dots,L_{X_j(r)}\xi_{r*}Y_\ell,\dots,\xi_{r*}Y_q)\circ dB_r^j\}.$$

Noting the relation (13), we wee that the right hand side of (18) is

$$\sum_{j=0}^{m}\int_s^t \xi_r^* L_{X_j(r)}K(\theta^1,\dots,\theta^p,Y_1,\dots,Y_q)\circ dB_r^j .$$

We have thus proved (16).

Now the martingale part of $\xi_{s,r}^* L_{X_j(r)}K$ is

$$\sum_{k=1}^{m}\int_s^r \xi_{s,u}^* L_{X_k(u)}L_{X_j(u)}K\,dB_u^k$$

by the above formula (16). Therfore we have

$$\int_s^t \xi_{s,r}^* L_{X_j(r)}K\circ dB_r^j = \int_s^t \xi_{s,r}^* L_{X_j(r)}K\,dB_r^j + \frac{1}{2}\int_s^t \xi_{s,r}^* L_{X_j(r)}^2 K\,dr.$$

The formula (14) follows from this.

It is possible to prove (15) under a weaker condition as is stated in the theorem, by applying Theorem 2.1 and 4.1. Details are omitted.

5. SUPPORTS OF STOCHASTIC FLOW OF DIFFEOMORPHISMS

The stochastic flow of diffeomorphisms generated by SDE on the manifold M does not take all the diffeomorphism of M in general. The possible subset of diffeomorphisms that the flow can take depends on the structure of the Lie group generated by vector fields $X_0,\dots,X_m$ defining the SDE.

Assume that $X_0,\dots,X_m$ defining the SDE do not depend on t and that each is a complete vector field. In case that $X_0,\dots,X_m$ are commutative, i.e., $[X_i,X_j] = 0$, $i,j=0,\dots,m$, we have seen in Example 3.5 that the solution $\xi_{s,t}$ takes values in the commutative Lie group consisting of diffeomorphisms $\mathrm{Exp}\ t_0X_0\circ\dots\circ\mathrm{Exp}\ t_mX_m$. In case that the Lie algebra $\underline{\underline{L}}$ generated by $X_0,\dots,X_m$ is solvable, and finite dimensional, then $\xi_{s,t}$ takes values in the finite dimensional Lie group whose element is written as $\mathrm{Exp}\ s_1Y_1\circ\dots\circ\mathrm{Exp}\ s_nY_n$, where $\{Y_1,\dots,Y_n\}$ is a basis of $\underline{\underline{L}}$ introduced in Section 3. Generally, if the Lie algebra generated by $X_0,\dots,X_m$ is finite dimensional, then we can describe the possible diffeomorphisms that $\xi_{s,t}$ can take.

In order to see this, we need a fact from the differential geometry. Let $\underline{\underline{L}}$ be a finite dimensional Lie algebra whose elements are complete vector fields. Then there exists a Lie group G with properties (i) - (iii) below.

(i) G is a Lie transformation group of M, i.e., there exists a C^∞-map ϕ from the product manifold $G\times M$ into M such that

(a) for each g, $\phi(g,\cdot)$ is a diffeomorphism of M and

(b) $\phi(e,\cdot) = \text{identity}$, $\phi(gh,\cdot) = \phi(g,\phi(h,\cdot))$ for any g,h of G.

(ii) The map $g \longrightarrow \phi(g,\cdot)$ is an isomorphism from G into the group of all diffeomorphisms of M.

(iii) Let $\underline{\underline{G}}$ be the Lie algebra of G (= right invariant vector fields on G). For any X of $\underline{\underline{L}}$ there exists $\hat{X}$ of $\underline{\underline{G}}$ such that

$$(1) \qquad \hat{X}(f\circ\phi_x)(g) = Xf(\phi(g,x))$$

holds for any C^∞-function f on M. Here $f\circ\phi_x$ is a C^∞-function on G such that $f\circ\phi_x(g) = f\circ\phi(g,x)$. We will call G the <u>Lie(transformation) group associated with</u> $\underline{\underline{L}}$.

5.1. <u>Theorem.</u> Suppose that $X_0,\dots,X_m$ defining the SDE are complete C^∞-vector fields not depending on t and that the Lie algebra $\underline{\underline{L}}$ generated by $X_0,\dots,X_m$ is of finite dimension. Let G be the Lie group associated with the Lie algebra $\underline{\underline{L}}$. Then the solution $\xi_{s,t}$ takes values in the set of diffeomorphisms $\{\phi(g,\cdot)\ ;\ g\in G\}$.

<u>Proof.</u> We shall define an SDE on G. Let $\hat{X}_0,\dots,\hat{X}_m$ be right invariant vector fields of G related to $X_0,\dots,X_m$ by the formula (1), respectively. Consider an SDE on G:

$$d\hat{\xi}_t = \sum_{k=0}^{m} \hat{X}_k(\hat{\xi}_t)\circ dB_t^k .$$

The solution $\hat{\xi}_t(e)$ starting at e at time 0 is called a Brownian motion on G. It is actually conservative. Indeed, for an open neighborhood of e, define a sequence of stopping times T_n by induction:

$$T_1 = \inf\{t > 0\ ;\ \hat{\xi}_t(e) \notin U\} \quad (= \infty \ \text{ if } \{\dots\} = \phi)$$

$$T_n = \inf \{t > T_{n-1};\ \hat{\xi}^{-1}_{T_{n-1}}(e)\hat{\xi}_t(e) \notin U\} \quad (= \infty \text{ if } \{\ldots\} = \phi).$$

Then $T_1, T_2 - T_1, \ldots, T_n - T_{n-1}$ are independent identically distributed random variables such that $E[T_1] > 0$. Therefore T_n diverges to $+\infty$ as $n \to \infty$ by the law of large numbers. This shows that $\hat{\xi}_t(e)$ is conservative.

Set now $\xi_t(x,\omega) = \phi(\hat{\xi}_t(e,\omega), x)$. Then we have

$$\begin{aligned} f(\xi_t(x)) &= f\circ\phi(\hat{\xi}_t(e), x) \\ &= f\circ\phi(e,x) + \sum_{k=0}^{m} \int_0^t \hat{X}_k(f\circ\phi_x)(\hat{\xi}_s(e))\circ dB_s^k \\ &= f(x) + \sum_{k=0}^{m} \int_0^t X_k f(\xi_s(x))\circ dB_s^k. \end{aligned}$$

Therefore $\xi_t(x)$ is the solution of the given SDE. The proof is complete.

If the Lie algebra $\underline{\underline{L}}$ is of infinite dimension, the Lie transformation group associated with $\underline{\underline{L}}$ is not well defined except the case that the underlying manifold M is compact. So we shall look the support of diffeomorphisms from another aspect making use of Itô's formula for tensor fields. We first establish the following.

5.2. Theorem. Let K be a C^2 tensor field of type (p,q). Under the same smoothness condition for $X_0,\ldots,X_m$ as that of Theorem 4.3, it holds $\xi^*_{s,t}K = K$ a.s. if and only if $L_{X_k(r)}K = 0$, $r \in [s,t]$, $k=0,\ldots,m$.

Proof. If $L_{X_k(r)}K = 0$, $r \in [s,t]$, $k=0,\ldots,m$, the relation $\xi^*_{s,t}K = K$ is clear from Theorem 4.3. Conversely suppose $\xi^*_{s,t}K = K$ is satisfied. Then it holds from (14) of Section 4,

$$(2)\quad \sum_{k=1}^{m}\int_s^t (\xi^*_{s,r}L_{X_k(r)}K)(\theta^1,\dots,\theta^p,Y_1,\dots,Y_q)dB_r^k$$

$$= -\int_s^t \xi^*_{s,r}(\frac{1}{2}\sum_{k=1}^{m} L^2_{X_k(r)} + L_{X_0(r)})K(\theta^1,\dots,\theta^p,Y_1,\dots,Y_q)dr.$$

Since the left hand side is a continuous (local) martingale and the right hand side is a process of the bounded variation, both should be 0. This implies that the quadratic variation of the right hand side is 0, i.e.,

$$\sum_{k=1}^{m}\int_s^t |\xi^*_{s,r}L_{X_k(r)}K(\theta^1,\dots,\theta^p,Y_1,\dots,Y_q)|^2 dr = 0$$

holds for any $\theta^1,\dots,\theta^p \in T_x(M)^*$ and $Y_1,\dots,Y_q \in T_x(M)$. This implies $\xi^*_{s,r}L_{X_k(r)}K = 0$, so that we have $L_{X_k(r)}K = 0$, $r \in [s,t]$, $k=1,\dots,m$. Then $L_{X_0(r)}K$ should be 0, too, since the right hand side of (2) is also 0.

As an application of the above theorem, we give here four examples, characterizing the support of diffeomorphisms.

5.3. Example. Let M be a Riemannian manifold with the metric g. A vector field X on M is called a Killing vector field or an infinitesimal motion if

$$X(g(Y,Z)) = g([X,Y],Z) + g(Y,[X,Z])$$

is satisfied for any vector fields Y, Z. Considering that g is a tensor of type (0,2), the above is equivalent to $L_X g = 0$ by the formula (13) in Section 4. Now a diffeomorphism ϕ of M is called a motion of M if $g(Y,Z) = g(\phi_*Y,\phi_*Z)$ (isometry) is satisfied for any vector fields Y and Z. It is equivalent to $\phi^*g = g$.

Now let $\xi_{s,t}$ be a stochastic flow of diffeomorphisms determined by an SDE. Then it follows from Theorem 5.2 that $\xi_{s,t}$ is a motion for any $s < t$ a.s., if and only if $X_j(r)$, $r \in [s,t]$, $j=0,\dots,m$ are all Killing vector fields.

5.4. Example. Let Ω be a positive differential form of order d. (volume element). A diffeomorphism ϕ ; $M \longrightarrow M$ is called volume preserving if $\phi^*\Omega = \Omega$. The solution $\xi_{s,t}$ is volume preserving if and only if $L_{X_k(r)}\Omega = 0$ for any $r \in [s,t]$ and $k=0,\dots,m$.

If $X_0(r),\dots,X_m(r)$ are Killing vector fields, then the solution $\xi_{s,t}$ is volume preserving since it is a motion.

5.5. Example. Let M be a manifold where a linear connection is defined. A diffeomorphism ϕ is called affine if it maps each parallel vector field along each curve τ of M into a parallel vector field along the curve $\phi(\tau)$. A vector field X is called an infinitesimal affine transformation of M if for each $x \in M$, a local one parameter group of local transformations ϕ_t of a neighborhood U is affine for each t. We can prove similarly as the above that $\xi_{s,t}$ is an affine transformation for any $s < t$ a.s. if and only if $X_0(r),\dots,X_m(r)$, $r \in [s,t]$, $k=0$, $\dots,m$ are all infinitesimal affine transformations.

5.6. Example. Let M be a complex manifold. A map $\phi : M \longrightarrow M$ is called holomorphic if $f \circ \phi$ is a holomorphic function on M for any holomorphic function f on M. Let J be the almost complex structure. Then a map $\phi : M \longrightarrow M$ is holomorphic if and only if $J(\phi_* Y) = \phi_* JY$ holds for any vector field Y.

Let X be a real vector field. It is called <u>analytic</u> if it is written as

$$X = \sum_{i=1}^{d} X^i \frac{\partial}{\partial z_i} + \sum_{i=1}^{d} \bar{X}^i \frac{\partial}{\partial \bar{z}_i},$$

where $X^i(z)$, $i=1,\dots,d$ are holomorphic functions and $(z_1,\dots,z_d)$, $z_i = x_i + \sqrt{-1}y_i$ is a holomorphic coordinate. Here $\bar{X}^i$ means the complex conjugate of X^i. Then X is analytic if and only if $JL_X = L_X J$.

Let X be a complete vector field and let ϕ_t be the one parameter group of transformations generated by X. Then it is known that ϕ_t is holomorphic for each t if and only if X is an analytic vector field. We will show the similar fact for SDE.

5.7. <u>Theorem.</u> Suppose that the solution $\xi_{s,t}$ defines a flow of diffeomorphisms. It is holomorphic for any $s < t$ a.s. if and only if $X_0,\dots,X_m$ are analytic vector fields.

<u>Proof.</u> By Theorem 2.2, it holds

$$(3)\quad J(\xi_{s,t*}Y) = JY - \sum_{k=0}^{m}\int_s^t JL_{X_k(r)}\xi_{s,r*}Y\circ dB_r^k$$

$$(4)\quad \xi_{s,t*}JY = JY - \sum_{k=0}^{m}\int_s^t L_{X_k(r)}\xi_{s,r*}JY\circ dB_r^k .$$

If $\xi_{s,t}$ is holomorphic, then $\xi_{s,t*}JY = J\xi_{s,t*}Y$, so that we have

$$\sum_{k=0}^{m}\int_s^t (JL_{X_k(r)} - L_{X_k(r)}J)\xi_{s,r*}Y\circ dB_r^k = 0.$$

This implies $JL_{X_k(r)} = L_{X_k(r)}J$, $r \in [s,t]$, $k=0,\dots,m$. (See the proof of Theorem 5.2). Therefore $X_k(r)$, $r \in [s,t]$, $k=0,\dots,m$ are analytic vector fields.

Conversely suppose that $X_k(r)$ are analytic. Set $\Phi_{s,t}(Y) =$

$J(\xi_{s,t*}Y) - \xi_{s,t*}JY$. Then (3), (4) and the relation $L_{X_k(r)}J = JL_{X_k(r)}$ implies

$$\Phi_{s,t}(Y) = -\sum_{k=0}^{m}\int_s^t L_{X_k(r)}\Phi_{s,r}(Y)\circ dB_r^k .$$

The above linear SDE has a unique solution $\Phi_{s,t}(Y)$, which should be identically 0. This proves that $\xi_{s,t}$ is holomorphic.

6. ITÔ'S FORMULA FOR STOCHASTIC PARALLEL DISPLACEMENT OF TENSOR FIELDS

This section concerns an Itô's formula for stochastic parallel displacement of tensor field along curves governed by a stochastic differential equation. Stochastic parallel displacement along Brownian curves on Riemannian manifold was introduced by K. Itô [17], [18]. Our definition is close to [18].

We shall prepare some facts on parallel displacement needed later. Let M be connected, paracompact C^∞-manifold of dimension d where an affine connection is defined. Suppose we are given a one parameter family of into homeomorphisms ϕ_t ; $M \longrightarrow M$, $t \in [0,a]$, such that $\lim_{t\downarrow 0} \phi_t(x) = x$ for any x. Let u_t be a tangent vector belonging to $T_{\phi_t(x)}(M)$. We denote by u_0 the parallel displacement of u_t along the curve $\phi_s(x)$, $0 \le s \le t$ from the point $\phi_t(x)$ to x. Then the map $\pi_{t,x} : u_t \longrightarrow u_0$ defines an isomorphism from $T_{\phi_t(x)}(M)$ to $T_x(M)$. Now, given a vector field Y on M, we define a vector field $\pi_t Y$ by $(\pi_t Y)_x = \pi_{t,x} Y_{\phi_t(x)}$, $\forall x \in M$. The one parameter family of vector fields $\pi_t Y$, $t \ge 0$ satisfies

$$(1) \quad \frac{d}{dt}(\pi_t Y)_x = (\pi_t \nabla_{\dot\phi} Y)_x, \qquad \forall x \in M,$$

where $\nabla_{\dot\phi} Y$ is the covariant derivative of Y along the curve ϕ_t. If $\phi_t(x)$ is a solution of an ordinary differential equation:

$$\dot\phi_t = \sum_{j=1}^{r} X_j(\phi_t)v_j(t), \qquad \phi_0 = x,$$

where $X_1,\dots,X_r$ are vector fields on M and $v_1(t),\dots,v_r(t)$ are smooth scalar functions, then equation (1) becomes

$$(2)\qquad \frac{d}{dt}(\pi_t Y)_x = \sum_{j=1}^{r} (\pi_t \nabla_{X_j} Y)_x v_j(t).$$

The inverse map $\pi_{t,x}^{-1}$ defines another vector field $\pi_t^{-1}Y$ by $(\pi_t^{-1}Y)_x = \pi^{-1}_{t,\phi_t^{-1}(x)} Y_{\phi_t^{-1}(x)}$, which is the parallel displacement of $Y_{\phi_t^{-1}(x)}$ along the curve ϕ_s, $0 \le s \le t$ from $\phi_t^{-1}(x)$ to x. It holds

$$(3)\qquad \frac{d}{dt}(\pi_t^{-1}Y)_x = -\sum_{j=1}^{r} (\nabla_{X_j} \pi_t^{-1}Y)_x v_j(t).$$

Let $T_x(M)^*$ be the cotangent space at x (dual of $T_x(M)$). The dual $\pi^*_{t,x}$ is an isomorphism from $T_x(M)^*$ to $T_{\phi_t(x)}(M)^*$ such that $\langle\pi^*_{t,x}\theta, Y\rangle = \langle\theta, \pi_{t,x}Y\rangle$ holds for any $\theta \in T_x(M)^*$ and $Y \in T_{\phi_t(x)}(M)$. Given a 1-form θ (covariant vector field), $\pi_t^*\theta$ is a 1-form defined by $(\pi_t^*\theta)_x = \pi^*_{t,\phi_t^{-1}(x)}\theta_{\phi_t^{-1}(x)}$. The 1-form $\pi_t^{*-1}\theta$ is defined similarly.

The <u>parallel displacement</u> $\pi_t K$ <u>of the tensor field</u> K <u>of type</u> (p,q) <u>along the curve</u> ϕ_s is defined by the relation

$$(4)\qquad (\pi_t K)_x(\theta^1,\dots,\theta^p,Y_1,\dots,Y_q) = K_{\phi_t(x)}(\pi_t^*\theta^1,\dots,\pi_t^*\theta^p,\pi_t^{-1}Y_1,\dots,\pi_t^{-1}Y_q).$$

If K is a vector field, it coincides clearly with the parallel displacement mentioned above. If K is a 1-form, it coincides with $\pi_t^{*-1}K$. Hence we can write the above relation as

$$(4')\qquad (\pi_t K)_x(\theta^1,\dots,\theta^p,Y_1,\dots,Y_q) = K_{\phi_t(x)}(\pi_t^{-1}\theta^1,\dots,\pi_t^{-1}\theta^p,\pi_t^{-1}Y_1,\dots,\pi_t^{-1}Y_q).$$

Let X be a complete vector field and ϕ_t, the one parameter group of transformations generated by X. Then the <u>covariant derivative of the tensor field</u> K is defined by

$$(5) \qquad (\nabla_X K)_x(\theta^1,\dots,\theta^p,Y_1,\dots,Y_q) = \frac{d}{dt}(\pi_t K)_x(\theta^1,\dots,\theta^p,Y_1,\dots,Y_q)\Big|_{t=0}.$$

The following relation is easily checked.

$$(6) \qquad (\nabla_X K)_x(\theta^1,\dots,\theta^p,Y_1,\dots,Y_q)$$
$$= X(K_x(\theta^1,\dots,\theta^p,Y_1,\dots,Y_q))$$
$$- \sum_{k=1}^{p} K_x(\theta^1,\dots,\nabla_X\theta^k,\dots,\theta^p,Y_1,\dots,Y_q)$$
$$- \sum_{\ell=1}^{q} K_x(\theta^1,\dots,\theta^p,Y_1,\dots,\nabla_X Y_\ell,\dots,Y_q).$$

Now let $\xi_{s,t}(x)$ be the solution of the SDE

$$d\xi_t = \sum_{k=0}^{m} X_k(t,\xi_t)\circ dB_t^k,$$

where $X_1(t),\dots,X_m(t)$ are $C^{4,\alpha}$ vector fields for some $\alpha > 0$ and $X_0(t)$ is a $C^{3,\alpha}$ vector field. We shall assume that the solution is conservative. In the sequel we shall define the parallel displacement of a vector field from $\xi_{s,t}(x)$ to x along the curve $\xi_{s,r}(x)$, $r \in [s,t]$. Since the curve $\xi_{s,r}(x)$, $s \le r \le t$ are not smooth a.s., the definition of the parallel displacement mentioned above is not applied directly. We shall define the stochastic parallel displacement following the idea of Itô [18].

A stochastic analogue of equation (2) is this:

$$(7) \qquad (\pi_{s,t}Y)_x = Y_x + \sum_{k=0}^{m}\int_s^t (\pi_{s,r}\nabla_{X_k(r)}Y)_x \circ dB_r^k, \qquad \forall x \in M.$$

Here, $\pi_{s,t}$ is a stochastic linear map acting on the space of vector fields such that $\pi_{s,t}(fY)_x = f(\xi_{s,t}(x))(\pi_{s,t}Y)_x$ for scalar function f. The existence and uniqueness of the solution $\pi_{s,t}$ can be shown as follows. Let $(x_1,\ldots,x_d)$ be a local coordinate. Set $\partial_j = \frac{\partial}{\partial x_j}$. Then equation (7) is written as

$$(8)\quad (\pi_{s,t}\partial_j)_x = (\partial_j)_x + \sum_k \sum_{\alpha,\ell} \int_s^t X_k^\alpha(\xi_{s,r}(x))\Gamma_{\alpha j}^\ell(\xi_{s,r}(x))(\pi_{s,r}\partial_\ell)_x \circ dB_r^k$$

$$j=1,\ldots,d,$$

where $X_k = \sum_\alpha X_k^\alpha \partial_\alpha$ and $\Gamma_{\alpha j}^\ell$ is the Christoffel symbol. It may be considered as an equation on the tangent space $T_x(M)$. It has clearly a unique solution $(\pi_{s,t}\partial_j)_x$, $j=1,\ldots,d$ for each x. Define $(\pi_{s,t}Y)_x = \sum_j Y^j(\xi_{s,t}(x))(\pi_{s,t}\partial_j)_x$ if $Y = \sum Y^j \partial_j$. Then it is the desired solution of (7).

We shall call $(\pi_{s,t}Y)_x$ the <u>parallel displacement of</u> $Y_{\xi_{s,t}(x)}$ <u>along the curve</u> $\xi_{s,r}(x)$, $s \le r \le t$. We can show similarly as the proof of Theorem 3.1, Chapter II that $(\pi_{s,t}Y)_x$ is a C^3-vector field provided that Y is a C^3 vector field.

6.1. <u>Theorem.</u> The solution $\pi_{s,t}$ of (7) satisfies the following backward equation

$$(9)\quad (\pi_{s,t}Y)_x = Y_x + \sum_{k=0}^m \int_s^t (\nabla_{X_k(r)}\pi_{r,t}Y)_x \circ \hat{d}B_r^k .$$

<u>Proof.</u> Let us introduce the matrix valued process $\Pi_{s,t} = (\pi_{s,t}^{ij})$ and $P_k = (P_k^{j\ell})$ by

$$(\pi_{s,t}\partial_i)_x = \sum_j \pi_{s,t}^{ij}(\partial_j)_x, \qquad P_k^{j\ell}(t,x) = \sum_\alpha X_k^\alpha(t,x)\Gamma_{\alpha j}^\ell(x).$$

From (8), the matrix $\Pi_{s,t}$ satisfies

$$(10) \qquad \Pi_{s,t}(x) = I + \sum_{k=0}^{m}\int_s^t P_k(r,\xi_{s,r}(x))\Pi_{s,r}(x)\circ dB_r^k .$$

We denote by $\Pi_{s,t}(x,A)$ the solution of (10) where the initial state I is replaced by the matrix A. Then the joint process $(\xi_{s,t}(x),\Pi_{s,t}(x,A))$ is governed by a system of SDE on the product manifold $M\times GL(d)$, where $GL(d)$ is the Lie group consisting of all non-singular $d\times d$-matrices. Set $F(x,A) = A$ and apply Itô's backward formula to $\Pi_{s,t}(x,A)$. Then after a simple computation, we arrive at

$$\Pi_{s,t}(x,A) = A + \sum_{k=0}^{m}\int_s^t \{\Pi_{r,t}(x,A)P_k(r,x) + (X_k(r)\Pi_{r,t})(x,A)\}\circ \hat{d}B_r^k .$$

In case $A = I$, $\Pi_{s,t}(x) \equiv \Pi_{s,t}(x,I)$ satisfies

$$\Pi_{s,t}(x) = I + \sum_{k=0}^{m}\int_s^t \{\Pi_{r,t}(x)P_k(r,x) + (X_k(r)\Pi_{r,t})(x)\}\circ \hat{d}B_r^k .$$

Then the component $\pi_{s,t}^{ij}$ satisfies

$$\pi_{s,t}^{ij} = \delta_{ij} + \sum_{k=0}^{m}\int_s^t \sum_\alpha X_k^\alpha(r)\{\sum_\ell \Gamma_{\alpha\ell}^j \pi_{r,t}^{i\ell} + \partial_\alpha(\pi_{r,t}^{ij})\}\circ \hat{d}B_r^k .$$

This proves

$$(\pi_{s,t}\partial_i)_x = (\partial_i)_x + \sum_{k=0}^{m}\int_s^t (\nabla_{X_k(r)}\pi_{r,t}\partial_i)_x \circ \hat{d}B_r^k .$$

The formula (9) follows immediately.

We will denote the linear map $Y_{\xi_{s,t}(x)} \longrightarrow (\pi_{s,t}Y)_x$ as $\pi_{s,t,x}$. Then it is an isomorphism from $T_{\xi_{s,t}(x)}(M)$ to $T_x(M)$. Given a vector field Y, we denote by $(\pi_{s,t}^{-1}Y)_x$ the stochastic parallel displacement of Y along $\xi_{s,r}$, $s \le r \le t$ from $\xi_{s,t}^{-1}(x)$ to x, which is defined by $(\pi_{s,t}^{-1}Y)_x = \pi^{-1}_{s,t,\xi_{s,t}^{-1}(x)} Y_{\xi_{s,t}^{-1}(x)}$.

6.2. <u>Theorem.</u> It holds

$$(11)\qquad (\pi_{s,t}^{-1}Y)_x = Y_x - \sum_{k=0}^{m}\int_s^t (\nabla_{X_k(r)}\pi_{s,r}^{-1}Y)_x \circ dB_r^k$$

$$(12)\qquad = Y_x - \sum_{k=0}^{m}\int_s^t (\pi_{r,t}^{-1}\nabla_{X_k(r)}Y)_x \circ d\hat{B}_r^k .$$

<u>Proof.</u> Let $\Sigma_{s,t}(x) = (\sigma_{s,t}^{ij}(x))$ be the inverse matrix of $\Pi_{s,t}(x)$ of (10). Then it holds

$$\pi_{s,t}^{-1}(\partial_j)_x = \sum_{\ell} \sigma_{s,t}^{j\ell}(\xi_{s,t}^{-1}(x))(\partial_\ell)_x .$$

We shall compute $\sigma_{s,t}^{j\ell}(\xi_{s,t}^{-1}(x))$ using Itô's formula. It is easy to see that the inverse $\Sigma_{s,t}$ satisfies

$$\Sigma_{s,t}(x) = I - \sum_{k=0}^{m}\int_s^t \Sigma_{s,r}(x)P_k(r,\xi_{s,r}(x))\circ dB_r^k$$

(c.f. Proof of Theorem 4.4, Chapter II). On the other hand, it holds

$$d_t\xi_{s,t}^{-1} = -\sum_{k=0}^{m} \xi_{s,t*}^{-1}X_k(t)(\xi_{s,t}^{-1})\circ dB_t^k .$$

Consequently,

$$d_t(\Sigma_{s,t}(\xi_{s,t}^{-1})) = (d_t\Sigma_{s,t})(\xi_{s,t}^{-1}) - \sum_{\alpha} \frac{\partial\Sigma_{s,t}}{\partial x_\alpha}(\xi_{s,t}^{-1})(\xi_{s,t*}^{-1}X_k(t))^{\alpha}\circ dB_t^k$$

$$= -\sum_{k=0}^{m} \Sigma_{s,t}(\xi_{s,t}^{-1})P_k(t)\circ dB_t^k$$

$$- \sum_{k=0}^{m} (\xi_{s,t*}^{-1}X_k(t)\Sigma_{s,t})(\xi_{s,t}^{-1})\circ dB_t^k .$$

Therefore $K_{s,t}(x) = \Sigma_{s,t}\circ\xi_{s,t}^{-1}(x)$ satisfies

$$d_tK_{s,t} = -\sum_{k=0}^{m} \{K_{s,t}P_k(t) + X_k(t)K_{s,t}\}\circ dB_t^k .$$

The component $\kappa_{s,t}^{ij} = \sigma_{s,t}^{ij}\circ\xi_{s,t}^{-1}$ is then written as

$$\kappa_{s,t}^{ij} = \delta_{ij} - \sum_{k=0}^{m}\int_s^t \sum_{\alpha} X_k^{\alpha}(\sum_{\beta} \Gamma_{\alpha\beta}^{j}\kappa_{s,r}^{i\beta} + \partial_\alpha(\kappa_{s,r}^{ij}))\circ dB_r^k .$$

This proves (11). The formula (12) can be proved similarly as that of Theorem 6.1.

The dual $\pi^*_{s,t}$ of $\pi_{s,t}$ is defined as before. It is acting on 1-forms. It holds $<\pi^*_{s,t}\theta, Y>_{\xi_{s,t}(x)} = <\theta, \pi_{s,t}Y>_x$ for any 1-form θ and vector field Y. We shall obtain equations for $\pi^*_{s,t}\theta$ and $\pi^{*-1}_{s,t}\theta$.

6.3. Theorem. It holds

$$(13)\qquad (\pi^*_{s,t}\theta)_x = \theta_x - \sum_{k=0}^{m}\int_s^t (\nabla_{X_k(r)}\pi^*_{s,r}\theta)_x \circ dB_r^k ,$$

$$(14)\qquad = \theta_x - \sum_{k=0}^{m}\int_s^t (\pi^*_{r,t}\nabla_{X_k(r)}\theta)_x \hat{\circ} dB_r^k .$$

$$(15)\qquad (\pi^{*-1}_{s,t}\theta)_x = \theta_x + \sum_{k=0}^{m}\int_s^t (\pi^{*-1}_{s,r}\nabla_{X_k(r)}\theta)_x \circ dB_r^k ,$$

$$(16)\qquad = \theta_x + \sum_{k=0}^{m}\int_s^t (\nabla_{X_k(r)}\pi^{*-1}_{r,t}\theta)_x \hat{\circ} dB_r^k .$$

Proof is similar to that of Theorem 4.2. It is omitted.

The stochastic parallel displacement of tensor field K of type (p,q) is defined by

$$(17)\qquad (\pi_{s,t}K)_x(\theta^1,\dots,\theta^p,Y_1,\dots,Y_q)$$

$$= K_{\xi_{s,t}(x)}(\pi^*_{s,t}\theta^1,\dots,\pi^*_{s,t}\theta^p,\pi^{-1}_{s,t}Y_1,\dots,\pi^{-1}_{s,t}Y_q).$$

6.4. Theorem. It holds

$$(18)\qquad \pi_{s,t}K = K + \sum_{k=0}^{m}\int_s^t \pi_{s,r}\nabla_{X_k(r)}K \circ dB_r^k$$

$$(19)\qquad = K + \sum_{k=1}^{m}\int_s^t \pi_{s,r}\nabla_{X_k(r)}K dB_r^k + \int_s^t \pi_{s,r}(\frac{1}{2}\sum_{k=1}^{m}\nabla^2_{X_k(r)} + \nabla_{X_0(r)})K dr$$

$$(20) \qquad = K + \sum_{k=0}^{m} \int_s^t \nabla_{X_k(r)} \pi_{r,t} K \circ \hat{d}B_r^k$$

$$(21) \qquad = K + \sum_{k=1}^{m} \int_s^t \nabla_{X_k(r)} \pi_{r,t} K \hat{d}B_r^k + \int_s^t \left(\frac{1}{2} \sum_{k=1}^{m} \nabla^2_{X_k(r)} + \nabla_{X_0(r)}\right) \pi_{r,t} K dr$$

<u>Proof</u> can be carried over similarly to that of Theorem 4.3, making use of the formula (6) instead of (13) in Section 4. We omit the details.

7. APPLICATION TO PARABOLIC PARTIAL DIFFERENTIAL EQUATIONS

It is a well known fact that some parabolic partial differential equations of the second order can be solved via stochastic method: The solution can be expressed by the expected value of some function of the solution of a suitable SDE. In this section, we will see that Itô's backward formula is helpful for solving some parabolic partial differential equations.

Consider the Cauchy problem of the second order parabolic partial differential equation on $[0,a] \times R^d$:

$$(1) \qquad \frac{\partial u_t(x)}{\partial t} = \frac{1}{2} \sum_{i,j=1}^{d} a_{ij}(t,x) \frac{\partial^2}{\partial x_i \partial x_j} u_t(x) + \sum_{i=1}^{d} b^i(t,x) \frac{\partial}{\partial x_i} u_t(x),$$

where $a_{ij}(t,x)$ and $b_i(t,x)$ are written as

$$(2) \qquad a_{ij}(t,x) = \sum_{k=1}^{m} X_k^i(t,x) X_k^j(t,x),$$

$$b_i(t,x) = X_0^i(t,x),$$

making use of d-vector functions $X_k(t,x) = (X_k^1(t,x),\dots,X_k^d(t,x))$. We shall construct the solution of equation (1) by a probabilistic method.

Consider Itô's backward SDE on R^d:

$$(3) \qquad \hat{d}\hat{\xi}_s = -X_0(s,\hat{\xi}_s)ds - \sum_{k=1}^{m} X_k(s,\hat{\xi}_s)\hat{d}B_s^k$$

We denote by $\hat{\xi}_{s,t}(x)$, $s < t$ the solution of equation (3) with the terminal condition $\hat{\xi}_t = x$, i.e.

$$\hat{\xi}_{s,t}(x) = x - \int_s^t X_0(r,\hat{\xi}_{r,t}(x))dr - \sum_{k=1}^m \int_s^t X_k(r,\hat{\xi}_{r,t}(x))\hat{d}B_r^k .$$

7.1. <u>Theorem.</u> Suppose that coefficients $X_0,\ldots,X_m$ of equation (3) are $C_g^{2,\alpha}$ functions for some $\alpha > 0$ and their derivatives up to the second order are bounded. Let f be a $C_g^{2,\alpha}$ function such that $|f|$, $|f'_{x_i}|$, $|f''_{x_i x_j}|$ are bounded. For a given t_0, define $u_t(x)$, $t \geq t_0$ by

(4) $\quad u_t(x) = E[f(\hat{\xi}_{t_0,t}(x))]$.

Then it is a C^1 function of t, $C_g^{2,\alpha}$ function of x. Further it is a solution of (2) with the initial condition $\lim_{t \downarrow t_0} u_t(x) = f(x)$.

<u>Proof.</u> The map $\hat{\xi}_{s,t}$; $R^d \longrightarrow R^d$ is $C^{2,\beta}$ for any $\beta < \alpha$. Hence $E[f(\hat{\xi}_{t_0,t}(x))]$ is a C^2 function of x. It holds

(5) $$\partial_i\partial_j E[f(\hat{\xi}_{s,t}(x))] = \sum_{k,\ell} E[\partial_\ell \partial_k f(\hat{\xi}_{s,t}(x))\partial_i \hat{\xi}^\ell_{s,t}(x)\partial_j \hat{\xi}^k_{s,t}(x)] + \sum_k E[\partial_k f(\hat{\xi}_{s,t}(x))\partial_i\partial_j \hat{\xi}^k_{s,t}(x)].$$

We shall estimate $\partial_i\partial_j E[f(\hat{\xi}_{s,t}(x))] - \partial_i\partial_j E[f(\hat{\xi}_{s,t}(y)]$. It holds

(6) $$|E[\partial_k f(\hat{\xi}_{s,t}(x))\partial_i\partial_j \hat{\xi}^k_{s,t}(x)] - E[\partial_k f(\hat{\xi}_{s,t}(y))\partial_i\partial_j \hat{\xi}^k_{s,t}(y)]|$$

$$\leq E[|\partial_k f(\hat{\xi}_{s,t}(x)) - \partial_k f(\hat{\xi}_{s,t}(y))||\partial_i\partial_j \hat{\xi}^k_{s,t}(x)|] + E[|\partial_i\partial_j \hat{\xi}^k_{s,t}(x) - \partial_i\partial_j \hat{\xi}^k_{s,t}(y)||\partial_k f(\hat{\xi}_{s,t}(y))|]$$

$$\leq LE[|\hat{\xi}_{s,t}(x) - \hat{\xi}_{s,t}(y)|^2]^{\frac{1}{2}} E[|\partial_i\partial_j \hat{\xi}^k_{s,t}(x)|^2]^{\frac{1}{2}} + E[|\partial_i\partial_j \hat{\xi}^k_{s,t}(x) - \partial_i\partial_j \hat{\xi}^k_{s,t}(y)|^2]^{\frac{1}{2}} E[|\partial_k f(\hat{\xi}_{s,t}(y))|^2]^{\frac{1}{2}},$$

where L is a Lipschitz constant of $\partial_k f$. We have the estimation

$$E[|\hat{\xi}_{s,t}(x) - \hat{\xi}_{s,t}(y)|^2]^{\frac{1}{2}} \leq C_1|x-y|$$

$$E[|\partial_i\partial_j\hat{\xi}^k_{s,t}(x) - \partial_i\partial_j\hat{\xi}^k_{s,t}(y)|^2]^{\frac{1}{2}} \leq C_2|x-y|^\alpha.$$

See Sections 2 and 3 of Chapter II. Therefore (6) is dominated by $C_3|x-y|^\alpha$ with some positive C_3. We can prove the similar estimation for the first term of the right hand side of (5). Consequently, $\partial_i\partial_j E[f(\hat{\xi}_{s,t}(x))]$ is Hölder continuous of order α.

Now, let us apply Itô's backward formula to the function $f(x)$ and the process $\hat{\xi}_{s,t}(x)$, interchanging the forward and backward variables. Then,

$$f(\hat{\xi}_{s,t}(x)) - f(x) = \sum_{k=1}^m \int_s^t X_k(r)(f\circ\hat{\xi}_{s,r})(x)dB^k_r + \int_s^t L(r)(f\circ\hat{\xi}_{s,r})(x)dr.$$

Take the expectation of each term. The expectation of the first term of the right hand side is 0. We then have

$$u_t(x) - u(x) = \int_{t_0}^t E[L(r)(f\circ\hat{\xi}_{t_0,r})(x)]dr.$$

Interchanging the order of the integral E and the differential $L(r)$, we get

$$u_t(x) = u(x) + \int_{t_0}^t L(r)u_r(x)dr.$$

This proves immediately that $u_t(x)$ is a C^1 function of t and $\frac{\partial u_t}{\partial t} = L(t)u_t(x)$. The proof is complete.

We can solve the parabolic partial differential equation on the manifold by the same way, obviously. Even a suitable parabolic equation for tensor field can be handled similarly. Let $X_0(t),\dots,X_m(t)$ be $C^{4,\alpha}$ vector fields on M for some $\alpha > 0$. Let $X_0(t)$ be a $C^{3,\alpha}$ vector field on M. Let $\hat{\xi}_{s,t}(x)$ be the solution of the backward equation of the form

$$d\hat{\xi}_s = -\sum_{k=0}^{m} X_k(X,\hat{\xi}_s)\circ d\hat{B}_s^k$$

with the terminal condition $\hat{\xi}_t = x$. For a C^2 tensor field of type (p,q), we define tensor fields K_t with the parameter t;

$$K_{t,x}(\theta^1,\ldots,\theta^p,Y_1,\ldots,Y_q) = E[(\hat{\xi}^*_{t_0,t}K)_x(\theta^1,\ldots,\theta^p,Y_1,\ldots,Y_q)].$$

Then we have the following.

7.2. <u>Theorem.</u> If K has a compact support, then K_t defined above is the solution of the equation.

$$\begin{cases} \dfrac{\partial K_t}{\partial t} = (\frac{1}{2}\sum_{k=1}^{m} L^2_{X_k(t)} + L_{X_0(t)})K_t \\ \lim\limits_{t\downarrow t_0} K_t = K \end{cases}$$

<u>Proof.</u> We shall omit $\theta^1,\ldots,\theta^p,Y_1,\ldots,Y_q$ for simplicity. Apply Itô's formula for tensor field (15) in Theorem 4.3, interchanging the forward and backward variables. Then

$$\hat{\xi}^*_{t_0,t}K - K = \sum_{k=1}^{m}\int_{t_0}^{t} L_{X_k(r)}\hat{\xi}^*_{t_0,r}K\,dB_r^k + \int_{t_0}^{t}(\frac{1}{2}\sum_{k=1}^{m} L^2_{X_k(r)} + L_{X_0})\xi^*_{t_0,r}K\,dr.$$

Taking the expectation of each of the above, we get

$$\begin{aligned} K_t - K &= \int_{t_0}^{t} E[(\frac{1}{2}\sum_{k=1}^{m} L^2_{X_k(r)} + L_{X_0})\xi^*_{t_0,r}K)]dr \\ &= \int_{t_0}^{t} (\frac{1}{2}\sum_{k=1}^{m} L^2_{X_k(r)} + L_{X_0})K_r\,dr. \end{aligned}$$

This proves the theorem.

Let $\hat{\pi}_{s,t}K$ be the parallel displacement of the tensor field K along the path $\{\hat{\xi}_{r,t}, s \le r \le t\}$. We shall define an another tensor field $\tilde{K}_t$ with parameter t by $\tilde{K}_t = E[\hat{\pi}_{t_0,t}K]$, i.e.,

$$\tilde{K}_t(\theta^1,\dots,\theta^p,Y_1,\dots,Y_q) = E[\hat{\pi}_{t_0,t}K(\theta^1,\dots,\theta^p,Y_1,\dots,Y_q)].$$

Then we have the following.

7.3. Theorem. If K is of compact support, then $\tilde{K}_t$ defined above is the solution of the following equation.

$$\begin{cases} \dfrac{\partial \tilde{K}_t}{\partial t} = (\frac{1}{2}\sum_{k=1}^m \nabla^2_{X_k(t)} + \nabla_{X_0(t)})\tilde{K}_t \\ \lim_{t \downarrow t_0} \tilde{K}_t = K. \end{cases}$$

Proof is carried over similarly as Theorem 7.2, applying Itô's formula (Theorem 6.4) for the parallel displacement.

BIBLIOGRAPHICAL NOTES

Itô's backward formula is announced in S. Watanabe [43]. The proof presented at Section 1 is adapted from Kunita [25]. Itô's formula for tensor fields is studied in Watanabe [43], Kunita [24] and Bismut [3] in various settings. The present proof is adapted from [24].

Most materials of Section 3 are adapted from Kunita [23]. The proof of Theorem 3.7 contained errors in [23], which is corrected here. A result analogous to Theorem 3.7 was first obtained by Yamato [45] when the Lie algebra generated by coefficients of SDE is nilpotent. Problems of representing solutions of SDE are studied in Kunita [22], Krener-Lobry [20] and Fliess-Normand-Cyrot [9] discuss similar problems from different stand points, employing Baker-Campbell-Hausdorff formula or formal series of K. T. Chen.

Bismut [3] and Ikeda-Watanabe [14] contain some other interesting examples where the supports of stochastic flows are characterized.

Itô's formula for stochastic parallel displacement is adapted from Kunita [24]. Shigekawa [39] proposes a different approach to the formula.

REFERENCES

[1] P. Baxendale, Wiener processes on manifolds of maps, Proc. Royal Soc. Edinburgh, 87A (1980), 127-152.

[2] J. M. Bismut, A generalized formula of Itô and some other properties of stochastic flows, Z. W. 55 (1981), 331-350.

[3] J. M. Bismut, Mécanique aléatoire, Lecture Notes in Math. 866 (1981).

[4] Yu. N. Blagovescenskii-M. I. Freidlin, Certain properties of diffusion processes depending on a parameter, Soviet Math. Dokl. 2 (1961), 633-636.

[5] K. T. Chen, Decomposition of differential equations, Math. Annalen 146 (1962), 263-278.

[6] J. L. Doob, Stochastic processes, John Wiley and Sons, New York, 1953.

[7] K. D. Elworthy, Stochastic dynamical system and their flows, Stochastic analysis ed. by A. Friedman and M. Pinsky, 79-95, Academic Press, New York, 1978.

[8] M. Emery, Une topologie sur l'espace des semimartingales, Séminaire de Prob. XIII, Lecture Notes in Math. 721 (1979), 260-280.

[9] M. Fliess-D. Normand-Cyrot, Algébres de Lie nilpotents, formule de Baker-Campbell-Hausdorff et intégrales iterées de K. T. Chen, Séminaire de Prob. XVI, Lecture Notes in Math., to appear.

[10] T. Funaki, Construction of a solution of random transport equation with boundary condition, J. Math. Soc. Japan 31 (1979), 719-744.

[11] N. Ikeda-S. Manabe, Stochastic integral of differential forms and its applications, Stochastic Analysis ed. by A. Friedman and M. Pinsky, 175-185, New York, 1978.

[12] N. Ikeda-S. Manabe, Integral of differential forms along the path of diffusion processes, Publ. RIMS, Kyoto Univ. 15 (1979), 827-852.

[13] N. Ikeda-S. Watanabe, Stochastic differential equations and diffusion processes, North Holland-Kodansha, 1981.

[14] N. Ikeda-S. Watanabe, Stochastic flow of diffeomorphisms, to appear.

[15] K. Itô, Stochastic differential equations in a differentiable manifold (1), Nagoya Math. J. 1 (1950), 35-37, (2), Mem. Coll. Sci. Univ. Kyoto Math. 28 (1953), 81-85.

[16] K. Itô, Lectures on stochastic processes, Tata Institute of Fundamental Research, Bombay, 1960.

[17] K. Itô, The Brownian motion and tensor fields on Riemannian manifold, Proc. Intern. Congr. Math. Stockholm, 536-539, 1963.

[18] K. Itô, Stochastic parallel displacement, Probabilistic methods in differential equations, Lecture Notes in Math. 451 (1975), 1-7.

[19] S. Kobayashi-K. Nomizu, Fundations of differential geometry I, John Wiley and Sons, New York, 1963.

[20] A. J. Krener-C. Lobry, The complexity of solutions of stochastic differential equations, Stochastics 4 (1981), 193-203.

[21] N. V. Krylov-B. L. Rozovsky, On the first integrals and Liouville equations for diffusion processes, Proc. Third Conf. Stoch. Diff. System, Lecture Notes in Control and Information Science, to appear.

[22] H. Kunita, On the representation of solutions of stochastic differential equations, Séminaire des Probabilités XIV, Lecture Notes in Math. 784 (1980), 282-303.

[23] H. Kunita, On the decomposition of solutions of stochastic differential equations, Proc. Durham Conf. Stoch. Integrals, Lecture Notes in Math. 851 (1981), 213-255.

[24] H. Kunita, Some extensions of Itô's formula, Séminaire de Probabilités, XV, Lecture Notes in Math. 850 (1981), 118-141.

[25] H. Kunita, On backward stochastic differential equations, Stochastics 6 (1982), 293-313.

[26] H. Kunita, Stochastic differential equations and stochastic flows of homeomorphisms, to appear.

[27] H. Kunita, Stochastic partial differential equations connected with non-linear filtering, to appear in the Proceedings of C.I.M.E. Session on Stochastic control and filtering, Cortona, 1981.

[28] H. Kunita-S. Watanabe, On square integrable martingales, Nagoya Math. J. 30 (1967), 209-245.

[29] P. Malliavin, Un principe de transfert et son application au calcul de variations, C. R. Acad. Sci. Paris, 284, Serie A (1977), 187-189.

[30] P. Malliavin, Stochastic calculus of variation and hypoelliptic operators, Proc. Intern Symp. SDE Kyoto 1976 (ed. by K. Itô) 195-263, Kinokuniya, Tokyo.

[31] P. Malliavin, Géométrie differentielle stochastique, Les Presses de l'Université de Montréal, Montréal, 1978.

[32] Y. Matsushima, Differentiable manifolds, Marcel Dekker, New York, 1972.

[33] P. A. Meyer, Probability and potentials, Blaisdel, Waltham, Massachusetts, 1966.

[34] P. A. Meyer, Integrales stochastiques I-IV, Séminaire de Prob. I, Lecture Notes in Math. 39 (1967), 72-162.

[35] P. A. Meyer, Geometrie stochastique sans larmes, Séminaire de Prob. XV, Lecture Notes in Math. 850 (1981), 44-102.

[36] P. A. Meyer, Flot d'une equation differentielle stochastique, Séminaire de Prob. XV, Lecture Notes in Math. 850 (1981), 103-117.

[37] J. Neveu, Bases mathématiques du calcul des probabilités, Masson et Cie., Paris, 1964.

[38] B. L. Rozovsky, On the Itô-Ventzel formula, Vestnik of Moscow University, N. 1 (1973), 26-32. (In Russian).

[39] I. Shigekawa, On stochastic horizontal lifts, Z. W. 59 (1982), 211-222.

[40] D. W. Stroock-S. R. S. Varadhan, On the support of diffusion processes with application to the strong maximum principle, Proc. Sixth Berkeley Symp. Math. Statist. Prob. III, 333-359, Univ. California Press, Berkeley, 1972.

[41] D. W. Stroock-S. R. S. Varadhan, Multidimensional diffusion processes, Springer-Verlag, Berlin, 1979.

[42] A. D. Ventcel', On equations of the theory of conditional Markov processes, Theory of Prob. Appl. 10 (1965), 357-361.

[43] S. Watanabe, Flow of diffeomorphisms difined by stochastic differential equation on manifolds and their differentials and variations (in Japanese), Suriken Kokyuroku 391 (1980), 1-23.

[44] T. Yamada-Y. Ogura, On the strong comparison theorems for solutions of stochastic differetial equations, Z. W. 56 (1981), 3-19.

[45] Y. Yamato, Stochastic differential equations and nilpotent Lie algebra, Z. W. 47 (1979), 213-229.

QUELQUES PROPRIETES DES EXPOSANTS CARACTERISTIQUES

PAR F. LEDRAPPIER

PREFACE

L'objet de ce cours est de définir les exposants caractéristiques d'un produit de matrices aléatoires stationnaires et d'en étudier quelques propriétés. Les exposants sont définis simplement par sous-additivité. Le théorème ergodique sous-additif montre en fait un résultat de convergence presque sûre. Le théorème ergodique multiplicatif d'Osseledets qui s'en déduit est une loi des grands nombres qui précise le comportement exponentiel des images des vecteurs de $\mathbb{R}^d$ par le produit de matrices. Dans la suite ce théorème va nous servir à expliciter les relations entre les exposants et d'autres quantités qui apparaissent.

Un exemple naturel d'application pour le théorème d'Osseledets est les applications différentiables d'une variété dans elle-même. En effet d'après les règles de composition pour les applications différentielles, la différentielle de l'itérée de la transformation apparaît comme un produit d'opérateurs linéaires. Une mesure de probabilité invariante détermine alors une loi stationnaire sur ces opérateurs. Les exposants qu'on peut ainsi définir sont des quantités dynamiques liées à la transformation et à la mesure invariante. Nous les comparerons avec d'autres quantités dynamiques : dans le cas ergodique, l'entropie est plus petite que la somme des exposants positifs. Le défaut d'égalité dans cette formule peut-être expliqué par une propriété géométrique de la mesure : nous montrons en dimension 2 la formule de Young liant les exposants, l'entropie et la dimension de la mesure. Le théorème d'Osseledets permet en effet de montrer une estimation (presque partout) de la mesure des petites boules. La démonstration introduit plusieurs techniques de théorie ergodique "à la Pesin".

Un autre exemple important est le cas des produits de matrices indépendantes : nous montrons comment le théorème d'Osseledets permet de retrouver des résultats connus comme la formule de Fürstenberg donnant le plus grand exposant. Nous montrons également le critère de Fürstenberg pour assurer que les exposants ne sont pas tous égaux. Pour cela, en suivant la démonstration originale de Fürstenberg, nous introduisons une entropie associée à la marche aléatoire et nous montrons que cette entropie est plus petite que le plus grand exposant. Nous avons également ici en dimension 2 une formule liant cette entropie, l'exposant et une dimension de la mesure invariante.

Nous donnons enfin deux autres exemples de problèmes où le théorème d'Osseledets intervient dans l'étude d'équations aux différences à coefficients aléatoires. Par exemple Pastour a montré que l'opérateur de Schrödinger a un spectre singulier presque partout dès que l'exposant d'une certaine famille de matrices est positif. Nous reprenons son argument et donnons deux exemples. Nous mentionnons également un résultat récent de Key qui caractérise la récurrence de certaines marches aléatoires en milieu aléatoire sur Z par le fait que deux certains exposants sont nuls pour un produit de matrices naturellement associé au modèle.

En général, les exposants apparaissent aujourd'hui souvent difficiles à calculer explicitement (voir pourtant le joli exemple que Gérard Letac a exposé lors de l'Ecole d'Eté), mais le théorème d'Osseledets permet d'obtenir quelques relations, comme nous essayons de le montrer ici. La préparation de ce cours a largement bénéficié de nombreuses conversations avec mes collègues du Laboratoire de Probabilités que je voudrais remercier. L'attention et les remarques de participants à l'Ecole d'Eté m'ont également été précieux. Je voudrais encore remercier P.L. Hennequin de m'avoir invité à faire ce cours et pour la générosité discrète avec laquelle il a organisé cette session de l'Ecole d'Eté.

I - THEOREME ERGODIQUE MULTIPLICATIF D'OSSELEDETS

1. Exposants caractéristiques

Nous considérons une suite stationnaire $\{A_n, n \geq 0\}$ de matrices réelles carrées d x d et nous formons le produit $A^{(n)} = A_{n-1} \ldots A_0$. On peut toujours se ramener au modèle suivant : soit $(X, \mathcal{A}, m)$ un espace probabilisé et θ une application mesurable de X dans X qui laisse m invariante ($m(\theta^{-1}B) = m(B)$ pour tout B de $\mathcal{A}$) et soit A une application mesurable de X dans les matrices d x d. On obtient une suite stationnaire en posant $A_n = A.\theta^n$ et alors on note :

$$A^{(n)}(x) = A(\theta^{n-1} x).\ A(\theta^{n-2} x) \ldots .\ A(x).$$

Notons E l'espace euclidien $\mathbb{R}^d$ et pour p entier naturel, $\overset{p}{\Lambda} E$ les espaces puissance extérieure de E. (On peut considérer $\overset{p}{\Lambda} E$ comme l'espace des formes p-linéaires alternées sur le dual E^*).

Si A désigne une matrice d x d, identifiée à l'opérateur de E dans lui-même, nous notons $\overset{p}{\Lambda} A$ l'opérateur de $\overset{p}{\Lambda} E$ dans lui-même canoniquement associé à A (par exemple par la formule :

$$(\overset{p}{\Lambda} A)\ f(a_1^*, \ldots, a_p^*) = f(A^* a_1^*, \ldots, A^* a_p^*)).$$

Nous noterons $\|\ \|$ une norme quelconque sur chaque espace de matrices. La proposition suivante définit les exposants caractéristiques de la suite $\{A_n, n \geq 0\}$ ou plutôt de la famille $(X, \mathcal{A}, m, \theta, A)$.

1.1 - Proposition - définition

Avec les notations ci-dessus, supposons que :

$$\int \log^+ \|A(x)\|\ m(dx) < +\infty.$$

Alors pour tout $1 \leqq p \leqq d$, la limite de la suite

$$\frac{1}{n} \int \log \|\overset{p}{\Lambda} A^{(n)}(x)\|\quad m(dx) \quad \text{existe dans } \mathbb{R} \cup \{-\infty\}.$$

On appelle exposants caractéristiques $\lambda_1 \ldots \lambda_d$ les nombres réels tels que :

$$\lambda_1 + \lambda_2 + \ldots + \lambda_p = \lim_n \frac{1}{n} \int \log \|\overset{p}{\Lambda} A^{(n)}(x)\|\quad m(dx),$$

en posant $\lambda_p = -\infty$ si la limite est $-\infty$.

Démonstration :

Il est d'abord clair que ni l'hypothèse, ni la conclusion, ni la valeur de la limite ne dépendent de la norme choisie sur l'espace des matrices.

Si, par exemple pour p = 1, on choisit $|| \ ||_0$ une norme d'opérateur de $\mathbb{R}^d$ dans $\mathbb{R}^d$, on a l'égalité :

$$||A^{(n+m)}||_0 = ||A_{n+m-1} \dots A_n A_{n-1} \dots A_0||_0$$
$$\leq ||A_{n+m-1} \dots A_n||_0 \, ||A_{n-1} \dots A_0||_0 .$$

Par stationnarité, il s'ensuit que la suite $C_n = \int \log ||A^{(n)}(x)||_0 \, m(dx)$ vérifie $C_{n+m} \leq C_n + C_m$, et donc que la suite $\frac{C_n}{n}$ converge.

La convergence de $\frac{1}{n} \int \log ||\overset{p}{\Lambda} A^{(n)}(x)|| m(dx)$ pour $p > 1$ s'établit de la même manière ■

Rappelons que l'on peut écrire toute matrice carrée A comme un produit $A = K_1 \, \Delta \, K_2$ avec K_1 et K_2 unitaires, Δ diagonale à éléments positifs.

Notons $\delta_1(A) \geq \delta_2(A) \geq \dots \geq \delta_d(A)$ les termes diagonaux de Δ rangés en ordre décroissant.

1.2 - Proposition

Sous les conditions de 1.1, les exposants caractéristiques λ_i sont donnés par $\lambda_i = \lim_n \frac{1}{n} \int \log \delta_i(A^{(n)}(x)) \, m(dx)$.

En particulier, on a $\lambda_1 \geq \lambda_2 \geq \dots \geq \lambda_d$.

Démonstration :

En notant encore $|| \ ||_0$ la norme d'opérateur dans l'espace euclidien, on voit que $\delta_1(A) = ||A||_0$ et donc que :

$$\lambda_1 = \lim_n \frac{1}{n} \int \log \delta_1(A^{(n)}(x)) \, m(dx) \text{ par définition.}$$

D'autre part puisque $\overset{p}{\Lambda} A = (\overset{p}{\Lambda} K_1) \, (\overset{p}{\Lambda} \Delta) \, (\overset{p}{\Lambda} K_2)$ constitue une décomposition pour l'opérateur $\overset{p}{\Lambda} A$, nous avons $\delta_1(\overset{p}{\Lambda} A) = \delta_1(A) \dots \delta_p(A)$, et donc :

$$\lambda_1 + \dots + \lambda_p = \lim \frac{1}{n} \int (\log \delta_1(A^{(n)}(x)) + \dots + \log \delta_p(A^{(n)}(x)) \, m(dx).$$

Comme $\delta_i \leq \delta_{i-1}$, on en déduit de proche en proche que :

$$\lambda_i = \lim_n \frac{1}{n} \int \log \delta_i(A^{(n)}(x))\, m(dx) \text{ si } \lambda_i > -\infty ,$$

et que λ_i ne peut valoir $-\infty$ que si

$$\lim_n \frac{1}{n} \int \log \delta_i(A^{(n)}(x))\, m(dx) = -\infty \quad \blacksquare$$

1.3 - Corollaire

Sous les conditions de 1.1, on a :

$$\lambda_1 + \dots + \lambda_d = \int \log |\det A(x)| \; m(dx).$$

On peut en effet écrire, que les termes soient finis ou infinis :

$$\lambda_1 + \dots + \lambda_d = \lim_n \frac{1}{n} \int \log \Big(\prod_{i=1}^{d} \delta_i (A^{(n)}(x)) \Big) m(dx)$$

$$= \lim_n \frac{1}{n} \int \log |\det A^{(n)}(x)| \, m(dx)$$

$$= \int \log |\det \; A(x)| \, m(dx) \quad \blacksquare$$

Remarques et références :

Un système $(X, \mathcal{A}, m, \theta)$ est dit ergodique si tout ensemble mesurable invariant est négligeable ou de complémentaire négligeable. Nous supposerons le plus souvent dans la suite que le système de base du modèle est ergodique.

Remarquons que si la mesure m est un mélange de mesures invariantes, $m = \int m_\rho \, \mu(d\rho)$, nous avons pour les exposants $\lambda_i(m) = \int \lambda_i(m_\rho) \, \mu(d\rho)$ et que l'on peut toujours représenter toute mesure stationnaire sur $(\mathbb{R}^{dxd})^{\mathbb{N}}$ comme mélange de mesures ergodiques.

La notion d'exposant caractéristique, non pas en moyenne mais trajectoire par trajectoire est naturelle dès que l'on étudie la stabilité des solutions d'une équation différentielle ordinaire. Une bonne datation est alors l'année 1892, où paraissent les mémoires de :

A.M. LYAPOUNOV : Problème général de la stabilité du mouvement. En traduction française dans Ann. Fac. Sci. Univ. Toulouse 9 (1907) p. 203-475,

et de H. POINCARE : Méthodes nouvelles de la mécanique céleste. Paris, Gauthier-Villars (1892)

2° Théorème ergodique sous-additif

Nous allons obtenir des théorèmes de convergence presque sûre vers les exposants caractéristiques. Pour cela, nous utiliserons le théorème ergodique sous-additif sous la forme suivante :

2.1 - Théorème (Kingman 1968)

Soient $(X, \mathcal{Q}, m, \theta)$ un système ergodique et f_n une suite de fonctions telles que $f_1^+ \in L^1$, $f_{n+m} \leq f_n + f_m \circ \theta^n$ et $\inf_n \frac{1}{n} \int f_n = c \geq -\infty$.

Alors $\frac{1}{n} f_n$ converge presque partout vers c.

Démonstration :

Remarquons tout d'abord que la suite $\int f_n$ est sous-additive et donc la suite $\frac{1}{n} \int f_n$ converge vers $\inf_n \frac{1}{n} \int f_n = c$.

Définissons les fonctions $\overline{f}(x)$ et $\underline{f}(x)$ par :

$$\overline{f}(x) = \limsup_n \frac{1}{n} f_n(x) \quad , \quad \underline{f}(x) = \liminf_n \frac{1}{n} f_n(x).$$

Par sous-additivité on a $\overline{f}(x) \leq \overline{f}(\theta x)$, $\underline{f}(x) \leq \underline{f}(\theta x)$ et donc ces fonctions sont presque partout égales à une constante $\overline{f}$ ou $\underline{f}$.

Le théorème est alors la conséquence immédiate des propositions 2.2 et 2.5.

2.2 - Proposition

On a $\underline{f} \geq c$.

Démonstration :

Supposons $\underline{f} > -\infty$ et choisissons $\varepsilon > 0$.

On peut trouver une fonction $n(x)$ mesurable telle que pour presque tout x, on ait :

$$(1) \qquad f_{n(x)} \leq n(x)\ (\underline{f} + \varepsilon).$$

Pour $N > 1$, posons $A_N = \{n(x) \geq N\} \cap \{(1) \text{ n'est pas vraie}\}$ et définissons :

$$\tilde{f}(x) = \underline{f} \text{ sur } X \setminus A_N \qquad\qquad \tilde{n}(x) = n(x) \text{ sur } X \setminus A_N$$

$$\qquad = \max(\underline{f}, f_1) \text{ sur } A_N \qquad\qquad\qquad = 1 \quad \text{sur } A_N$$

Nous avons maintenant :

$$(2) \qquad f_{\tilde{n}(\ )} \leq \sum_{i=0}^{\tilde{n}(x)-1} (\tilde{f} + \varepsilon)\ (\theta^i x) \text{ partout.}$$

Nous posons alors pour tout x de X :

$$n_o(x) = 0 \qquad n_1(x) = \tilde{n}(x) \text{ et par récurrence}$$

$$n_j(x) = n_{j-1}(x) + \tilde{n}(\theta^{n_{j-1}(x)} x).$$

Pour P entier, P > N posons enfin

$$J_P(x) = \inf \{j | n_j(x) \geq P-N\} .$$

Par sous-additivité nous pouvons alors écrire en tout point x,

$$f_P(x) \leq \sum_{j=0}^{J_P(x)-1} f_{\tilde{n}(\theta^{n_j(x)}x)} (\theta^{n_j(x)} x) + \sum_{j=n_{J_P(x)}}^{P-1} f_1(\theta^j_x).$$

En sommant les inégalités (2) appliquées en chaque $\theta^{n_j(x)}(x)$, nous avons :

$$f_P(x) \leq \sum_{j=0}^{n_{J_P(x)}-1} (\tilde{f}+\varepsilon)(\theta^j x) + \sum_{j=n_{J_P(x)}}^{P-1} f_1(\theta^j x).$$

Comme $n_{J_P(x)} \geq P-N$ nous pouvons encore majorer :

$$f_P(x) \leq \sum_{j=0}^{P-N-1} (\tilde{f}+\varepsilon)(\theta^j x) + \sum_{j=P-N}^{P-1} (\tilde{f}^+ + f_1^+ + \varepsilon)(\theta^j x).$$

Cette relation étant vraie partout, nous avons :

$$c \leq \frac{1}{P} \int f_P \leq \frac{P-N}{P} \int (\tilde{f}+\varepsilon) + \frac{N}{P} \int (f^+ + f_1^+ + \varepsilon).$$

En faisant tendre P vers l'infini, puis N vers l'infini (et alors $\int \tilde{f} \searrow \underline{f}$ car $\max(f_1, \underline{f}) \in L^1$) et enfin ε vers zéro, nous obtenons la proposition 2.2, dans le cas où $\underline{f} > -\infty$.

Supposons maintenant $\underline{f} = -\infty$. Pour tout M on peut trouver n'(x) mesurable telle que on ait :

(1') $$f_{n'(x)} \leq n'(x) M.$$

La même démonstration permet d'établir que c ≤ M ; et ceci étant vrai pour tout M montre que $c = -\infty \leq \underline{f}$. c.q.f.d.

2.3 - Lemme

> Si θ est une transformation préservant la mesure sur un espace de probabilité $(X, \mathcal{A}, m)$ et si $f^+ \in L^1$, $\limsup_n \frac{1}{n} f \circ \theta^n \leq 0$ partout.

Démonstration du lemme :

Soit $\delta > 0$. Nous avons :

$$\sum_{n>0} m(\frac{1}{n} f \circ \theta^n \geq \delta) = \sum_{n>0} m(f \geq n\delta) \leq \frac{1}{\delta} \int f^+ < +\infty.$$

En presque tout point, nous avons donc :

$$\limsup_n \frac{1}{n} f \circ \theta^n \leq \delta .$$

2.4 - Proposition

$$\bar{f} \leq \int f_1 .$$

Démonstration :

La démonstration est analogue à celle de la proposition 2.2 : Supposons d'abord $\bar{f} < +\infty$ et choisissons $\varepsilon > 0$.

On peut trouver une fonction $n(x)$ mesurable telle que pour presque tout x, on ait :

$$n(x)\bar{f} \leq f_{n(x)} + \varepsilon n(x).$$

Par sous-additivité nous aurons alors presque partout :

$$(3) \qquad n(x)\bar{f} \leq \sum_{i=0}^{n(x)-1} (f_1 + \varepsilon)\ (\theta^i x).$$

Posons encore $A_N = \{n(x) \geq N\} \cap \{(3) \text{ n'est pas vraie}\}$ et :

$$\tilde{f}(x) = f_1(x) \text{ sur } X \setminus A_N \qquad \tilde{n}(x) = x(x) \text{ sur } X \setminus A_N$$

$$= \max(f_1(x), \bar{f}) \text{ sur } A_N \qquad = 1 \text{ sur } A_N.$$

Nous avons maintenant :

$$\tilde{n}(x)\bar{f} \leq \sum_{i=0}^{n(x)-1} (\tilde{f}+\varepsilon)(\theta^i x).$$

Pour P entier, P > N, nous obtenons par le même procédé de sommation que plus haut :

$$P\bar{f} \leq \sum_{j=0}^{P-N-1} (\tilde{f}+\varepsilon)\ (\theta^j x) + \sum_{j=P-N}^{P-1} (\bar{f}^+ + \tilde{f}^+ + \varepsilon)\ (\theta^j x).$$

Nous obtenons encore en intégrant, divisant par P et passant à la limite, $\bar{f} \leq \int \tilde{f} + \varepsilon$. De nouveau, quand N tend vers l'infini $\int \tilde{f}$ décroît vers $\int f_1$, ce qui établit la proposition quand $\bar{f} < +\infty$.

Si on pouvait avoir $\overline{f} = +\infty$, cela entraînerait :

(3') $\qquad n'(x)\, M \leq f_{n'(x)} + \varepsilon n'(k)$

pour un $M > \int f_1$ et une fonction $n'(x)$ mesurable.

La même démonstration donne encore $M \leq \int f_1$, ce qui montre que $\overline{f} = +\infty$ est impossible.

2.5 - Proposition

Pour tout j entier > 0, on a $\overline{f} \leq \frac{f_j}{j}$.

Démonstration :

Posons $\overline{f}_j = \limsup_n \frac{1}{n} f_{jn}$.

Montrons d'abord que $\overline{f}_j$ est constant et égal à $j\overline{f}$, la proposition 2.4 appliquée à f_{jn} et θ^j donne alors le résultat. Nous savons que $\overline{f}_j \leq j\overline{f}$ car la limite est prise sur une sous-suite.

D'autre part, la sous-suite qui réalise $\overline{f}$ a une infinité d'éléments dans l'un des $\mathbb{Z} j + k$, $0 \leq k < j$ et donc :

$$j\overline{f} \leq \sup_{0 \leq k < j} \overline{f}_j \circ \theta^k .$$

Par sous-additivité nous savons encore que :

$$\frac{1}{n} f_{(n+1)j} \leq \frac{1}{n} f_1 + \frac{1}{n} (f_{jn} \circ \theta) + \frac{1}{n}(f_{j-1} \circ \theta^{jn+1})$$

et donc en passant à la limite supérieure et utilisant le lemme 2.3, $\overline{f}_j \leq \overline{f}_j \circ \theta$.

Ces relations ne sont compatibles que si $\overline{f}_j$ est constante et égale à $j\overline{f}$ et ceci achève la démonstration de la proposition 2.5 et du théorème.

En rassemblant les résultats des deux premiers paragraphes, nous obtenons :

2.6 - Théorème

Soient $(X, \mathcal{A}, m, \theta)$ un système ergodique, A une application mesurable dans les matrices réelles dxd telle que $\int \log^+ ||A(x)||\, m(dx) < \infty$.

Formons $A^{(n)}(x) = A(\theta^{n-1}x) \ldots A(x)$.

Si $\lambda_1 \geq \lambda_2 \geq \ldots \geq \lambda_d$ désignent les exposants caractéristiques de la suite $A_n(x) = A(\theta^n x)$

$$\left\lfloor \begin{array}{l} \frac{1}{n} \log \|\overset{p}{\Lambda} A^{(n)}(x)\| \to \lambda_1 + \ldots + \lambda_p \quad m \quad \text{p.s.} \\ \text{et } \frac{1}{n} \log \delta_i(A^{(n)}(x)) \to \lambda_i \quad m \quad \text{p.s.} \end{array} \right.$$

Remarques et références :

Le théorème 2.1 est une application du théorème de Kingman.

J.F.C. Kingman : The ergodic theory of subadditive processes. J. Royal Statist. Soc. B.30 (1968) p. 499-510.

J.F.C. Kingman : Subadditive processes. Ecole d'été de Probabilités de Saint-Flour Lect. Notes in maths n° 539 - Springer - Berlin (1976).

Sous la forme donnée ici,une autre démonstration est dans :

Y. Derriennic : Sur le théorème ergodique sous-additif. C.R.A.S. Paris 281 A(1975) 985-988.

Nous avons suivi un article récent :

Y. Katznelson and B. Weiss : A simple proof of some ergodic theorems. Israël J. of maths (1982).

Le théorème 2.6 de Fürstenberg et Kesten, présenté ici comme une conséquence du théorème sous-additif, est en fait antérieur :

H. Fürstenberg and H. Kesten : Products of random matrices. Ann. Math. Stat. 31 (1960) 457-469.

Dans le cas d'une système non ergodique, 2.1 dit que l'ensemble de convergence presque sûre est de mesure 1 pour toute mesure de probabilité ergodique, et donc pour toute mesure invariante. (On ne considère que la loi de la famille dénombrable $f_n \circ \theta^m$ $m, n \geq 0$ de variables aléatoires réelles).

Le théorème ergodique ponctuel correspond bien sûr au cas particulier d'une suite f_n additive, où à la fois les suites f_n et $-f_n$ sont sous-additives.

3° Théorème ergodique multiplicatif

En fait, les exposants caractéristiques introduits en 1° décrivent des taux de croissance de normes de $A^{(n)}(x)v$ pour certains vecteurs v de $\mathbb{R}^d$. Le théorème ergodique multiplicatif précise ce comportement.

3.1 - Théorème

Soient $(X, \mathcal{Q}, m ; \theta)$ un système ergodique, A une application mesurable de X dans $\mathcal{G}L(d,\mathbb{R})$, telle que les variables $\log||A(x)||$ et $\log||A^{-1}(x)||$ sont intégrables.

Il existe un ensemble B, mesurable, tel que $\theta^{-1}B \subseteq B$, $m(B)=1$ et si x appartient à B, il existe une filtration de $\mathbb{R}^d$ en sous-espaces vectoriels :

$$\mathbb{R}^d = V_x^r \supset V_x^{r-1} \supset \ldots\ldots \supset V_x^1 \supset V_x^o = \{0\} \text{ telle que :}$$

i) l'application $x \to V_x^j$ est mesurable

ii) $A(x)V_x^j = V_{\theta x}^j$

iii) Il existe des nombres positifs $\mu_1 < \mu_2 < \ldots < \mu_r$ tels que

pour $1 \le j \le r$

$v \in V_x^j \setminus V_x^{j-1}$ si et seulement si

$$\lim_{n\to+\infty} \frac{1}{n} \log ||A_x^{(n)} v|| = \log \mu_j .$$

La suite des $\log \mu_j$, répétés chacun avec la multiplicité $\dim V_x^j - \dim V_x^{j-1}$, et ordonnée en décroissant est la suite des exposants caractéristiques.

Ce paragraphe est consacré à la démonstration du théorème 3.1.

Ecrivons la décomposition de $A^{(n)}(x)$ sous la forme

$A^{(n)}(x) = L_n(x)\, \Delta_n(x)\, K_n(x)$ avec $\Delta_n(x)$ matrice diagonale dont les éléments non nuls $\delta_i(A^{(n)}(x))$ sont rangés en ordre décroissant.

La démonstration consiste à montrer que si x appartient à un certain ensemble mesurable B, les matrices $(A^{(n)*}(x)\, A^{(n)}(x))^{1/2n}$ convergent vers une matrice symétrique Λ_x.

Les μ_j seront alors les valeurs propres de Λ_x et les espaces V_x^j la somme des sous-espaces propres de Λ_x correspondant aux valeurs propres inférieures à μ_j. Les estimations sont alors suffisantes pour établir iii).

Soit B_0 l'ensemble de convergence dans le théorème 2.6. B_0 est un ensemble invariant mesurable de mesure 1 et si x appartient à B_0,

(1) $\frac{1}{n} \log \delta_i(A^{(n)}(x)) \to \lambda_i$ quand n tend vers l'infini.

Soit B l'ensemble suivant :

$$B = B_0 \cap \{\lim_n \frac{1}{n} \log||A(\theta^n(x)|| = 0\}$$

$$\cap \{\lim_n \frac{1}{n} \log||A^{-1}(\theta^n x)|| = 0\} .$$

L'ensemble B est mesurable, invariant.

On a m(B) = 1 par le lemme 2.3 et :

3.2 - <u>Proposition</u>

Soient x dans B et $\delta > 0$. Il existe un entier N tel que si $n \geq N$, $k \geq 0$ et si $u_{i,j}^{n,k}$ désigne le terme général de la matrice $K_n K_{n+k}^*$, on a :

$$|u_{i,j}^{n,k}| \leq e^{-n|\lambda_i - \lambda_j|} e^{2n\delta} .$$

Nous montrons d'abord que la proposition 3.2 entraîne le théorème 3.1. D'abord comme $|\det A| \geq \frac{1}{||A^{-1}||^d}$, les exposants caractéristiques sont tous finis et d'après (1) si $x \in B$, nous avons $\Delta_n^{\frac{1}{n}}(x)$ converge vers une matrice fixe Δ quand n tend vers l'infini. Fixons donc x dans B. Nous voulons montrer que $(A^{(n)*}(x) A^{(n)}(x))^{\frac{1}{2n}}$, qui est donnée par $K_n^*(x) \Delta_n^{\frac{1}{n}}(x) K_n(x)$ converge.

Soit K_x un point d'accumulation de la suite de matrices unitaires $K_n(x)$ et posons $\Lambda_x = K_x^* \Delta K_x$. La matrice Λ_x est un point d'accumulation de la suite $(A^{(n)*}(x) A^{(n)}(x))^{\frac{1}{2n}}$.

D'après 3.2 si K'_x est un autre point d'accumulation de la suite $K_n(x)$, le terme général $U_{i,j}$ de $K_x K'^*_x$ est nul si $\lambda_i \neq \lambda_j$. Cela entraîne que $K_x K'^*_x$ commutte avec Δ et donc que :

$$\Lambda_x = K_x^* \Delta K_x = K'^*_x \Delta K'_x .$$

La suite $(A^{(n)*}(x)\,A^n(x))^{\frac{1}{2n}}$ est donc telle que toute sous suite admet une sous-suite convergeant vers Λ_x. Il y a bien la convergence annoncée.

Les valeurs propres μ_j de Λ_x et leurs multiplicités sont celles de Δ et sont donc bien données par la suite des exponentielles des exposants caractéristiques. Les espaces V_x^j annoncés sont alors donnés par $K_x^* U^j$, où U^j désigne le sous-espace de $\mathbb{R}^d$ engendré par les sous-espaces propres de Δ de valeurs propres inférieures à μ_j.

Donc tout vecteur v de $V_x^j \ominus V_x^{j-1}$ est de la forme $v = K_x^* u$ pour un u vérifiant :

$$\begin{cases} u = u_s & s = 1,\dots,d \qquad \sum |u_s|^2 \neq 0 \\ u_s = 0 & \text{si} \quad \lambda_s \neq \log \mu_j . \end{cases}$$

Montrons alors que pour v dans $V_x^j \ominus V_x^{j-1}$ $\frac{1}{n}\log ||A^{(n)}(x)\,v||$ tend vers $\log \mu_j$.

Nous avons :

$$||A^{(n)}(x)\,v||^2 = (\Delta_n^2(x)\,K_n(x)\,K_x^*\,u,\ K_n(x)\,K_x^*\,u)$$

$$= \sum_k \delta_k^2\,(A^{(n)}(x))\,|\sum_s u_{k,s}^n\ \ u_s|^2$$

en notant $u_{k,s}^n$ le terme général de $K_n(x)\,K_x^*$.

Comme $u_s = 0$ sauf si $\lambda_s = \log \mu_j$, et comme d'après 3.2 et (1), nous avons, pour n grand :

$$\text{si } \lambda_k > \lambda_s \quad |u_{k,s}^n| \le e^{n(\lambda_s - \lambda_k)}\ e^{2n\delta} \text{ et } \delta_k(A^{(n)}(x)) \le e^{n\lambda_k}\ e^{n\delta}$$

$$\text{et si } \lambda_k \le \lambda_s \quad \delta_k(A^{(n)}(x)) \le e^{n\lambda_s}\ e^{n\delta} ,$$

il vient :

$$\limsup_n \frac{1}{n}\ \log ||A^{(n)}(x)\ v|| \ \le \lambda_s = \log \mu_j .$$

D'autre part en ne comptant que les valeurs de k avec $\lambda_k = \lambda_s = \log \mu_j$,

$$||A^{(n)}(x)v||^2 \ge e^{2n\lambda_s}\ e^{-2n\delta} \sum_k (|\sum_{\lambda_s=\lambda_k} u_{k,s}^n\ u_s|^2)$$

et $\sum_k \left| \sum_{s,\, \lambda_s = \lambda_k} u^n_{k,s}\, u_s \right|^2 \geq ||u||^2 - \sum_k \left| \sum_{s\, \lambda_s \neq \lambda_k} u^n_{k,s}\, u_s \right|^2$

$$\geq \frac{1}{2}||u||^2 \text{ pour n assez grand.}$$

D'où $\liminf_n \frac{1}{n} \log ||A^{(n)}(k)\, v|| \geq \log \mu_j$.

La propriété étant vraie pour tout j $j=1,..,r$ établit la caractérisation iii/ des sous-espaces V^j_x.

Les propriétés i/ et ii/ suivent immédiatement de cette caractérisation. Reste à montrer la proposition 3.2.

Notons $\rho = \inf\{|\lambda_i - \lambda_j|,\ \lambda_i \neq \lambda_j\}$ et si $\rho > 0$, choisissons un entier M tel que $M\rho \geq 2 \log 4d$. (Si $\rho = 0$, la proposition 3.2 est immédiate).

3.3 - Lemme

Soient x un point de B et $\delta > 0$. Il existe N_1 tel que si $n \geq N_1$ et $u^{n,k}_{i,j}$ désigne le terme général de la matrice $K_n K^*_{n+k}$, on a , si $k \leq M$,

(3) $$|u^{n,k}_{i,j}| \leq e^{-n|\lambda_i - \lambda_j|}\, e^{n\delta}.$$

Démonstration :

Choisissons N_1 assez grand pour que pour tout $1 \leq k \leq M$ et $n \geq N_1$, on ait à la fois :

$$||A^{(k)}(\theta^n x)|| \leq e^{\frac{n\delta}{5}}, \quad ||A^{(k)}(\theta^n x)^{-1}|| \leq e^{\frac{n\delta}{5}}$$

$$|\lambda_i| \leq \frac{n\delta}{5M}, \quad |\log \delta_i(A^{(n)}(x)) - n\,\lambda_i| \leq \frac{n\delta}{5} \quad \text{pour } i=1,\dots,d.$$

Nous pouvons écrire l'identité $A^{(n+k)}(x) = A^{(k)}(\theta^n x)\, A^{(n)}(x)$ sous la forme suivante, en omettant la variable x :

$$A^{(k)}.\, \theta^n L_n\ \Delta_n K_n = L_{n+k}\ \Delta_{n+k}\ K_{n+k}.$$

D'où l'on déduit :

$$K_{n+k} K^*_n = \Delta^{-1}_{n+k} (L^*_{n+k}\ A^{(k)}.\, \theta^n L_n)\ \Delta_n$$

et $$K_n K^*_{n+k} = \Delta^{-1}_n\ (L^*_n (A^{(k)}.\theta^n)^{-1}\ L_{n+k})\ \Delta_{n+k}\ .$$

Le terme général de $L^*_{n+k} A^{(k)} . \theta^n L_n$ est majoré par $||A^{(k)} \circ \theta^n||$, donc par $e^{\frac{n\delta}{5}}$ pour $n \geq N_1$. Il vient donc, en transposant,

$$|u^{n,k}_{i,j}| \leq (\delta_j(A^{(n+k)}(x))^{-1} \delta_i(A^{(n)}(x)) e^{\frac{n\delta}{5}}$$

ce qui donne (3) quand $\lambda_i \leq \lambda_j$.

La deuxième relation donne de même (3) pour les i,j avec $\lambda_i \geq \lambda_j$

Pour $1 \leq i \leq d$, notons r_i le nombre d'exposants distincts supérieurs ou égaux à λ_i.

3.4 - Lemme

Soient x un point de B et δ , $0 < \delta \leq \frac{\rho}{4d}$.

Il existe N_2 tel que si $n \geq N_2$, on a pour tout $k \geq 0$

$$|u^{n,k}_{i,j}| \leq e^{-n|\lambda_i - \lambda_j|} e^{2|r_i - r_j| n\delta} . \tag{4}$$

Démonstration :

Choisissons N_2 plus grand que N_1 donné par le lemme 3.3 et tel que $4d\, e^{-N_2 \delta} \leq 1$. Nous allons établir (4) par récurrence sur la partie entière de $\frac{k-1}{M}$

En effet le lemme 3.3 montre (4) pour tout $n \geq N_2$ et $1 \leq k \leq M$.

En écrivant que :

$$K_n K^*_{n+k+M} = K_n K^*_{n+M} K_{n+M} K^*_{n+k+M},$$

nous allons montrer que si (4) est vraie pour tout $n \geq N_2$ et k, (4) est encore vraie pour tout $n \geq N_2$ et k+M.

Nous avons en effet :

$$u^{n,k+M}_{p,q} = \sum_i u^{n,M}_{p,i} u^{n+M,k}_{i,q} . \tag{5}$$

Reportons dans (5) les inégalités en séparant les valeurs de i pour lesquelles respectivement $\lambda_i = \lambda_p$, $\lambda_i = \lambda_q$ et les autres :

$$\begin{aligned}|u^{n,k+M}_{p,q}| \leq\ & d\, e^{-(n+M)|\lambda_p - \lambda_q|} e^{2(n+M)|r_p - r_q|\delta} \\ & + d\, e^{-n|\lambda_p - \lambda_q|} e^{n\delta} \\ & + \sum_{i\ \lambda_i \neq \lambda_p, \lambda_q} e^{-n|\lambda_i - \lambda_p|} e^{-(n+M)|\lambda_i - \lambda_q|} e^{n\delta} e^{2(n+M)|r_i - r_q|\delta}\end{aligned}$$

Si λ_i est compris entre λ_p et λ_q, nous pouvons écrire :

$$|\lambda_i - \lambda_p| + |\lambda_i - \lambda_q| = |\lambda_p - \lambda_q| \quad \text{et} \quad |r_i - r_q| \leq |r_p - r_q| - 1 .$$

Si λ_i n'est pas compris entre λ_p et λ_q, alors

$$|\lambda_i - \lambda_p| + |\lambda_i - \lambda_q| \geq |\lambda_p - \lambda_q| + \rho \quad \text{et} \quad |r_i - r_q| \leq |r_p - r_q| + d - 1.$$

Nous trouvons finalement :

$$|u_{p,q}^{n,k+M}| \leq e^{-n|\lambda_p - \lambda_q|} e^{2n\delta |r_p - r_q|} . C$$

avec une estimation pour C donnée par :

$$C \leq d\, e^{-M\rho} e^{2Md\delta} + d\, e^{-N_2\delta} + d\, e^{-M\rho} e^{2Md\delta}$$
$$+ d\, e^{-n\rho} e^{-M\rho} e^{2Md\delta} e^{2nd\delta} .$$

Nous avons $C \leq 3d\, e^{-\frac{M\rho}{2}} + \frac{1}{4}$ par notre choix de δ et de N_2, et finalement $C \leq 1$ par notre choix de M.

Ceci montre donc le lemme 3.4 et la proposition 3.2 est une reformulation de 3.4.

Nous utiliserons parfois du théorème 3.1 seulement le résultat partiel suivant :

3.5 - Corollaire

Sous les conditions du théorème 3.1, il existe un ensemble B invariant de mesure 1, tel que si x appartient à B, $\frac{1}{n} \log ||A^{(n)}(x)\, v||$ converge pour tout v non nul de $\mathbb{R}^d$ et les limites possibles sont exactement les exposants caractéristiques.

Remarques et références :

Le théorème 3.1 constitue une partie du théorème d'Osseledets :

V.I. Oseledeč : A multiplicative ergodic theorem, Ljapunov characteristic numbers for dynamical systems, Trans. Moscow Math. Soc. 19 (1968) 197 - 231.

La démonstration donnée ici suit :

M.S. Raghunathan : A proof of Oseledec' multiplicative ergodic theorem. Israël J. Maths 32 (1979) 356.
D. Ruelle : Ergodic theory of differentiable dynamical systems. Publ. math. de l'IHES, 50 (1979) 27-58.

En fait l'hypothèse $\log ||A^{-1}(x)||$ intégrable n'est pas nécessaire. Si en effet nous avons seulement $\log ||A(x)||$ intégrable, nous ne pouvons utiliser que la relation :

$$K_{n+k} K_n^* = \Delta_{n+k}^{-1} (L_{n+k}^* A^{(k)} \circ \theta^n L_n) \Delta_n.$$

La même démonstration nous donne alors que sous les conditions de 3.2, nous avons pour n assez grand et pour tout k :

$$|u_{i,j}^{n,k}| \leq e^{-n(\lambda_i - \lambda_j)} e^{2n\delta} \text{ pour } i \geq j$$

autrement dit : $|u_{i,j}^{n,k}| \leq e^{-n(\lambda_i - \lambda_j)^+} e^{2n\delta}$

(En remplaçant éventuellement les λ_j valant $-\infty$ par $-\frac{1}{\delta}$).

On en déduit la proposition 3.2 en remarquant que comme $K_n K_{n+k}^*$ est unitaire, les termes $u_{i,j}^{n,k}$ $i < j$ sont donnés par leurs cofacteurs B_{ij} et que

$$|B_{ij}| \leq \sum_{i_k, j_k} \prod_k |u_{i_k, j_k}|$$

où la somme porte sur toutes les suites (i_k, j_k) avec les $i_k (j_k)$ deux à deux distincts et différents de $i(j)$. En particulier $\sum_k \lambda_{i_k} + \lambda_i = \sum_k \lambda_{j_k} + \lambda_j$ et $\sum_k (\lambda_{i_k} - \lambda_{j_k}) = \lambda_j - \lambda_i$.

Nous avons bien

$$|B_{ij}| \leq \sum_{i_k, j_k} e^{2n(d-1)\delta} e^{-n \sum_k (\lambda_{i_k} - \lambda_{j_k})^+}$$

$$\leq (d-1)!\, e^{2n(d-1)\delta} e^{-n(\lambda_j - \lambda_i)}.$$

D'où la proposition 3.2 suit.

Cet énoncé peut se généraliser alors en dimension infinie. Il est possible de le démontrer si on suppose qu'il n'y a qu'un nombre fini d'exposants positifs ou nuls et certaines hypothèses :

D. Ruelle : Characteristic exponents and invariant manifolds in Hilbert space, Annals of Math, 115, (1982) 243-290.

R. Mañé : Lyapunov exponents and stable manifolds for compact transformation, prétirage.

Enfin l'extension de 3.1 au cas non ergodique est immédiate. Remarquons en effet que l'ensemble B du théorème 3.1 est défini comme un certain ensemble de convergence, qui est de mesure 1 pour toute mesure ergodique.

L'ensemble B pour lequel il existe une filtration avec les propriétés 3.1 i/ ii/ iii/ est donc de mesure un pour toute mesure invariante. Les quantités $r(x)$, $\mu_j(x)$ $1 \leq j \leq r(x)$ sont alors des fonctions invariantes.

4° Cas inversible

Dans ce paragraphe, nous supposerons que la transformation θ est inversible et que son inverse θ^{-1} est mesurable. La transformation θ^{-1} préserve encore la mesure m.

Si A est une application mesurable de K dans $\mathcal{G}L(d, \mathbb{R})$, nous allons considérer la famille $A^{(-1)}(x) = A^{-1}(\theta^{-1}x)$ et nous noterons :

(1) $$A^{(-n)}(x) = A^{(-1)(n)} = A^{-1}(\theta^{-n}x)\dots A^{-1}(\theta^{-1}x).$$

Remarquons qu'avec cette notation, si $A^{(0)} = \mathrm{Id}$ nous avons :

$$A^{(m+n)}(x) = A^{(m)}(\theta^n x)\, A^{(n)}(x),$$

pour tous m, n de $\mathbb{Z}$.

4.1 - Proposition

Si $\lambda_1 \geq \lambda_2 \geq \dots \geq \lambda_d$ sont les exposants caractéristiques de la famille $(X, \mathcal{Q}, m, \theta, A)$, les exposants caractéristiques de la famille $(X, \mathcal{Q}, m, \theta^{-1}, A^{(-1)})$ sont $-\lambda_d \geq -\lambda_{d-1} \geq \dots \geq -\lambda_1$.

En effet d'après la formule (1) $A^{(-n)}(x) = (A^{(n)}(\theta^{-n}x))^{-1}$ et donc

$$\delta_i(A^{(-n)}(x)) = \delta^{-1}_{d-i+1}(A^{(n)}(\theta^{-n}x))$$

et 4.1 suit alors de la proposition 1.2 ■

4.2 - Théorème

Soient $(X, \mathcal{Q}, m)$ un espace probabilisé, θ une transformation mesurable ainsi que son inverse θ^{-1}, préservant la mesure m et ergodique, A une application mesurable de X dans $\mathcal{G}L(d, \mathbb{R})$ avec $\log\|A(x)\|$ et $\log\|A^{-1}(x)\|$ intégrables.

Il existe un ensemble $B \subset X$, mesurable invariant $m(B) = 1$ et pour tout x dans B une décomposition de $\mathbb{R}^d = \bigoplus_{i=1}^{r} W_x^i$ avec :

i) $x \to W_x^i$ est mesurable

ii) $A(x)\, W_x^i = W_{\theta x}^i$

iii) $v \in W_x^i \iff \frac{1}{n}\log\|A^{(n)}(x)v\| \to \log\mu_i$

quand $n \to +\infty$ et quand $n \to -\infty$.

$$\text{où } A^{(n)}(x) = A(\theta^{n-1}x) \;\ldots.\; A(\theta x)\, A(x) \qquad \text{pour } n > 0$$
$$= A^{-1}(\theta^{-n}x)\ldots\; A^{-1}(\theta^{-1}x) \qquad \text{pour } n < 0.$$

Démonstration :

En appliquant le théorème 3.1 aux deux familles $(X, \mathcal{Q}, m, \theta, A)$ et $(X, \mathcal{Q}, m, \theta^{-1}, A^{(-1)})$, nous obtenons deux ensembles B_1 et B_2 mesurables invariants et sur $B_1 \cap B_2$ deux décompositions mesurables invariantes de $\mathbb{R}^d$:

$$\{0\} \subset V_x^1 \subset V_x^2 \subset \ldots.. \subset V_x^r = \mathbb{R}^d$$

et
$$\{0\} \subset U_x^r \subset U_x^{r-1} \subset \qquad \subset U_x^1 = \mathbb{R}^d$$

telles que :

$$v \in V_x^j \Longleftrightarrow \limsup_{n\to+\infty} \frac{1}{n} \log ||A^{(n)}(x)\, v|| \leq \log \mu_j$$

$$v \in U_x^j \Longleftrightarrow \limsup_{n\to-\infty} \frac{1}{|n|} \log ||A^{(n)}(x)\, v|| \leq - \log \mu_j .$$

Nous avons alors les deux propriétés suivantes :

1° $\dim U_x^{j+1} + \dim V_x^j = d$ pour tout j, $0 \leq j < d$ car par définition et d'après la proposition 4.1, $\dim U_x^{j+1}$ est le nombre d'exposants $> \log \mu_j$ et $\dim V_x^j$ est le nombre d'exposants $\leq \log \mu_j$.

2° $U_x^{j+1} \cap V_x^j = \{0\}$ pour tout $0 \leq j < d$ car c'est un sous-espace invariant et si v appartient à $U_x^{j+1} \cap V_x^j$, $v = A^{(-n)}(\theta^n x)\, A^{(n)}(x) v$ et donc :

$$||v|| \leq \mu_{j+1}^{-n}\, e^{n\varepsilon_n(\theta^n x)}\, \mu_j^n\, e^{n\,\varepsilon'_n(x)}\, ||v||$$

avec $\varepsilon_n(\theta^n x)$ et $\varepsilon'_n(x)$ tendant vers zéro en probabilité ; ce qui est impossible si $||v|| \neq 0$.

Il s'ensuit que en posant $W_x^j = V_x^j \cap U_x^j$, nous définissons une décomposition mesurable invariante de $\mathbb{R}^d$ et si v est un vecteur non nul de W_x^j, nous avons

$$\lim_{n\to+\infty} \frac{1}{n} \log ||A^{(n)}(x)\, v|| = \log \mu_j$$

car v est un vecteur de V_x^j mais pas de V_x^{j-1} par la propriété 2° et de même :

$$\lim_{n\to-\infty} \frac{1}{-n} \log ||A^{(n)}(x)\, v|| = - \log \mu_j \quad \blacksquare$$

Soient une famille $(X, \mathcal{Q}, m, \theta, A)$ avec la fonction $\log^+ ||A(x)||$ intégrable et $1 \leq p \leq d$.

Si nous considérons la famille $(X, \mathcal{Q}, m, \theta, \overset{p}{\Lambda} A)$ nous avons

$$\overset{p}{\Lambda} A^{(n)} = (\overset{p}{\Lambda} L_n)(\overset{p}{\Lambda} \Delta_n)(\overset{p}{\Lambda} K_n).$$

Les exposants caractéristiques de la famille $\overset{p}{\Lambda} A$ sont donc donnés par les sommes $\sum_{j=1}^{p} \lambda_{i_j}$ pour chaque suite i_j $1 \leq i_1 < i_2 < \ldots < i_p \leq d$.

Un même exposant peut être obtenu ainsi par plusieurs combinaisons. La limite des $((\overset{p}{\Lambda} A^{(n)})^* (\overset{p}{\Lambda} A^{(n)}))^{\frac{1}{2n}}$ considérée dans la démonstration de 3.1 est l'opérateur $\overset{p}{\Lambda} \Lambda$. Comme les sous-espaces V_x de la filtration définie au théorème 3.1 sont construits à partir de cette limite, il est facile de vérifier que si $\mu_{j,p}$ est un exposant de la famille $\overset{p}{\Lambda} A$, le sous espace $V_x^{j,p}$ de $\overset{p}{\Lambda} E$ associé à $\mu_{j,p}$ par le théorème 3.1 est le sous-espace engendré par les éléments de la forme $v_1 \wedge \ldots \wedge v_p$, $v_i \in V_x^{j_i}$, $\prod_{i=1}^{p} \mu_{j_i} \leq \mu_{j,p}$.

Supposons maintenant θ inversible et soit $\mu_{j,p}$ un exposant de la famille $(X, \mathcal{Q}, m, \theta, \overset{p}{\Lambda} A)$.

Montrons que le sous-espace $W_x^{j,p}$ correspondant est le sous-espace engendrée par les :

$$u_1 \wedge u_2 \wedge \ldots \wedge u_p \quad u_i \in W_x^{j_i}, \quad \prod_{i=1}^{p} \mu_{j_i} = \mu_{j,p}.$$

En effet d'après ce qui précède, nous savons que $u_1 \wedge u_2 \wedge \ldots \wedge u_p$ appartient à $V_x^{j,p}$ et à $U_x^{j,p}$ et donc à $W_x^{j,p}$ par définition.

Mais d'autre part la dimension de $W_x^{j,p}$ est donnée par la multiplicité de $\mu_{j,p}$ et nous avons suffisamment de vecteurs indépendants de la forme annoncée.

Cette propriété a de multiples conséquences. Par exemple :

4.3 - <u>Proposition</u>

Sous les hypothèses de 4.2 soient $\log \mu_j$ un exposant caractéristique, W_x^j le sous-espace associé et $\det_j(x)$ le déterminant de la restriction de A à W_x^j. Alors :

$$\dim W_x^j \log \mu_j = \int \log |\det_j(x)| \; m(dx).$$

Posons $p_j = \dim W_x^j$. Nous savons que $\Lambda^{p_j} W_x^j$ est un vecteur unidimensionnel invariant de la décomposition 3.1 de $\Lambda^{p_j} A(x)$ et correspondant à l'exposant $p_j \log \mu_j$.

Nous avons $\lim_{n\to\pm\infty} \frac{1}{n} \log ||(\overset{p_j}{\Lambda} A^{(n)}(x))(\overset{p_j}{\Lambda} W_x^j)|| = p_j \log \mu_j$

et l'action de $\overset{p_j}{\Lambda} A(x)$ sur $(\overset{p}{\Lambda} W_x^j)$ est la multiplication par $\det_j(x)$.

Nous avons donc :

$$p_j \log \mu_j = \lim_{n\to\pm\infty} \frac{1}{n} \log \prod_{i=0}^{n-1} |\det_j(\theta^i x)| = \int \log |\det_j(x)| \, m(dx)$$

par le théorème ergodique.

De même :

4.4 - Proposition

Sous les hypothèses de 4.2, notons $E_x^u(E_x^s)$ le sous-espace de R^d engendré par les W_x^j associés à des $\mu_j > 1$ (< 1) et $\mathcal{J}_u(x)$ $\mathcal{J}_s(x)$ le déterminant de $A(x)$ restreint à $E_x^u(E_x^s)$. Nous avons :

$$\sum_i \lambda_i^+ = \log |\mathcal{J}_u(x)| \, m(dx)$$

$$\sum_i \lambda_i^- = \log |\mathcal{J}_s(x)| \, m(dx).$$

(en prenant $\mathcal{J}_u(\mathcal{J}_s) = 1$ si $E^u(E^s)$ est réduit à $\{0\}$).

C'est la même vérification que 4.3.

4.5 - Proposition

Sous les hypothèses de 4.2, soient u un vecteur de W_x^i et v un vecteur de W_x^j, $i \neq j$:

$$\frac{1}{n} \log \frac{||A^{(n)}(x) u \wedge A^{(n)}(x) v||}{||A^{(n)}(x)u|| \, ||A^{(n)}(x)v||} \to 0$$

m presque sûrementquand n tend vers l'infini.

En effet $u_\Lambda v$ appartient à $\omega_x^i \wedge \omega_x^j$ qui est inclus dans le sous-espace invariant associé à $\log \mu_i + \log \mu_j$ et :

$$\lim_n \frac{1}{n} \log ||A^{(n)}(x) \, u \wedge A^{(n)}(x) \, v|| =$$

$$\lim_n \frac{1}{n} \log ||\overset{2}{\Lambda} A^{(n)}(x) \, (u_\Lambda v)|| = \log \mu_i + \log \mu_j. \text{ etc...}$$

Remarques et références :

Le théorème 4.2 se généralise facilement à un "cocycle"

$$(X, \mathcal{Q}, m, \theta_t, t \in \mathbb{R}, A^{(t)})$$

où θ_t est une action de $\mathbb{R}$ sur $(X, \mathcal{Q}, m)$ telle que l'application $(t,x) \to \theta_t x$ est mesurable et préserve m et $A^{(t)}$ est une application mesurable de $X \times \mathbb{R}$ dans $\mathcal{G}L(d,\mathbb{R})$ telle que :

$$A^{(t+s)}(x) = A^{(t)}(\theta_x^s)\, A^{(s)}(x).$$

On interprète parfois 4.5 et les résultats analogues de la manière suivante.

Pour deux sous-espaces E_1 et E_2 de $\mathbb{R}^2$, on pose :

$$\alpha(E_1, E_2) = \inf\left\{ \frac{||u_1 \wedge u_2||}{||u_1||\ ||u_2||} \,\Big|\, u_1 \in E_1,\ u_2 \in E_2 \right\} \ ;$$

$\alpha(E_1, E_2)$ représente le "sinus de l'angle" entre les deux sous-espaces.

On déduit de 4.5 que sous les hypothèses de 4.2, si $j_1 \neq j_2$, $\frac{1}{n} \log \alpha(W^{j_1}_{\theta^n x}, W^{j_2}_{\theta^n x})$ tend vers 0 quand n tend vers l'infini : l'angle entre les sous-espaces invariants ne décroît pas exponentiellement vers 0 le long des orbites.

L'ensemble de ces résultats constitue le théorème d'Osseledets (voir référence paragraphe 3).

5° Mesures invariantes

Considérons l'espace $\mathbb{P}^{d-1}$ des directions de $\mathbb{R}^d$. C'est l'espace quotient de $R^d \setminus \{0\}$ par la relation d'équivalence : $u \sim v$ s'il existe λ réel tel que $u = \lambda v$. Si A est un opérateur linéaire de $\mathbb{R}^d$ dans $\mathbb{R}^d$, il préserve la relation $\sim$ et nous noterons encore A l'action quotient de cet opérateur sur $\mathbb{P}^{d-1}$.

Soient $(X, \mathcal{Q}, m, \theta)$ un système dynamique ergodique, θ inversible et A une application mesurable de X dans $\mathcal{G}L(d,\mathbb{R})$ telle que les fonctions $\log ||A(x)||$ et $\log ||A^{-1}(x)||$ soient intégrables. Définissons sur $X \times P^{d-1}$ la transformation $\hat{\theta}$ par :

$$\hat{\theta}(x, \dot{u}) = (\theta x, A(x)\dot{u}).$$

5.1 - Proposition

Soit F la fonction sur $X \times P^{d-1}$ définie par $F(x,\dot{u}) = \log \frac{||A(x)u||}{||u||}$.

Pour toute mesure de probabilité $\hat{m}$ $\hat{\theta}$ invariante et ergodique sur $X \times P^{d-1}$, se projetant en m sur X, il existe j $1 \leq j \leq r$ tel que :

i/ $\int F(x,\dot{u})\, \hat{m}(dx,d\dot{u}) = \log \mu_j$

ii/ la mesure $\hat{m}$ est portée par l'ensemble des $(x,\dot{u})$ vérifiant $u \in W_x^j$.

Inversement pour tout j, $1 \leq j \leq r$, il existe au moins une telle mesure.

Démonstration :

Remarquons tout d'abord que :

$$\sum_{i=0}^{n-1} F \circ \hat{\theta}^i(x,\dot{u}) = \log \frac{||A^{(n)}(x)u||}{||u||}$$ et que la mesure $\hat{m}$ étant ergodique et la fonction $F(x,\dot{u})$ intégrable nous avons :

$$\frac{1}{n} \sum_{i=0}^{n-1} F \circ \hat{\theta}^i(x,\dot{u}) \to \int F\, d\hat{m}$$ $\hat{m}$ presque sûrement quand n tend vers l'infini.

D'après 3.5 cela impose qu'il existe j avec $1 \leq j \leq r$, $\int F\, d\hat{m} = \log \mu_j$ et $u \in V_x^j \setminus V_x^{j-1}$ $\hat{m}$ presque sûrement.

De même en remarquant que

$$\sum_{i=1}^{n} F \circ \hat{\theta}^{-i}(x,\dot{u}) = \log \frac{||u||}{||A^{(-n)}(x)u||},$$

nous avons encore la convergence de :

$$\frac{1}{n} \log ||A_x^{(-n)} u|| \text{ vers } - \int F \, d\hat{m} = - \log \mu_j$$

quand n tend vers l'infini. Nous avons donc $u \in U_x^j \setminus U_x^{j+1}$ $\hat{m}$ presque sûrement. Par définition de W_x^j, $u \in W_x^j$ presque sûrement.

Inversement, nous considérons l'espace de Banach séparable $C(\mathbb{P}^{d-1})$ des fonctions continues sur $\mathbb{P}^{d-1}$, muni de la norme uniforme. Le dual fort de l'espace $L^1(X, m, C(\mathbb{P}^{d-1}))$ est l'espace $L^\infty(X, m, M(\mathbb{P}^{d-1}))$ des classes d'applications scalairement mesurables de X dans l'espace vectoriel topologique des mesures sur $\mathbb{P}^{d-1}$ muni de la convergence faible. (cf Bourbaki L VI Intégration chapitre 6 § 2 proposition 10. Hermann Paris (1959)).

Le sous-ensemble M_j de $L^\infty(X, m, M(P^{d-1}))$ formé des classes d'applications à valeurs mesures m_x telles que pour m presque tout x, la mesure m_x est une mesure de probabilité de support inclus dans W_x^j est un fermé faible de la boule unité de $L^\infty(X, m, M(P^{d-1}))$. L'ensemble M_j est donc un ensemble convexe compact pour la topologie faible.

L'application $\hat{\theta}$ induit une application linéaire continue de $L^1(X, m, C(\mathbb{P}^{d-1}))$ dans lui-même et donc par dualité une application affine $\hat{\theta}$ faiblement continue. Enfin, d'après 4.2 ii/ θ préserve M_j.

Il existe donc un point fixe par $\hat{\theta}$ dans M_j autrement dit une famille de mesures de probabilités $\hat{m}_j(x,.)$ sur $\mathbb{P}^{d-1}$ telles que :

$\hat{m}_j(x, W_x^j) = 1$ m-presque sûrement et

$\hat{\theta}(\hat{m}_j(x,.)) = \hat{m}_j(x,.)$ ce qui signifie exactement que la mesure :

$\hat{m}_j(dx,du) = \hat{m}_j(x,du)\, m(dx)$ est $\hat{\theta}$ invariante.

II - ENTROPIE ET EXPOSANTS

1° Entropie des systèmes dynamiques

Nous rappelons dans ce paragraphe la définition et les propriétés de l'entropie des systèmes dynamiques. Soient P et Q deux partitions dénombrables mesurables d'un espace probabilisé $(X, \mathcal{A}, m)$. On appelle entropie conditionnelle de P sachant Q la quantité $H(P|Q)$ suivante :

$$H(P|Q) = -\sum_{i,j} m(P_i \cap Q_j) \log m(P_i|Q_j)$$

où P_i et Q_j décrivent les éléments de P et de Q. Pour retenir les propriétés élémentaires de l'entropie, on peut interpréter $H(P|Q)$ comme la quantité d'information qu'apporte la connaissance de P à celui qui connaît déjà Q. Nous les résumons dans :

1.1 - Proposition

Soient P,Q,R trois partitions dénombrables mesurables d'un espace probabilisé $(X, \mathcal{A}, m)$

i/ $0 \leq H(P|Q) \leq +\infty$ et $H(P|Q)$ est nul si et seulement si P est une sous-partition de Q

ii/ $H(P \vee Q|R) = H(Q|R) + H(P|Q \vee R)$

iii/ si R est une sous-partition de Q

$$H(P|R) \geq H(P|Q)$$

iv/ $H(P|Q) \leq \log \operatorname{card} P$.

Les démonstrations sont élémentaires et laissées au lecteur. Si on note Q_o la partition réduite au seul élément X, on appelle entropie de la partition P la quantité $H(P) = H(P|Q_o)$.

Soit maintenant θ une transformation mesurable de $(X, \mathcal{A})$ laissant la mesure m invariante. Si H est une partition d'entropie finie, nous appelons entropie moyenne de P et notons $h_m(P, \theta)$ la quantité :

$$h_m(P,\theta) = \inf_n \frac{1}{n} H(P \vee \theta^{-1} P \vee \ldots \vee \theta^{-n+1} P).$$

1.2 - Proposition :

$$|h_m(P,\theta) - h_m(Q,\theta)| \leq H(P|Q) + H(Q|P).$$

La proposition 1.2 se déduit facilement de (1.1 ii et iii). On définit l'entropie de la transformation θ par :

$$h_m(\theta) = \sup h_m(P,\theta)\ H(P) < +\infty .$$

La proposition 1.2 permet de calculer $h_m(\theta)$ comme limite de $h_m(P_n,\theta)$ où les P_n sont des partitions de plus en plus fines et engendrant finalement la σ-algèbre $\mathcal{A}$. L'entropie d'une transformation est un nombre réel positif, fini ou infini. Nous avons les propriétés et exemples fondamentaux suivants.

1.3 - Proposition

$h_m(\theta^n) = n\, h_m(\theta)$ et si θ est inversible $h_m(\theta^{-1}) = h_m(\theta)$.

Il suit en effet immédiatement de la définition que pour toute partition P d'entropie finie,

$$h_m(P,\theta^{-1}) = h_m(P) \text{ et}$$

$$h_m\Big(\bigvee_{i=0}^{n-1} \theta^{-i} P,\ \theta^n\Big) = n\, h_m(P,\theta).$$

En considérant une famille P_k engendrant $\mathcal{A}$, 1.3 suit ∎

1.4 - Exemple

Soit $X = \mathbb{R}/\mathbb{Z}$, m la mesure de Haar sur X, θ la translation de α sur X ; $h_m(\theta) = 0$.

Si en effet la partition P_k est une partition en k intervalles de longueur 1/k, il est clair que la partition $P_k \vee \theta^{-1} P_k \vee .. \vee \theta^{-n+1} P_k$ a moins de kn éléments et donc que

$$h_m(P_k,\theta) \leq \inf_n \frac{1}{n} \log kn = 0.$$

On conclut en remarquant que de telles partitions engendrent la σ-algèbre des boréliens de X.

1.5 - Exemple

Soient $(A,p = \{p_a, a\in A\})$ un espace probabilisé fini, $(X,m) = (A,p)^{\mathbb{Z}}$ le produit indexé par $\mathbb{Z}$ de copies de (A,p), θ le décalage des coordonnées $((\theta x)_n = x_{n+1})$ $h_m(\theta) = - \sum_a p_a \log p_a$.

En effet la partition P définie par la coordonnée 0 est indépendante de ses images et

$$h_m(P,\theta) = \frac{1}{n} H(P \vee \theta^{-1} P \vee \dots \vee \theta^{-n+1} P) = - \sum_a p_a \log p_a .$$

Pour la partition $P_k = \theta^{-k} P \vee \dots \vee \theta^k P$, nous avons de même

$$h_m(P_k,\theta) = \inf_n \frac{n+2k}{n} H(P) = H(P).$$

1.6 - Exemple

Soient $X = C_0(\mathbb{R}^+, \mathbb{R})$ l'espace des fonctions continues de $\mathbb{R}^+$ dans $\mathbb{R}$, nulles en 0, m la mesure de Wiener et θ définie par

$$\theta X(t) = X(t+1) - X(1) \;;\; h_m(\theta) = +\infty .$$

Les σ-algèbres engendrées par les variables $X(t+s) - X(s)$, $0 \leq s < t < 1$, sont en effet continues et indépendantes de leurs images par θ^n, $n \in \mathbb{N}$.

Références, démonstrations, exemples, compléments, etc... se trouvent dans le livre de P. Billingsley : Ergodic theory and information. J.Wiley and Sons (1965). Pour les développements ultérieurs les plus importants, voir :

D.S. Ornstein : Ergodic theory, Randomness and Dynamical Systems. Yale University Press, New-Haven(1974).

2° Applications différentiables

Le cadre naturel du théorème d'Osseledets est celui d'une application différentiable d'une variété. Soient X une variété riemannienne compacte de dimension d, θ une application différentiable de X dans X, m une mesure de probabilité invariante. Considérons la différentielle $d\theta$, application du fibré tangeant TX dans lui-même. Le théorème d'Osseledets donne des renseignements sur le comportement des applications $d_x\theta^n$, en m-presque tout point x.

Choisissons en effet un isomorphisme fibré mesurable τ entre TX et $X \times \mathbb{R}^d$ tel que, pour tout x, τ définit un isomorphisme d'espaces euclidiens τ_x entre $T_x X$ muni du produit scalaire de la métrique riemannienne et l'espace $\mathbb{R}^d$, muni du produit scalaire canonique.

Posons $A(x) = \tau_{\theta x}\, d_x\theta\, \tau_x^{-1}$.

Nous avons bien défini une application mesurable de $(X, \mathcal{A})$ dans les opérateurs de $\mathbb{R}^d$. Les règles de composition de la différentielle donnent :

$$A^{(n)}(x) = \tau_{\theta^n x}\, d_x\theta^n\, \tau_x^{-1} .$$

Remarquons que si θ est inversible et son inverse différentiable, la formule ci-dessus est encore valable pour n négatif. (cf. formule (1) du paragraphe I.4).

Les résultats de la première partie se traduisent alors sur $d\theta$ de la manière suivante :

2.1 - Théorème

Soient X une variable riemannienne compacte de dimension d, θ une application différentiable de X dans X, m une mesure de probabilité invariante ergodique. Il existe des nombres $\lambda_1 \geq \lambda_2 \ldots \geq \lambda_d$ tels que les suites $\frac{1}{n} \log ||\overset{p}{\Lambda}\, d_x\, \theta^n||$ convergent m presque partout vers $\lambda_1 + \lambda_2 + \ldots + \lambda_p$, $1 \leq p \leq d$.

Il existe un ensemble B invariant de mesure 1, et une filtration mesurable de la restriction TB de TX à B,

$$TB = V^r \supset \ldots \supset V^1 \supset V^0 = \{0_x, x \in B\},$$

invariante par $d\theta$ telle que le vecteur v de $T_x X$ appartient à $V_x^j \setminus V_x^{j-1}$ si et seulement si la suite $\frac{1}{n} \log \|d_x \theta^n v\|_{\theta^n x}$ converge vers un certain exposant $\log \mu_j$.

Si l'application θ est un difféomorphisme, il existe un ensemble B' invariant de mesure 1 et une décomposition mesurable invariante de la restriction TB' de TX à B'

$$TB = \bigoplus_{j=1}^{r} W^j$$ telle que le vecteur v de $T_x X$ appartient à un W_x^j si et seulement si les limites de la suite $\frac{1}{n} \log\|d_x\theta^n v\|_{\theta^n x}$ existent et sont égales quand n tend vers plus l'infini et vers moins l'infini.

Dans ce paragraphe nous voulons établir :

2.2 - Théorème

Soient (X, θ, m) un système dynamique différentiable ergodique ;

$$h_m(\theta) \le \sum_{i=1}^{d} \lambda_i^+ .$$

Démonstration :

La démonstration consiste à se ramener à une estimation sur les différentielles $d\theta^n$ et à utiliser alors un argument géométrique. L'argument géométrique est donné par le lemme 2.3.

Dans un espace métrique, notons $B_d(F,\alpha)$ l'ensemble des points situés à une distance de la partie F plus petite que α.

2.3 - Lemme

Soit A une application linéaire de l'espace euclidien $\mathbb{R}^d$. Il existe une constante $C_1(d)$ telle que le nombre de boules disjointes de rayon $\alpha/2$ que peut rencontrer $B(A(B(0,\alpha)), 2\alpha)$ est majoré par

$$C_1(d) \prod_{i=1}^{d} \max(\delta_i(A), 1).$$

Démonstration de 2.3

Notons $A = K_1 \Delta K_2$ la décomposition de A ; les nombres $\delta_i(A)$ sont les coëfficients diagonaux de la matrice diagonale Δ.

Les opérateurs K_1 et K_2 étant unitaires conservent les distances et le nombre cherché est donc le même pour la matrice A et la matrice Δ.

Notons d' la distance $d'(x,y) = \max_i (x_i - y_i)$ le nombre cherché est plus petit que le nombre de d'-boules disjointes de rayon $\alpha/2\sqrt{d}$ que peut rencontrer $B_{d'}(\Delta(B_{d'}(0,\alpha)), 3\alpha)$.

Le lemme revient alors à compter le nombre de cubes disjoints de côté $1/2\sqrt{d}$ que peut rencontrer un paralléllépipède de côtés $2(\delta_i+3)$, $i=1,\ldots, d$. Ce nombre est majoré par $(4\sqrt{d})^d \prod_{i=1}^{d} (\delta_i + 4)$. Nous obtenons $C_1 \leq (20\sqrt{d})^d$ en remplaçant $\delta_i + 4$ par $5 \max(\delta_i, 1)$■

L'étape suivante consiste à linéariser. Notons en effet $\exp_x$ l'application exponentielle de $T_x X$ dans X ; par compacité, il existe un nombre δ tel que $\exp_x$ soit un difféomorphisme entre la boule de centre 0 et de rayon δ dans $T_x X$ et la boule de centre x et de rayon δ sur X. Fixons alors $n > 0$. Par continuité uniforme de $d_x\theta^n$ nous pouvons choisir ε_0 assez petit pour que $\varepsilon_0 \leq \delta/4 \max \{||d_x\theta^n|| \; x \in X\}$ et que si $\varepsilon < \varepsilon_0$, on a dès que $d(x,y) \leq \varepsilon$,

(1) $d(\theta^n y, \exp_{\theta^n x} d_x\theta^n \exp_x^{-1} y) \leq \varepsilon$ et

(2) $1/2 \leq \delta_i(d_x\theta^n)/ \delta_i(d_y \theta^n) \leq 2.$

La démonstration de 2.2 s'achève alors par un calcul d'entropie : un sous-ensemble fini E de X est dit ε-séparé si deux points distincts de E sont toujours distants de plus de ε. Pour tout ε, $0 < \varepsilon < \varepsilon_0$, choisissons E_ε, ε séparé maximal et P_ε une partition $P_\varepsilon = \{P_x, x \in E_\varepsilon\}$ telle que pour chaque x de E_ε, P_x est inclus dans la fermeture de son intérieur, et l'intérieur de P_x est l'ensemble des y qui vérifient pour tout $x' \neq x$ de E_ε $d(y,x) < d(y,x')$. Nous allons estimer le nombre d'éléments de P_ε que rencontre chaque $\theta^n P_x$. Nous avons :

$$\theta^n P_x \subset \theta^n(\exp_x (B(0,\varepsilon))) \text{ par maximalité de } E_\varepsilon$$

$$\subset \exp_{\theta^n x} (B(d_x \theta^n(B(0,\varepsilon)),\varepsilon)) \text{ d'après (1).}$$

Les éléments x' de E_ε tels que $P_{x'}$ rencontre $\theta^n P_x$ sont donc les centres de $\varepsilon/2$-boules disjointes (car E_ε est ε-séparé) qui rencontrent

$$\exp_{\theta^n x} (B(d_x \theta^n (B(0,\varepsilon)), 2\varepsilon)).$$

D'après le lemme 2.3, il y a donc moins de $C_1(d) \prod_{i=1}^{d} \max(\delta_i(d_x\theta^n),1)$ éléments de la partition P_ε qui rencontrent $\theta^n P_x$; il y a donc moins de $C_1(d) \prod_{i=1}^{d} \max(\delta_i(d_x\theta^n),1)$ éléments de la partition $\theta^{-n} P_\varepsilon$ qui rencontrent P_x.

Nous avons donc :

$$H(\theta^{-n}P_\varepsilon|P_\varepsilon) = \sum_{x\in E} m(P_x)\left(\sum_{y\in E} m(\theta^{-n} P_y|P_x) \log \frac{1}{m(\theta^{-n} P_y|P_x)}\right)$$

$$\leq \sum_{x\in E} m(P_x) \log\left(C_1(d) \prod_{i=1}^{d} \max(\delta_i (d_x\theta^n),1)\right).$$

D'après (2) nous avons si y appartient à P_x,

$$\log^+ \delta_i(d_x\theta^n) \leq \log 2 + \log^+\delta_i(d_y\theta^n) .$$

D'où finalement :

$$H(\theta^{-n} P_\varepsilon|P_\varepsilon) \leq \log C_1(d).\ 2^d + \sum_{i=1}^{d} \int \log^+(\delta_i(d_y\theta^n))m(dy).$$

Si nous faisons tendre ε vers zéro, nous avons d'après 1.2 et 1.3

$$n\, h_m(\theta) = h_m(\theta^n) = \lim_{\varepsilon\to 0} h_m(P_\varepsilon, \theta^n) \leq \lim_{\varepsilon\to 0} H(\theta^{-n} P_\varepsilon | P_\varepsilon)$$

$$\leq \log C_1(d) 2^d + \sum_{i=1}^{d} \int \log^+ (\delta_i\, (d_y \theta^n))\, m(dy).$$

Nous obtenons la majoration annoncée en faisant tendre n vers l'infini et en remarquant que par la convergence presque sûre I.2.6 , $\frac{1}{n} \int \log^+ \delta_i (d_y \theta^n) m(dy)$ converge vers λ_i^+ ■

Remarques et références :

La formule analogue à 2.2 dans le cas non ergodique s'obtient facilement par mélange de mesures ergodiques

$$h_m(\theta) = \int h_{m_\rho}(\theta)\, d\rho \leq \sum_i \int \lambda_i^+(m_\rho) d\rho .$$

La démonstration n'a utilisé que la convergence presque sûre I.2.6 et non pas les filtrations du théorème II.2.1.

Le résultat est en général une inégalité stricte. Soit μ la mesure de Riemann sur X. Si la mesure m admet une densité par rapport à μ, le théorème 2.2 est attribué à Margulis (sans référence). Il a été établi dans le cas général par Ruelle, dont nous avons suivi la démonstration :

D. RUELLE : An inequality for the Entropy of Differentiable Maps.Bol Soc. Bras Mat.9 (1978) p. 83-87.

En fait si la mesure m est absolument continue par rapport à la mesure μ et si θ est un C^1 difféomorphisme tel que les applications $x \to d_x\theta$ et $x \to d_x \theta^{-1}$ satisfont à une condition de Hölder, il y a égalité. C'est la formule de Pesin voir :

Ya. B PESIN : Lyapunov Characteristic Exponents and Smooth Ergodic Theory, Russ. Mat. Surveys, 32 ; 4 (1977) 55-114.

Une démonstration plus simple a été donnée par Mañé :

R. Mañé : A proof of Pesin's formula, Ergod Th. & Dynam. Sys. 1 (1981) 77-93.

Remarquons que dans ce cas les relations :

$h_m(\theta) = h_m(\theta^{-1})$ (1.3), $h_m(\theta) = \sum \lambda_i^+$

$h_m(\theta^{-1}) = - \sum \lambda_i^-$ (par I.4.1)sont compatibles : d'après I.1.3 nous avons en effet :

$$\sum \lambda_i = \int \log|\det d_x\theta| \quad m(dx) = 0$$

car si $m = h\mu$, $|\det d_x \theta| = \frac{h}{h.\theta}$.

En général, l'inégalité de 2.2 est stricte. Il est même possible de montrer une réciproque :

Si les différentielles $d_x\,\theta$ et $d_x\,\theta^{-1}$ sont höldériennes, si la mesure invariante m est ergodique, si 0 n'est pas un exposant du système et si la formule de Pesin est satisfaite :

$$h_m(\theta) = \sum \lambda_i^+$$

alors l'ensemble des points x de X tels que pour toute fonction continue f sur X, $\frac{1}{n}\sum_{i=0}^{n-1} f(\theta^i x)$ converge vers m(f) quand n tend vers l'infini est de μ mesure positive.

Physiquement on a envie d'interpréter ce résultat en disant que les mesures que l'on "voit" avec une probabilité positive sont les mesures satisfaisant un certain principe variationnel.

Le langage de la mécanique statistique et des probabilités seraient alors pertinents pour décrire les phénomènes déterministes avec sensibilité forte aux conditions initiales. Cf.

D.RUELLE : Attracteurs étranges. La Recherche 108 (1980).

3° Théorème de Shannon - Mc. Millan - Breiman

Pour utiliser le théorème d'Osseledets et obtenir des comparaisons plus précises que 2.2, il nous faut un résultat de convergence presque sûre concernant l'entropie.

Nous rappelons dans ce paragraphe le résultat et la démonstration classiques de la "propriété d'équipartition". Nous avons d'abord :

3.1 - Lemme

Soient $\mathcal{B}_n$ $n \geq 0$ une suite croissante de σ-algèbres, $\mathcal{B}_\infty$ la σ-algèbre engendrée par les $\mathcal{B}_n$, P une partition d'entropie finie. Alors :

$$\int \sup_n \sum_i 1_{P_i} \log \frac{1}{E(1_{P_i}|\mathcal{B}_n)} . dm \leq H(P) + 1 .$$

Démonstration :

Pour $\alpha > 0$, nous avons :

$$\left\{ \sup_n \sum_i 1_{P_i} \log \frac{1}{E(1_{P_i}|\mathcal{B}_n)} > \alpha \right\} = \sum_i P_i \cap \left\{ \sup_n E((e^{-\alpha} - 1_{P_i})|\mathcal{B}_n) > 0 \right\}.$$

Mais par le lemme maximal, la mesure de l'ensemble $P_i \cap \{\sup_n E((e^{-\alpha} - 1_{P_i})|\mathcal{B}_n) > 0\}$ est plus petite que $e^{-\alpha}$. Nous avons donc la majoration suivante pour $f(\alpha) = m(\{\sup_n \sum_i 1_{P_i} \log \frac{1}{E(1_{P_i}|\mathcal{B}_n)} > \alpha\})$:

$$f(\alpha) \leq \sum_i \min(e^{-\alpha}, m(P_i)),$$

Le lemme suit en observant que par conséquent, $\int_0^\infty f(\alpha)d\alpha \leq \sum_i \int_0^\infty \min(e^{-\alpha}, m(P_i))d\alpha$ et que le calcul du deuxième membre donne H(P)+1.

3.2 - Théorème

Soient $(X, \mathcal{A}, m ; \theta)$ un système dynamique ergodique et P une partition d'entropie finie.

Notons $f_n(x)$ la mesure de l'atome de la partition $P \vee \theta^{-1} P \vee \ldots \vee \theta^{-n+1} P$ qui contient x. Alors $-\frac{1}{n} \log f_n(x)$ converge m presque sûrement vers $h_m(P, \theta)$.

Démonstration :

Nous avons en effet, en notant $\mathcal{B}_j$ la σ-algèbre engendrée par les partitions $\theta^{-1}P,\dots,\theta^{-j}P$ et P_i, $i\in I$, les éléments de P :

$$f_n(x) = \sum_{i_1\dots i_n \in I^n} 1_{P_{i_n}}(\theta^{n-1}x)\,1_{P_{i_{n-1}}}(\theta^{n-2}x)\dots 1_{P_{i_1}}(x)$$

$$m(P_{i_n})\,E(1_{P_{i_{n-1}}}|\mathcal{B}_1)(\theta^{n-2}x)\dots E(1_{P_{i_1}}|\mathcal{B}_{n-1})(x)$$

et donc :

$$-\log f_n(x) = \sum_{j=0}^{n-1}\Big(-\sum_i 1_{P_i}\log E(1_{P_i}|\mathcal{B}_{n-j-1})\Big)(\theta^j x).$$

D'après le lemme 3.1 si nous posons :

$$g_m = \sup_{n\geq m}\sum_i 1_{P_i}\left|\log\frac{E(1_{P_i}|\mathcal{B}_\infty)}{E(1_{P_i}|\mathcal{B}_n)}\right|,$$

les fonctions g_m forment une suite décroissante vers 0 de fonctions intégrables. Nous pouvons alors estimer :

$$\left|-\log f_n(x) + \sum_{j=0}^{n-1}\Big(-\sum_i 1_{P_i}\log E(1_{P_i}|\mathcal{B}_\infty)(\theta^j x)\Big)\right| \leq \sum_{j=0}^{n-1} g_{n-j-1}(\theta^j x).$$

Donc la suite $-\frac{1}{n}\log f_n(x)$ a, presque partout et en L^1, le même comportement que les moyennes ergodiques d'une fonction intégrable. La suite $-\frac{1}{n}\log f_n(x)$ converge donc vers une constante qui est égale à

$$\lim_n -\int\frac{1}{n}\log f_n\,dm = \lim_n\frac{1}{n}H(P\vee\dots\vee\theta^{-n+1}P) = h_m(P,\theta)\ \blacksquare$$

On peut obtenir de la même manière :

3.3 - Théorème

Soient $(X,\mathcal{a}, m\ ;\theta)$ un système ergodique avec θ inversible et P une partition mesurable d'entropie finie. Soient a et b deux constantes positives. Notons $f_n(x)$ la mesure de l'atome de la partition $\vee\theta^j P$, $-an \leq j\leq bn$ qui contient x. Alors m presque sûrement, $-\frac{1}{n}\log f_n(x)$ converge vers $(a+b)\,h_m(P,\theta)$ quand n tend vers l'infini.

On trouvera encore remarques, extensions, références etc... dans le livre de Billingsley section 13.

4° Minoration de la dimension

Nous avons donc des théorèmes de convergence presque sûre pour différentes quantités dynamiques les exposants (I.2.6) et l'entropie (3.2). Dans le reste de ce chapitre, nous allons en déduire des relations géométriques que doit satisfaire la mesure en termes de coëfficients de Lipschitz ou de dimension. Nous montrons dans ce paragraphe une minoration de la dimension de Hausdorff des ensembles de mesure positive.

4.1 - Proposition

Soit θ un C^1 difféomorphisme de la variété X, m une mesure θ invariante ergodique avec $\lambda_1 > 0 > \lambda_d$. Alors en m presque tout x,

$$\liminf_{\varepsilon\to 0} \frac{\log m(B(x,\varepsilon))}{\log \varepsilon} \geq h_m(\theta) \left(\frac{1}{\lambda_1} - \frac{1}{\lambda_d}\right).$$

Démonstration de 4.1 :

Nous avons d'abord un lemme de théorie de la mesure qui va nous permettre de construire de bonnes partitions.

4.2 - Lemme

Soient $r > 0$, ν une mesure de probabilité sur $\mathbb{R}$ concentrée sur $[0,r]$ et $0 < a < 1$. Alors pour Lebesgue presque tout x, la série $\sum_{k=0}^{\infty} \nu([x-a^k, x+a^k])$ converge.

Démonstration de 4.2 :

Appelons N_k l'ensemble des points x, $0 \leq x \leq r$ avec $\nu([x-a^k,x+a^k]) \geq \frac{1}{k^2}$, et recouvrons N_k par des intervalles $C_{i,k}$, $C_{i,k} = [x_{i,k}-a^k, x_{i,k}+a^k]$ $1 \leq i \leq S_k$ en choisissant les $x_{i,k}$ dans N_k et tels que tout point ne rencontre que deux intervalles $C_{i,k}$. Nous pouvons estimer S_k par $\frac{S_k}{k^2} \leq \sum_{i=1}^{S_k} \nu(C_{i,k}) \leq 2$.

D'autre part si λ désigne la mesure de Lebesgue,

$$\lambda(N_k) \leq S_k\ 2a^k \leq 4\ a^k\ k^2.$$

Par Borel-Cantelli, λ-presque tout point n'appartient qu'à un nombre fini d'ensembles N_k et donc en λ-presque tout point le terme général $\nu(x-a^k, x+a^k)$ est majoré par $\frac{1}{k^2}$ à partir d'un certain rang. q.e.d. ∎

Soient (X, d, m) un espace compact métrique mesuré. Un sous-ensemble E de X est dit surrégulier si pour tout a, $0 < a < 1$, la série

$$\sum_{k=0}^{\infty} m(B(E, a^k)) \text{ converge.}$$

4.3 - Corollaire

Tout point de X admet un système fondamental de voisinages dont la frontière est surrégulière.

Il suffit en effet d'appliquer 4.2 à la mesure image de m par l'application d(x,.) distance au point x, et à une suite a_n tendant vers 0.

4.4 - Corollaire

Pour tout $\varepsilon > 0$, il existe une partition de X dont les éléments sont de diamètre inférieur à ε et ont une frontière surrégulière.

Il suffit en effet de choisir la partition subordonnée à un sous-recouvrement fini d'un recouvrement en petits voisinages donnés par 4.3.

Nous pouvons maintenant démontrer 4.1. Soit $\delta > 0$. Nous commençons par choisir d'après 4.4 une partition P dont les éléments ont une frontière surrégulière et telle que :

$$h_m(P,\theta) \geq h_m(\theta) - \delta.$$

Notons ∂P la réunion des frontières des éléments de P.

Si deux points de X sont assez proches, leurs images par θ^i resteront assez proches pour que les deux points soient dans le même atome de la partition $\vee\theta^{-j} P, - an \leq j \leq bn$ et que l'on puisse appliquer 3.3.

Fixons en effet $t > 0$.

Par surrégularité de ∂P, il existe une fonction $C(x,t)$, $1 \geq C(x,t) > 0$ m presque partout, telle que :

(1) $$d(\theta^n x, \partial P) \geq C(x,t)\, e^{-|n|t} \text{ pour tout } n \text{ de } \mathbb{Z}.$$

Nous avons en effet

$$\sum_n m(\{x \mid d(\theta^n x, \partial P) < e^{-|n|t}\})$$

$$= \sum_n m(\{x \mid d(x, \partial P) < e^{-|n|t}\}) \text{ par invariance de } m$$

$$= \sum_n m(B(\partial P, e^{-|n|t})) < +\infty.$$

Par Borel-Cantelli, cela montre qu'il existe un ensemble X_1, $m(X_1) = 1$, tel que si x appartient à X_1, il n'existe qu'un nombre fini d'entiers n pour lesquels $d(\theta^n x, \partial P) < e^{-|n|t}$.

Définissons $C(x,t)$ par

$$C(x,t) = \inf(1, d(\theta^n x, \partial P) e^{|n|t}, n \in \mathbb{Z}).$$

Si x appartient à X_1 et si aucun $\theta^n x$, $n \in \mathbb{Z}$, n'appartient à ∂P, la fonction $C(x,t)$ est strictement positive. Nous avons bien $C(x,t) > 0$ m-presque partout.

Par continuité uniforme de $||d_x\theta||$ et de $||d_x \theta^{-1}||$ il existe d'autre part ε tel que, si $d(x,y) \leq \varepsilon$,

(2) $$e^{-t} \leq \frac{||d_x\theta||}{||d_y \theta||} \leq e^t \text{ et } e^{-t} \leq \frac{||d_x \theta^{-1}||}{||d_y \theta^{-1}||} \leq e^t.$$

Enfin, par le théorème ergodique ponctuel, pour presque tout x, il existe $N(x)$ tel que si $n \geq N(x)$,

(3) $$\sup_{0 \leq j \leq n} \prod_{k=0}^{j-1} ||d_{\theta^k x} \theta|| \leq e^{n \int \log||d_x \theta||\, m(dx)}\, e^{nt}$$

et

$$\sup_{0 \leq j \leq n} \prod_{k=0}^{j-1} ||d_{\theta^{-k} x} \theta^{-1}|| \leq e^{n \int \log||d_x \theta^{-1}||\, m(dx)}\, e^{nt}.$$

Nous disons alors que si $n \geq N(x)$ et si $d(x,y) \leq \varepsilon\, C(x,t)\, e^{-n \int \log||d\theta|| - 3nt}$, nous avons pour tout j, $1 \leq j \leq n$,

$$d(\theta^j x, \theta^j y) \leq \varepsilon\, C(x,t)\, e^{-n \int \log ||d\theta||}\, e^{-2nt} \cdot \prod_{k=0}^{j-1} ||d_{\theta^k x} \theta||\, e^{(j-n)t}.$$

Notre choix de ε et (2) entraînent facilement cette propriété de proche en proche. Il suit alors de (1) et (3) que si $n \geq N(x)$ et

$$d(x,y) \leq \varepsilon\, C(x,t)\, e^{-n \int \log\|dn\theta\|} e^{-3nt},$$

nous avons $d(\theta^j x, \theta^i y) \leq d(\theta^j x, \partial P)$ pour $0 \leq j \leq n$ et donc x et y sont dans le même élément de $\bigvee_{j=0}^{n} \theta^{-j}P$.

De même, si $n \geq N(x)$ et si

$d(x,y) \leq \varepsilon\, C(x,t)\, e^{-n \int \log\|d\,\theta^{-1}\|} e^{-3nt}$, x et y sont dans le même élément de la partition $\bigvee_{j=0}^{n-1} \theta^j P$.

En posant alors $a_1(t) = \dfrac{1}{\int \log\|d.\theta\|+3t}$, $b_1(t) = \dfrac{1}{\int \log\|d.\theta^{-1}\|+3t}$

nous venons d'établir que pour n assez grand, si $d(x,y) \leq \varepsilon\, C(x,t)e^{-n}$, les points x et y appartiennent au même élément de la partition $\bigvee_{-a_1 n}^{b_1 n} \theta^j P$. En appliquant le théorème 3.3, il vient :

$$\liminf_n -\frac{1}{n}\log m(B(x,e^{-n})) \geq (a_1(t)+b_1(t))\; h_m(P,\theta)$$

m-presque partout, et donc par l'arbitraire de δ et de t,

$$(4) \qquad \liminf_{\varepsilon} \frac{\log m(B(x,\varepsilon))}{\log \varepsilon} \geq h_m(\theta)\left(\frac{1}{\int \log\|d.\theta\|} + \frac{1}{\int \log\|d.\theta^{-1}\|}\right).$$

Remarquons alors que le premier membre de (4) est le même si on considère la transformation θ^n au lieu de θ , mais que le deuxième devient, d'après 1.3 :

$$h_m(\theta)\left(\frac{n}{\int \log\|d.\theta^n\|} + \frac{n}{\int \log\|d.\theta^{-n}\|}\right).$$

Quand n tend vers l'infini, $\frac{1}{n}\int \log\|d.\theta^n\|$ tend vers λ_1 et $\frac{1}{n}\int \log\|d.\theta^{-n}\|$ tend vers $-\lambda_d$ par définition de λ_1 et $-\lambda_d$, ce qui achève la démonstration de 4.1 ∎

Montrons que ce résultat est lié à la notion de dimension de Hausdorff. Soit $t > 0$ et A une partie d'un ensemble compact métrique (X,d).

Pour tout $\varepsilon > 0$ considérons les recouvrements U_i de A par des ensembles mesurables U_{ij} de diamètre $\delta_{ij} \leq \varepsilon$.

Posons :

$$\mu_t(A) = \lim_{\varepsilon \to 0} \left(\inf_{U_i} \left(\sum_j \delta_{ij}^t \right) \right).$$

Il est facile de voir qu'il existe t_0, $0 \leq t_0 \leq +\infty$ tel que :

$$\mu_t(A) = +\infty \quad \text{si } t < t_0$$

$$\mu_t(A) = 0 \quad \text{si } t > t_0.$$

Cette valeur t_0 s'appelle la dimension de Hausdorff de la partie A, notée dim A.

Si m est une mesure de probabilité, nous noterons dim m = inf {dim A, A mesurable et m(A) > 0} .

4.5 - Corollaire

Sous les hypothèses de 4.1, alors $\dim m \geq h_m(\theta) \left(\frac{1}{\lambda_1} - \frac{1}{\lambda_d}\right)$.

Si en effet A est un ensemble de mesure positive, en choisissant $t' < h_m(\theta) \left(\frac{1}{\lambda_1} - \frac{1}{\lambda_d}\right)$, il existe $A' \subset A$, $m(A') > 0$ et ε' tel que si $\varepsilon \leq \varepsilon'$ $m(B(x,\varepsilon)) \leq \varepsilon^{t'}$ sur A'.

Pour tout ensemble U_{ij} qui rencontre A', de diamètre δ_{ij}, $\delta_{ij} < \varepsilon'/2$ nous avons $m(U_{ij}) \leq m(B(x,2\delta_{ij})) \leq 2^{t'} \delta_{ij}^{t'}$ et donc :

$$\mu_{t'}(A) \geq 2^{-t'} \inf_{U_i} \sum_i m(U_{ij}) \geq 2^{-t'} m(A') > 0.$$

Nous avons bien $t' \leq \dim A$ pour tout $t' < h_m(\theta) \left(\frac{1}{\lambda_1} - \frac{1}{\lambda_d}\right)$, q.e.d.

Références et remarques :

La proposition 4.1 est due à Young.

L.S. YOUNG : Dimension, Entropy and Lyapunov Exponents. Ergod-Th. & Dynam. Syst. 2 (1982) p. 109-124.

Sa démonstration repose sur une version topologique du théorème de Shannon Mc. Millan - Breiman, due à Brin et Katok :

A.M. Brin and A. Katok : On local entropy - prétirage.

Dans le cas où θ est non inversible, le même raisonnement montre que :

$$(5) \qquad \liminf_{\varepsilon\to 0} \frac{\log m(B(x,\varepsilon))}{\log \varepsilon} \geq \frac{h_m(\theta)}{\lambda_1} .$$

On s'attend que les estimations 4.1 et (5) soient les meilleures possibles en petite dimension, en dimension 2 pour 4.1 et en dimension 1 pour (5). Nous verrons au paragraphe 5 que c'est le cas si $d_x\theta$ et $d_x\theta^{-1}$ satisfont à une condition de Hölder.

Si la convergence presque sûre est délicate à établir, ou n'a pas lieu, on peut s'intéresser à des convergences plus faibles des fonctions $\frac{\log m(B(x,\varepsilon))}{\log \varepsilon}$, en moyenne ou en probabilité ou etc... Il revient au même de donner d'autres définitions de la dimension d'un espace métrique probabilisé.

En général, ce sera de la forme :

$$\dim_i m = \limsup_{\varepsilon\to 0} \frac{1}{\log 1/\varepsilon} \; I_\varepsilon^i .$$

On peut par exemple définir I_ε^1 comme la plus petite entropie possible d'une partition de X dont les éléments ont un diamètre inférieur à ε . Nous obtenons ainsi la dimension de Renyi :

A. RENYI : Dimension, entropy and information. Transations of the second Prague Conférence on Information Theory, Statistical Decision Functions and Random Processes (1957) 545-556.

On peut également définir $I_\varepsilon^{2,\delta}$ comme le logarithme du nombre minimum de boules de rayon ε nécessaire pour obtenir une mesure totale supérieure à $1-\delta$. On a alors :

$$\dim_2 m = \lim_{\delta\to 0} \limsup_{\varepsilon\to 0} \frac{I_\varepsilon^{2,\delta}}{\log 1/\varepsilon}$$

F. LEDRAPPIER : Some relations Between Dimension and Lyapounov Exponents. Commun Math. Phys. 81 (1981) 229-238.

Ces différentes notions correspondent à différents procédés concrets d'évaluer la dimension "essentielle" d'une probabilité sur un espace métrique. Les relations entre elles sont des exercices plus ou moins délicats. (Cf. L.S. Young, paragraphe 4).

En dimension 1 on a alors si θ et $d\theta$ sont monotones par morceaux la formule : $\quad h_m(\theta) = \lambda_1 \dim_i m \qquad i = 1$ ou 2.

5° Exposants, entropie et dimension en dimension 2

Considérons le cas où la variété X est de dimension 2, et les exposants vérifient $\lambda_1 > 0 > \lambda_2$.

On s'attend dans ce cas que l'estimation 4.2 soit la meilleure possible. Pour le voir, nous devons, à l'inverse, montrer que si deux points x et y sont tels que leurs images $\theta^i x$ et $\theta^i y$ restent proches $n/\lambda_2 \leq i \leq n/\lambda_1$, alors nécessairement d(x,y) est plus petit que e^{-n}. C'est intuitif si on néglige et la non-linéarité et la non-uniformité.

Le théorème d'Osseledets nous donne en effet une décomposition de l'espace tangent en une direction dilatante et une direction contractante ; en notant (u_1,u_2) les coordonnées dans ce système,

l'ensemble des y tels que $d(\theta^i x,\theta^i y) \leq \varepsilon \quad 0 \leq i \leq n/\lambda_1$

est à peu près $\{\exp_x(u_1,u_2),\ |u_1| \leq \varepsilon\, e^{-n}\ ,\ |u_2| \leq \varepsilon\}$

et l'ensemble des y tels que $d(\theta^i x,\theta^i y) \leq \varepsilon \quad n/\lambda_2 \leq i \leq 0$

est à peu près $\{\exp_x(u_1,u_2),\ |u_1| \leq \varepsilon,\ |u_2| \leq \varepsilon e^{-n}\}$.

Cela nous donnerait le résultat cherché en appliquant le théorème 3.3 à une partition P dont les éléments ont un diamètre plus petit que ε .

Dans ce paragraphe, nous montrons comment contrôler le "à peu près" de la phrase précédente en supposant que les différentielles de θ et de θ^{-1} satisfont à une condition de Hölder. Nous avons :

5.1 - Théorème

Soient X une variété compacte de dimension 2, θ un difféomorphisme de X tel que les différentielles $d_x\theta$ et $d_x\theta^{-1}$ satisfont à une condition de Hölder en x et m une mesure de probabilité θ invariante ergodique avec les exposants λ_1 et λ_2, $\lambda_1 > 0 > \lambda_2$. Nous avons en m-presque tout point x :

$$\lim_{\varepsilon \to 0} \frac{\log m(B(x,\varepsilon))}{\log \varepsilon} = h_m(\theta)\ \left(\frac{1}{\lambda_1} - \frac{1}{\lambda_2}\right).$$

Démonstration :

Nous savons déjà par 4.1 que la $\liminf_{\varepsilon\to 0}$ convient. Il nous reste à estimer

$$\limsup_{\varepsilon\to 0} \frac{\log m(B(x,\varepsilon))}{\log \varepsilon} .$$

D'après le théorème I.4.2, il existe un ensemble invariant X_0, $m(X_0) = 1$ et sur X_0 deux directions E_x^1 et E_x^2 de T_xX dépendant mesurablement de x telles que

(1) $$v \in E_x^i \Longleftrightarrow \frac{1}{n} \log||d_x\theta^n v||_{\theta^n x} \to \lambda_i \quad \text{quand } n \to +\infty \text{ et quand } n \to -\infty .$$

Fixons $\chi > 0$ et posons pour $n \in \mathbb{Z}$, $x \in X_0$ et $i = 1,2$

$$\rho^i(n,x) = \frac{||d_x\theta^n v||_{\theta^n x}}{||v||_x} \quad \text{pour un } v \neq 0 \text{ de } E_x^i.$$

Posons alors :

(2) $$\begin{cases} B_1(x) = \sum_{n=0}^{\infty} \dfrac{e^{(\lambda_1-\chi)n}}{\rho^1(n,x)} \\ B_2(x) = \sum_{n=0}^{\infty} \rho^2(n,x)\, e^{-(\lambda_2+\chi)n} \end{cases} .$$

D'après (1) B_1 et B_2 sont finis sur X_0 ∎

5.2 - Lemme

Nous avons les relations :

$$\rho^1(1,x) \geq e^{(\lambda_1-\chi)} \frac{B_1(\theta x)}{B_1(x)} \quad \text{et} \quad \rho^2(1,x) \leq e^{\lambda_2+\chi} \frac{B_2(x)}{B_2(\theta x)} .$$

Démonstration :

Il suffit de remplacer, dans l'expression de $B_i(\theta x)$, $\rho^i(n,\theta x)$ par $\rho^i(n+1,x)/\rho^i(1,x)$.

5.3 - Lemme

Pour tout ε, $0 < \varepsilon < \chi$, il existe une fonction $A_\varepsilon(x)$ mesurable sur X_0 telle que :

i) $1 \leq B_i(x) \leq A_\varepsilon(x)$, $i=1,2$

ii) pour tout $n \geq 0$, $A_\varepsilon(\theta^n x) \leq e^{n\varepsilon} A_\varepsilon(x)$.

<u>Démonstration</u>

D'après (1), il existe une fonction $C_\varepsilon(x)$ finie sur X_0 vérifiant pour tout $n \geq 0$

$$\frac{1}{C_\varepsilon(x)}\, e^{n(\lambda_1 - \frac{\varepsilon}{2})} \leq \rho^1(n,x) \leq C_\varepsilon(x)\, e^{n(\lambda_1 + \frac{\varepsilon}{2})} .$$

Posons $A'_\varepsilon(x) = \sup\limits_{n\geq 0\, k\geq 0} \dfrac{\rho^1(k,x)}{\rho^1(n+k,x)}\, e^{\lambda_1 n}\, e^{-\varepsilon k}\, e^{-\frac{\varepsilon}{2} n}$.

On vérifie immédiatement que :

$$A'_\varepsilon(x) \leq C^2_\varepsilon(x) < +\infty \text{ sur } X_0$$

$$B_1(x) \leq A'_\varepsilon(x) \sum_{n=0}^{\infty} e^{-(\chi - \frac{\varepsilon}{2})n} \leq \frac{A'_\varepsilon(x)}{1 - e^{-(\chi - \frac{\varepsilon}{2})}}$$

$$A'_\varepsilon(\theta x) \leq e^{\varepsilon} A'_\varepsilon(x).$$

De même en posant :

$$A''_\varepsilon(x) = \sup_{n\geq 0\, k\geq 0} \frac{\rho^2(n+k,x)}{\rho^2(k,x)}\, e^{-\lambda_2 n}\, e^{-\varepsilon k}\, e^{-\frac{\varepsilon}{2} n}$$

nous obtenons $A''_\varepsilon < +\infty$ sur X_0, $A''_\varepsilon . \theta \leq e^{\varepsilon} A''_\varepsilon$ et $B_2 \leq A'' \dfrac{1}{1 - e^{-(\chi - \frac{\varepsilon}{2})}}$.

Le lemme est démontré en choisissant :

$$A_\varepsilon = \frac{1}{1 - e^{-(\chi - \frac{\varepsilon}{2})}} \max (A'_\varepsilon, A''_\varepsilon) \quad \blacksquare$$

Dans un voisinage O_x de 0 dans T_xX, définissons F_x par $F_x = \exp^{-1}_{\theta x}\, \theta\, \exp_x$. L'application F_x envoie O_x dans $T_{\theta x}X$. Repérons T_xX pour x dans X_0 dans les coordonnées définies par (E^1_x, E^2_x). Notons α l'exposant de Hölder de $d_x\theta$ et $d_x\theta^{-1}$. Le lemme 5.4 suivant contrôle la non-linéarité et la non-uniformité.

5.4 - <u>Lemme</u>

> Soient $0 < \varepsilon < \chi\, \alpha$, $K > 0$ fixés, x un point de X_0. Notons pour y dans T_xX $y = (u_1, u_2)$ ses coordonnées. Il existe une fonction mesurable sur X_0, $C(\varepsilon,K,x)$ telle que :

i) $C(\varepsilon,K,\theta x) \geq e^{-\varepsilon} C(\varepsilon,K,x)$

ii) Si on a à la fois $||y||_x \leq C(\varepsilon,K,x)$ et $B_1(x)B_2(x)|u_2| \leq K|u_1|$,

Alors $F_x y = (v_1,v_2)$ est bien défini et vérifie :

$$|v_1| \geq |u_1| \; e^{\lambda_1 - 2\chi} \; \frac{B_1(\theta x)}{B_1(x)}$$

et $B_1(\theta x)\, B_2(\theta x)\, |v_2| \leq K\, |v_1|$.

<u>Démonstration</u> :

Il existe δ tel que si $||y||_x \leq \delta$ l'application F_x est définie et il existe M tel que :

$$||F_{x,y} - d_x\theta(y)||_{\theta x} \leq M\, ||y||_x^{1+\alpha}.$$

Nous pouvons donc écrire :

$$|v_1| = |d_x\theta\, u_1 + (F_x y - d_x\theta(y))_1|$$

$$\geq |u_1| \; e^{\chi_1-\chi} \; \frac{B_1(\theta x)}{B_1(x)} - M||y||_x^{1+\alpha}$$

d'après 5.2.

En remarquant que $||y||_x \leq \sqrt{2}\max |u_1|$, $|u_2| \leq \sqrt{2}|u_1| \max\left(1, \frac{K}{B_1\, B_2}\right)$

la première conclusion de (5.4 ii) est vraie dès que $||y||_x \leq \delta$ et

$$(3) \qquad ||y||_x^{\alpha} \leq \frac{e^{\lambda-\chi} - e^{\lambda-2\chi}}{M\sqrt{2}} \quad \frac{B_1(\theta x)\, B_2(x)}{\max(B_1\, B_2, K)} .$$

De même nous pouvons écrire :

$$|v_2| = |d_x\theta \;\; u_2 + (F_x y - (d_x\theta(y))_2|$$

$$\leq \frac{e^{\lambda_2+\chi}\, B_2(x)}{B_2(\theta x)} \quad |u_2| + M||y||_x^{\alpha} .$$

Si la première conclusion de (5.4.ii) est vraie, nous pouvons majorer $||y||_x$ et $|u_2|$ en fonction de $|u_1|$, et $|u_1|$ en fonction de $|v_1|$.

Nous obtenons :

$$B_1(\theta x)\, B_2(\theta x)\, |v_2| \leq K|v_1|\,(e^{\lambda_2-\lambda_1+3\chi} + ||y||_x^{\alpha}\cdot M_2)$$

avec :
$$M_2 = M\sqrt{2}\, e^{-\lambda_1+2\chi} \max\left(\frac{B_2.\theta.B_1}{K}, \frac{B_2.\theta}{B_2}\right).$$

La deuxième conclusion de(5.4.ii) est vraie dès que :

(4) $$||y||_x^{\alpha} \leq \frac{1-e^{\lambda_2-\lambda_1+3\chi}}{M_2}.$$

En appelant D_1 et D_2 les constantes convenables et en appliquant le lemme 5.3.i avec $\varepsilon\alpha/2$, nous savons que (3) et (4) sont réalisés dès que :

$$||y||_x^{\alpha} \leq D_1 \quad \min\left(\frac{1}{K}, \frac{1}{A^2_{\frac{\varepsilon\alpha}{2}}(x)}\right) \quad \text{et}$$

$$||y||_x^{\alpha} \leq D_2 \quad \min\left(\frac{1}{K}, \frac{1}{A_{\frac{\varepsilon\alpha}{2}}(x)}\right) \cdot \frac{1}{A_{\frac{\varepsilon\alpha}{2}}(\theta x)}.$$

Si nous appelons alors $(C(\varepsilon,K,x))^{\alpha}$ la plus petite de ces quantités et de δ^{α}, la propriété ii) suit de (3) et (4) et la propriété i) de (5.3.ii).

En échangeant le rôle de θ et θ^{-1} nous avons de même :

5.4 (bis) - <u>Lemme</u>

Soient $0 < \varepsilon < \chi\alpha$, K fixés et x un point de X_0. Notons pour y dans T_xX, $y = (u_1,u_2)$ ses coordonnées.

Il existe des fonctions mesurables $B_1'(x)$, $B_2'(x)$, $C'(\varepsilon,K,x)$ définies sur X_0 telles que :

i) $C'(\varepsilon,K,\theta^{-1}x) \geq e^{-\varepsilon}\, C'(\varepsilon,K,x)$

ii) si à la fois $||y||_x \leq C'(\varepsilon,K,x)$ et $B_1'(x)B_2'(x)\,|u_1| \leq K\,|u_2|$ alors $F^{-1}_{\theta^{-1}x}\, y = (w_1,w_2)$ est bien défini et vérifie :

$$|w_2| \geq |u_2|\, e^{-\lambda_2-2\chi}\, \frac{B_2'(\theta x)}{B_2'(x)}$$

et $B_1'(\theta^{-1}x)\, B_2'(\theta^{-1}x)\, |w_1| \leq K\,|w_2|$.

Nous pouvons maintenant faire le raisonnement annoncé plus haut pour estimer $m(B(x,e^{n}))$: choisissons K assez grand pour que sur un ensemble A de mesure positive, à la fois B_1, B_2, B_1', B_2' sont inférieurs à K.

Alors 5.4 et 5.4 bis montrent que si x appartient à A et y vérifie :

$$d(\theta^i x, \theta^i y) \leq C(\varepsilon, K^2, \theta^i x) \text{ pour } 0 \leq i \leq \frac{n}{\lambda_1 - 2\chi}$$

$$d(\theta^i x, \theta^i y) \leq C^1(\varepsilon, K^2, \theta^i x) \text{ pour } \frac{n}{\lambda_2 + 2\chi} \leq i \leq 0$$

alors $d(x,y) \leq 2\, K\, e^{-n}$.

En effet, en considérant les coordonnées (u_1, u_2) de $\exp_x^{-1} y$, nous avons ou bien $|u_2| \leq |u_1|$, et donc $B_1\, B_2\, |u_2| \leq |u_1|\, K^2$ et nous pouvons appliquer 5.4 successivement en tous les points $\theta^i x$ aux images $\theta^i y$, $0 \leq i < \frac{n}{\lambda_1 - 2\chi}$, ou bien $|u_1| \leq |u_2|$, et donc $B_1'\, B_2'\, |u_1| \leq |u_2|\, K^2$ et nous pouvons appliquer 5.4 bis successivement en tous les points $\theta^i x$ aux images $\theta\, y$, $\frac{x}{\lambda_2 + 2\chi} < i \leq 0$.

Nous trouvons dans le premier cas :

$$\delta \geq d(\theta^{n/\lambda_1 - 2\chi} x, \theta^{n/\lambda_1 - 2\chi} y) \geq |u_1|\, e^n \frac{B_1(\theta^{n/\lambda_1 - 2\chi} x)}{B_1(x)}$$

$$\geq \frac{1}{2}\, d(x,y)\, e^n \frac{1}{K}$$

et de même dans le second cas.

Nous appliquons ensuite le lemme 5.5 suivant, dont la démonstration sera donnée en appendice (paragraphe 6).

5.5 - Lemme

> Soient X une variété compacte θ une application différentiable de X dans X et m une mesure de probabilité invariante. Soit f une fonction mesurable, $f > 0$ m-presque partout, telle qu'il existe $0 < A < 1$ avec $f(\theta x) \geq A\, f(x)$. Il existe un ensemble B de mesure arbitrairement proche de un et une partition P_B d'entropie finie tels que si x appartient à B et si y et x sont dans le même élément de chaque partition $\theta^{-i} P_B$, $0 \leq i < k$, alors $d(\theta^i x, \theta^i y) \leq f(\theta^i x)$ pour $0 \leq i < k$.

En appliquant le lemme 5.5 aux fonctions $C(\varepsilon, K^2, x)$ et $C'(\varepsilon, K^2, x)$, nous trouvons deux ensembles B et B', de mesure assez grande pour que $m(A \wedge B \wedge B') > 0$ et deux partitions P_B et $P'_{B'}$, tels que si x appartient à $A \cap B \cap B'$ et si y et x sont dans le même élément de chaque partition $\theta^i Q$, $-\frac{n}{\lambda_1 - 2\chi} \leq i \leq -\frac{n}{\lambda_2 + 2\chi}$ avec $Q = P_B \vee P'_{B'}$, alors $d(x,y) \leq 2\delta K e^{-n}$. D'après le théorème 3.3 cela implique que si x appartient à $A \cap B \cap B'$ nous avons :

$$\limsup_{\varepsilon \to 0} \frac{\log m(B(x,\varepsilon))}{\log \varepsilon} \leq h_m(Q,\theta)\left(\frac{1}{\lambda_1 - 2\chi} - \frac{1}{\lambda_2 + 2\chi}\right).$$

En remarquant que $h_m(Q,\theta)$ est majoré par $h_m(\theta)$ et que le premier membre est une fonction invariante, nous avons l'estimation suivante valable m presque partout ;

$$\limsup_{\varepsilon \to 0} \frac{\log m(B(x,\varepsilon))}{\log \varepsilon} \leq h_m(\theta)\left(\frac{1}{\lambda_1 - 2\chi} - \frac{1}{\lambda_2 + 2\chi}\right).$$

Le théorème 5.1 suit alors de l'arbitraire de χ.

Références et remarques :

Le théorème 5.1 est dû à Young. Les Lemmes 5.2 à 5.4 bis sont typiques de la théorie "à la Pesin" : nous utilisons le théorème d'Osseledets pour faire un changement de norme dans l'espace tangent tel que la différentielle est alors uniformément contractante ou dilatante. De plus, et c'est le point important, toutes les approximations sont contrôlées par des fonctions qui croissent ou décroissent lentement le long des orbites (propriété 5.3 i). La condition de Hölder intervient alors pour conserver cette propriété quand on délinéarise. Il est alors possible d'étendre beaucoup de résultats obtenus pour les systèmes "uniformément hyperboliques". Le mieux est de se reporter aux articles :

Ya.B.PESIN : Families of invariant manifolds corresponding to Non-zero characteristic Exponents. Math. of the USSR Izvestija, 10;6 (1978) p. 1261-1305.

A. FATHI, M.R. HERMAN, J.C. YOCCOZ : A proof of Pesin's stable manifold theorem. Prépublication - Orsay (1981).

A. KATOK : Lyapunov exponents, entropy and periodic orbits for diffeomorphisms Publ. Math. I.H.E.S. 51 (1980) 137-174.

Le Théorème 5.1 est également valable en dimension 1 :

Si θ est une application $C^{1+\alpha}$ d'un intervalle dans lui-même si m est une mesure ergodique telle que $\int \log |d\theta| \, dm > 0$, alors m-presque partout :

$$h_m(\theta) = \int \log |d\theta| \, dm \, . \, \lim_{\varepsilon \to 0} \frac{\log m(x-\varepsilon, x+\varepsilon)}{\log \varepsilon}$$

(Remarquons au contraire que par une modification du lemme 4.2, il est facile de voir que Lebesgue presque partout

$$\liminf_{\varepsilon \to 0} \frac{\log m(x-\varepsilon, x+\varepsilon)}{\log \varepsilon} \geq 1 \text{).}$$

En dimension plus grande que 2, il suffit de considérer les produits de deux systèmes pour se convaincre qu'il n'existe pas de formule aussi simple que 5.1 valable en toute généralité.

Dans un article qui est à l'origine de ce regain d'intérêt pour la dimension en systèmes dynamiques,

P. FREDIRCKSON, J. KAPLAN, E. YORKE et J. YORKE : (The Liapounov Dimension of Strange Attractors, à paraître dans J. Diff. Eq.)

introduisent la notion de dimension de Liapounov. Soient $(X, \mathcal{A}, m, \theta)$ un système différentiable, $\lambda_1 \geq \lambda_2 \geq .. \geq \lambda_d$ les exposants caractéristiques. Soit j le dernier entier tel que $\lambda_1 + \ldots + \lambda_j \geq 0$.

Posons $$\text{lia-dim } m = j + \frac{\lambda_1 + \ldots + \lambda_j}{|\lambda_{j+1}|}$$

(lia-dim m = 0 si $\lambda_1 < 0$, = d si $\lambda_1 + \ldots + \lambda_d > 0$).

On peut montrer que si $d_x\theta$ satisfait à une condition de Hölder et si le système est ergodique, $\dim_2 m \leq$ lia dim m . (Cf. notes du paragraphe 4 pour la définition de $\dim_2$).

Inversement dès la dimension 2, le théorème 5.1 montre que l'on ne peut avoir dim m = lia-dim m et dim X = 2 que si $h_m(\theta) = \lambda_1$, c'est-à-dire si on a égalité dans 2.2, ce qui n'est réalisé que pour certaines mesures (cf. notes du paragraphe 2).

6 - Appendice au chapitre II : démonstration du lemme 5.5

La démonstration du lemme 5.5 repose sur deux observations :

6.1 - Soit B un sous-ensemble de X de mesure positive, notons, pour $n \geq 1$,

$$B_n = \{x \in B / \theta^i x \notin B \quad 0 < i < n, \ \theta^n x \in B\} .$$

Alors B_n, $n \geq 1$ forme une partition de B et $\sum_{n \geq 1} n \, m(B_n) = 1$ (Formule de Kač).

6.2 - Soit X une variété compacte de dimension d. Il existe une constante C tel que si E est un ensemble ε séparé dans X, card $E \leq C \ \varepsilon^{-d}$.

Reprenons les notations de l'énoncé de 5.5. Pour tout $\delta > 0$ posons $B_\delta = \{f > \delta\}$. On peut choisir δ de manière à ce que la mesure de B_δ soit arbitrairement proche de 1. Supposons δ_0 choisi et notons $B = B_{\delta_0}$.

Soit L une constante de Lipschitz globale pour la transformation θ et pour chaque $n \geq 1$ choisissons E_n un ensemble $\frac{\delta A^n}{2L^n}$ séparé maximal dans X et Q_n une partition subordonnée. Par maximalité de E_n, les éléments $Q_{n,j}$ de Q_n sont de diamètre inférieur à $\frac{\delta A^n}{L^n}$.

Considérons alors P la partition dont les éléments sont $B_0 = X \setminus B$ et les $B_n \cap Q_{nj}$, $n \geq 1$, et vérifions les propriétés annoncées.

Si x appartient à B et si y appartient au même élément de P que x, x et y appartiennent au même ensemble B_{n_0} et donc $\theta^{n_0} x$ et $\theta^{n_0} y$ appartiennent à B.

D'autre part nous avons comme x est dans B_{n_0} et y dans le même ensemble $Q_{n_0,j}$,

$$d(x,y) \leq \frac{A^{n_0}\delta}{L^{n_0}} \leq \frac{A^{n_0} f(x)}{L^{n_0}} .$$

Pour $0 \leq i < n_0$, nous avons donc :

$$d(\theta^i x, \theta^i y) \leq L^i \ d(x,y) \leq \frac{L^i}{L^{n_0}} \ A^{n_0} \ f(x) \leq A^i f(x) \leq f(\theta^i x),$$

ce qui montre la propriété pour $0 \leq i < n_0$.

Si $k \leq n_0$ il n'y a rien à montrer de plus ; sinon nous pouvons recommencer à partir de $\theta^{n_0} x$ et $\theta^{n_0} y$ qui sont également dans le même élément de P. Et ainsi de suite jusqu'au premier instant n supérieur à k où $\theta^n x$ appartient à B.

Reste à vérifier que $H(P) < \infty$. Nous avons en notant R la partition de X en B_n, $n \geq 0$.

$$H(P) = H(R) + H(P \mid R).$$

Comme $\sum_{n=1}^{\infty} n\, m(B_n) < +\infty$, $H(R) < \infty$ et d'autre part :

$$H(P/R) \leq \sum_{n=1}^{\infty} m(B_n) \log(\text{card } Q_n)$$

$$\leq \sum_{n=1}^{\infty} m(B_n) \left(\log \frac{C2^d}{\delta^d} + n \log \frac{L^d}{A^d}\right)$$

$$< + \infty.$$

Ceci achève la démonstration du lemme.

Le lemme et sa démonstration sont dus à Mañé,

R. MAÑE : A proof of Pesin's formula. Ergod. Th. & Dynam. Sys. 1 (1981) 77-93.

III - FAMILLES INDEPENDANTES

1. La formule de Fürstenberg pour λ_1

Dans ce chapitre nous considérons une famille indépendante de matrices. Nous nous limiterons à des matrices de déterminant 1.

Plus précisément nous considérons l'ensemble M des mesures de probabilités sur SL(d, R) telles que :

i) $\int \log ||g|| \, \mu(dg) < +\infty$

ii) il n'existe pas de réunion finie de sous-espaces vectoriels propres de $\mathbb{R}^d$ invariante par μ-presque tout g.

La condition ii) est appelée l'irréductibilité. L'ensemble des mesures de probabilité sur SL(d, $\mathbb{R}$) vérifiant i) est muni de la topologie faible définie par les fonctions continues majorées en valeur absolue par k log $||g||$, $k \geq 0$. Il est clair alors que l'ensemble M est une intersection dénombrable d'ouverts dans cet espace. L'ensemble M est donc un espace polonais pour la topologie induite.

Soit μ une mesure de M . Pour décrire une famille indépendante de matrices de loi μ, nous considérerons l'espace (X,m), où $(X,m) = (SL(d, \mathbb{R}),\mu)^{\mathbb{N}}$ est l'espace produit de $\mathbb{N}$ copies de $(SL(d,\mathbb{R}), \mu)$, la transformation θ décalage des coordonnées $(\theta x)_n = x_{n+1}$ $n \geq 0$, et l'application A de X dans SL(d,$\mathbb{R}$) définie par la première coordonnée, $A(x) = x_0$.

D'après la condition i), nous pouvons appliquer I 1.1 et définir alors les exposants caractéristiques $\lambda_1,\ldots, \lambda_d$ d'une famille indépendante de matrices de loi μ.

Remarquons que si A est une matrice de SL(d, $\mathbb{R}$)
$1 \leq ||A^{-1}|| \leq C(d) \, ||A||^{d-1}$ et donc que nous avons également $\int \log ||A^{-1}(x)||m(dx)< \infty$ si bien que $\lambda_d > -\infty$.

Nous considérons encore comme au paragraphe I.5 l'espace $\mathbb{P}^{d-1}$ des directions de $\mathbb{R}^d$ et l'action naturelle de SL(d,$\mathbb{R}$) sur cet espace. Si ν est une mesure de probabilité sur $\mathbb{P}^{d-1}$, nous notons $\mu * \nu$ la mesure définie par :

$$\int f(t) \, \mu * \nu(dt) = \int f(gt) \, \nu(dt) \, \mu(dg).$$

Nous notons I_μ l'ensemble des mesures de probabilité sur $\mathbb{P}^{d-1}$ invariantes par μ, c'est-à-dire des mesures ν telles que $\mu * \nu = \nu$. L'ensemble I_μ est un convexe compact non vide de mesures de probabilité sur $\mathbb{P}^{d-1}$.

1.1 - Théorème

Soit F la fonction sur $(SL(d, \mathbb{R}) \times P^{d-1})$ définie par

$F(g,\dot{t}) = \log \frac{||gt||}{||t||}$ pour un t non nul de $\dot{t}$.

Quelle que soit la mesure ν de I_μ,

$$\lambda_1 = \int (F(g,\dot{t})\ \mu(dg)\ \nu(d\dot{t}).$$

Démonstration :

Considérons sur $X \times \mathbb{P}^{d-1}$ la transformation :

$$\hat{\theta} : \ \hat{\theta}(x,\dot{t}) = (\theta x,\ x_0\dot{t})$$

et remarquons que si ν est une mesure de I_μ, la mesure produit $m \times \nu$ est $\hat{\theta}$ invariante. Si nous notons encore $F(x,t) = \log \frac{||x_0\ t||}{||t||}$, nous avons immédiatement la formule

$$\sum_{i=0}^{n-1} F(\hat{\theta}^i(x,\dot{t})) = \log \frac{||x_{n-1} \dots x_0 t||}{||t||}.$$

En appliquant le théorème ergodique ponctuel, il vient que pour $m \times \nu$ -presque tout $(x,\dot{t})$, la suite $\frac{1}{n} \log \frac{||x_{n-1} \dots x_0 t||}{||t||}$ converge vers une fonction invariante, version de l'espérance conditionnelle de F par rapport à la σ-algèbre des invariants $E(F \mid \mathfrak{I})$. Autrement dit, il existe un ensemble X_1, $m(X_1) = 1$, tel que si x appartient à X_1, la convergence a lieu pour ν-presque tout t.

D'autre part, d'après le théorème I 3.1 il existe un ensemble X_0, X_0, $m(X_0) = 1$, tel que si x est dans X_0, il existe un sous-espace vectoriel propre de $\mathbb{R}^d$ V_x^{r-1} tel que si t n'est pas dans V_x^{r-1}, alors $\frac{1}{n} \log \frac{||x_{n-1} \dots x_0 t||}{||t||}$ converge vers λ_1.

Nous avons donc $E(F \mid \mathfrak{I}) = \lambda_1$ $(m \times \nu)$ p.s. et donc la conclusion du théorème sauf si :

(1) $\quad m(\{x \mid x \in X_0,\ \ \nu(V_x^{r-1}) > 0\}) > 0.$

Montrons que la condition d'irréductibilité interdit que (1) soit réalisé. Plus précisément, la mesure ν est "diffuse" dans le sens où il n'y a pas de sous-espace de dimension finie propre chargé par ν.

En effet, soit d_0 la plus petite dimension telle que il existe un sous-espace V de $\mathbb{R}^d$ de dimension d_0 chargé par ν , et ν_0 la plus grande mesure d'un sous-espace de dimension d_0. Soient $V_1, V_2, \ldots, V_k$ les sous-espaces tels que $\nu(\dot{V}_i) = \nu_0 \quad i=1,\ldots k$.

Si $d_0 < d$, la relation d'invariance :

$$\nu(\dot{V}_i) = \int \nu(g^{-1}\dot{V}_i)\, \mu(dg)$$

ne peut être satisfaite que si les $g^{-1}\ V_i$ sont des V_j μ-presque partout et donc si la réunion des V_j, $j=1,\ldots, k$ est invariante par μ-presque tout g, ce qui contredit l'irréductibilité de μ.

1.2 - Corollaire

L'application qui à μ associe $\lambda_1(\mu)$ est continue sur M .

Considérons en effet une suite μ_n convergeant vers μ dans M et ν_n une mesure de I_{μ_n}, $n \geq 0$. Pour toute sous-suite n_k, $k \geq 0$, la suite de mesures sur l'espace compact ν_{n_k} admet des points d'accumulation faible. Pour tout point d'accumulation ν_∞ d'une sous-suite (notée encore ν_n, $n \geq 0$) nous pouvons affirmer que ν_∞ appartient à I_μ car $\mu \times \nu_\infty$ est la limite faible des mesures $\mu_n \times \nu_n$ et en particulier $\mu * \nu_\infty$ est la limite faible des mesures $\mu_n * \nu_n$, i.e. $\mu * \nu_\infty = \nu_\infty$.
Donc d'après 1.1, nous avons :

$$\lambda_1(\mu) = \int F(g,t)\, \mu(dg)\, \nu_\infty(dt)$$

$$= \lim_n \int F(g,t)\, \mu_n(dg)\, \nu_n(dt),$$

car la fonction F(g,t) est continue et majorée par $\log||g||$; nous avons $\lambda_1(\mu) = \lim \lambda_1(\mu_n)$ d'après 1.1 encore. Ceci établit la continuité de λ_1.

1.3 - Corollaire

Si μ appartient à M , pour tout vecteur v de R^d, $v \neq 0$,
$\lim_n \frac{1}{n} \log||A^{(n)}(x)\ v|| = \lambda_1$ m presque partout quand n tend vers l'infini.

Nous avons en effet par I-3.5 que sur un ensemble X_0 de mesure un la limite existe pour tout v non nul de $\mathbb{R}^d$ et est un des exposants.

En particulier si $x \in X_0$, $\lim_n \frac{1}{n} \log ||A^{(n)}(x) v|| \leq \lambda_1$.

Soient alors v_0 dans $\mathbb{R}^d$ tel que :

$$m(\lim_n \frac{1}{n} \log ||A^{(n)}(x) v_0|| < \lambda_1) > 0 \quad \text{et } t_0 = \dot{v}_0$$

la direction de v_0.

Formons successivement $\mu_0 = \delta_{t_0}$, $\mu_n = \mu * \mu_{n-1}$, $\nu_n = \frac{1}{n} \sum_{i=0}^{n-1} \mu_i$ et choisissons ν un point d'accumulation faible de la suite ν_n de mesures sur $\mathbb{P}^{d-1}$.

La mesure ν est invariante, appartient à I_μ et nous avons :

$$\int F(g,t)\, \mu(dg)\nu(dt) \leq \overline{\lim}\ \frac{1}{n} \sum_{i=0}^{n-1} \int(\int F(g,t)\, \mu(dg))\, \mu_i(dt)$$

$$\leq \int \overline{\lim_n}\ \frac{1}{n} \sum_{i=0}^{n-1} F(g,\, x_{i-1}\, x_{i-2} \ldots x_0 v_0)\, \mu(dg)\mu(dx_{i-1}) \ldots \mu(dx_0)$$

$$\text{car } F(g.) \leq \log ||g||$$

$$= \int \overline{\lim_n}\ \frac{1}{n} \log \frac{||A^{(n)}(x)\ v_0||}{||v_0||}\ m(dx)$$

$$= \int \lim_n \frac{1}{n} \log ||A^{(n)}(x)\ v_0|| m(dx)$$

car nous savons que la limite existe presque partout. Notre hypothèse sur v_0 impose alors que $\int F(g.t)\, \mu(dg)\, \nu(dt) < \lambda_1$, ce qui contredit (1.1)■

Remarquons que ici, l'ensemble de convergence de (1.3) dépend de V_0, comme nous le verrons plus loin (théorème IV 1.2) sur un exemple.

Références :

Les résultats de ce paragraphe (et des suivants)sont dans :

H. FÜRSTENBERG : Non-commuting random products.Transactions. Amer. Math. Soc. 108 (1963) p. 377-428. (cf. en particulier le théorème 8.5).

Notre exposition ici n'est ni meilleure ni plus simple, nous voulons utiliser le lien entre le théorème d'Osseledets et ces résultats. cf. aussi :

Y. GUIVARC'H : Quelques propriétés asymptotiques des produits de matrices aléatoires. Ecole d'été de probabilités de Saint-Flour VIII. 1978. Springer L.N in maths 774 (1980).

Y. GUIVARC'H : Sur les exposants de Liapunoff des marches aléatoires à pas markovien. C.R.A.S. Paris.

La continuité 1.2, même en dehors de M est également discutée par :

Y. KIFER : Perturbations of random matrix products Z. Wahrscheinlichkeittheorie u. verw. Geb. 61(1982) 83-95.

Y. KIFER and E. SLUD : Perturbations of random matrix products in a reducible case prétirage.

2 - Entropie d'un produit indépendant de matrices

Nous introduisons dans ce paragraphe un nombre associé à une mesure μ de M.

Soient μ une mesure de M, ν une mesure sur $\mathbb{P}^{d-1}$, ν appartenant à I_μ.

Posons $a(\mu,\nu) = -\frac{1}{d}\int \log \frac{dg^{-1}\nu}{d\nu}(\dot{t})\ \nu(d\dot{t})\ \mu(dg)$, avec la convention que si $\frac{dg^{-1}\nu}{d\nu} = 0$, $-\log \frac{dg^{-1}\nu}{d\nu} = +\infty$ mais que l'on néglige cette contribution si elle n'intervient que sur un ensemble négligeable.

Remarquons que pour tout $(g,\dot{t})$, nous avons $-\log \frac{dg^{-1}\nu}{d\nu}(\dot{t}) \geq 1 - \frac{dg^{-1}\nu}{d\nu}(\dot{t})$ et que $(g,\dot{t}) \to \frac{dg^{-1}\nu}{d\nu}(\dot{t})$ est une version de la densité de la mesure $g^{-1}\nu(d\dot{t})\mu(dg)$ par rapport à la mesure $\nu \times \mu$. La fonction $-\log \frac{dg^{-1}\nu}{d\nu}(\dot{t})$ est donc minorée par une fonction intégrable, d'intégrale positive ou nulle. L'intégrale a bien un sens et définit un nombre positif, fini ou infini.

Remarquons encore que si $a(\mu,\nu)$ est fini, alors pour μ-presque tout g, nécessairement la mesure ν est absolument continue par rapport à $g^{-1}\nu$, ou encore $g\nu$ est absolument continue par rapport à ν.

2.1 - Définition

On appelle entropie de μ le nombre $\alpha(\mu)$

$$\alpha(\mu) = \inf \{a(\mu,\nu),\ \ \nu \in I_\mu\}.$$

2.2 - Théorème

La fonction réelle positive qui à μ dans M associe son entropie $\alpha(\mu)$ est semi-continue inférieurement sur M. De plus, il existe ν_0 dans I_μ telle que $\alpha(\mu) = a(\mu,\nu_0)$.

Démonstration :

Nous avons d'abord quelques lemmes pour contrôler le comportement des quantités $a(\mu,\nu)$.

2.3 - Lemme

Soient μ_n une suite convergeant vers μ dans M, ν_n des mesures de I_{μ_n} convergeant faiblement vers une mesure ν de I_μ, A un sous-ensemble ouvert de $\mathbb{P}^{d-1}$ dont la frontière est incluse dans une réunion finie de sous-espaces vectoriels propres de $\mathbb{R}^d$. Nous avons alors :

$$-\nu(A) \int \log \nu(gA) \leq \liminf_n - \nu_n(A) \int \log \nu_n(gA)\, \mu_n(dg).$$

Démonstration de 2.3 :

Observons d'abord que $-\log(1-t) = \sum_{m=1}^{\infty} \frac{t^m}{m}$ et donc qu'il suffit d'établir que pour tout $m \geq 1$

$$\int (1-\nu_n(gA))^m\, \mu_n(dg) \text{ converge vers } \int (1-\nu(gA))^m\, \mu(dg) \text{ quand}$$

n tend vers l'infini, et que $\nu_n(A)$ tend vers $\nu(A)$.

Or nous savons que ν_n tend vers ν faiblement, que les mesures $\mu_n^{(m)}$ sur $SL(d,\mathbb{R}) \times (\mathbb{P}^{d-1})^m$ définies par :

$$\mu_n^m = g^{-1}\nu_n(dt_1)\, g^{-1}\nu_n(dt_2) \ldots g^{-1}\nu_n(dt_n)\, \mu_n(dg)$$

convergent faiblement vers la mesure

$$\mu_\infty^m = g^{-1}\nu(dt_1)\, g^{-1}\nu(dt_2) \ldots g^{-1}\nu(dt_m)\, \mu(dg),$$

et enfin que les ensembles considérés $t_1 \notin A$, $t_2 \notin A$, ... $t_m \notin A$ sont des fermés de $SL(d,\mathbb{R}) \times (\mathbb{P}^{d-1})^m$ de frontière négligeable pour μ_∞^m.

Les convergences annoncées en découlent immédiatement ■

Notons $\mathcal{R}$ l'ensemble des recouvrements finis $\mathcal{C}$ de $\mathbb{P}^{d-1}$ par des ensembles A_i tels que la frontière de chaque A_i est incluse dans une réunion finie de sous-espaces vectoriels propres de $\mathbb{R}^d$, et les intérieurs des ensembles A_i sont disjoints. Posons pour μ dans M, ν dans I_μ et $\mathcal{C}$ dans $\mathcal{R}$:

$$\rho(\mathcal{C}, \mu, \nu) = -\int \Big(\sum_{A \in \mathcal{C}} \nu(A) \log \frac{\nu(gA)}{\nu(A)} \Big)\, \mu(dg).$$

Avec ces notations nous avons :

2.4 - Corollaire

Pour tout μ de M et $\mathcal{C}$ de $\mathcal{R}$, il existe ν_a dans I_μ tel que $\rho(\mathcal{C}, \mu, \nu_a) \leq \rho(\mathcal{C}, \mu, \nu)$ pour tout ν dans I_μ.
La fonction $\mu \to \rho(\mathcal{C}, \mu, \nu_a)$ est alors semi-continue inférieurement sur M.

Le corollaire suit clairement de 2.3 et du fait que I_μ est fermé dans le compact des mesures de probabilités sur P^{d-1}.

2.5 - Lemme

i) Si le recouvrement $\mathcal{G}$ raffine $\mathcal{G}'$ (i.e. chaque élément de $\mathcal{G}$ est inclus dans un élément de $\mathcal{G}'$)

$$\rho(\mathcal{G}', \mu, \nu) \leq \rho(\mathcal{G}, \mu, \nu)$$

ii) $$d\, da(\mu,\nu) = \sup \{\rho(\mathcal{G}, \mu,\nu) ; \mathcal{G} \in \mathcal{R}\} .$$

Démonstration de 2.5

Nous avons en effet :

$$\rho(\mathcal{G}', \mu, \nu) = -\int \log \Big(\sum_{A' \in \mathcal{G}'} 1_{A'}(t) \frac{\nu(gA')}{\nu(A')} \Big) \nu(dt)\, \mu(dg)$$

et

$$\sum_{A' \in \mathcal{G}'} 1_{A'}(t) \frac{\nu(gA')}{\nu(A')} = \nu \Big(\sum_{A \in \mathcal{G}} 1_A(t) \frac{\nu(gA)}{\nu(A)} \mid \sigma(\mathcal{G})\Big)$$

où $\sigma(\mathcal{G})$ désigne le σ-algèbre engendrée par la partition (aux négligeables près) $\mathcal{G}$. La propriété i) suit alors de l'inégalité de Jensen. Pour établir la propriété ii), rappelons que $\frac{dg^{-1}\nu}{d\nu}$ désigne la densité de la partie absolument continue de la mesure $g^{-1}\nu$ par rapport à la mesure ν . Nous avons alors :

$$\sum_{A \in \mathcal{G}} 1_A(t) \cdot \frac{\nu(gA)}{\nu(A)} \geq E_\nu\Big(\frac{dg^{-1}\nu}{d\nu} \mid \sigma(\mathcal{G})\Big)$$

et donc $\rho(\mathcal{G}, \mu, \nu) \leq d\, a(\mu,\nu)$ par Jensen .

D'autre part considérons une suite $\mathcal{G}_n$ d'éléments de $\mathcal{G}$, de plus en plus fins et tels que le diamètre maximal des éléments de $\mathcal{G}_n$ tende vers zéro. Alors la suite $\sum_{A \in \mathcal{G}_n} 1_A(t) \frac{\nu(gA)}{\nu(A)}$ converge pour tout g ν-presque partout vers la fonction $\frac{dg^{-1}\nu}{d\nu}$ et nous pouvons écrire :

$$d\, a(\mu,\nu) = - \int \log \frac{dg^{-1}\nu}{d\nu}(t) \quad \nu(dt)\ \mu(dg)$$

$$= \lim_{M,M' \to +\infty} - \int \log \Big[\max \Big(\min\Big(\frac{dg^{-1}\nu}{d\nu}(t), M'\Big), \frac{1}{M}\Big)\Big] \nu(dt)\mu(dg)$$

car la fonction $- \log \frac{dg^{-1}\nu}{d\nu}(t)$ est minorée par une fonction intégrable.

Nous avons alors, par convergence dominée :

$$d\, a(\mu,\nu) =$$

$$= \lim_{M,M' \to \infty} \lim_{n \to \infty} - \int \log \Big[\max \Big(\min\Big(\sum_{A \in \mathcal{G}_n} 1_A(t) \frac{\nu g(A)}{\nu(A)}, M'\Big), \frac{1}{M}\Big)\Big] \nu(dt)\mu(dg)$$

$$\leq \lim_{M' \to \infty} \sup_n - \int \log\Big(\min \Big(\sum_{A \in \mathcal{G}_n} 1_A(t) \frac{\nu(gA)}{\nu(A)}, M'\Big)\Big) \nu(dt)\ \mu(dg)$$

$$\leq \lim_{M'\to\infty} \sup_n \rho(\mathcal{A}_n, \mu,\nu) + \int \Big(\sum_{A\in \widehat{\mathcal{A}}_n(g,M')} \nu(A) \log \frac{\nu(gA)}{\nu(A)} \, \mu(dg)$$

en appelant $\widehat{\mathcal{A}}_n(g,M')$ l'ensemble des parties de $\mathcal{A}_n$ où $\frac{\nu(gA)}{\nu(A)} > M'$.

Remarqons que $\sum_{A\in \widehat{\mathcal{A}}_n(g,M')} \nu(A) \log \frac{\nu(gA)}{\nu(A)}$ s'écrit aussi $\sum_{A\in \widehat{\mathcal{A}}_n(g,M')} \frac{\nu(A)}{\nu(gA)} \log \frac{\nu(gA)}{\nu(A)} \nu(gA)$ et que sur $\widehat{\mathcal{A}}_n(g,M')$, $\frac{\nu(A)}{\nu(gA)} \log \frac{\nu(gA)}{\nu(A)} \leq \frac{\log M'}{M'}$.

Nous avons finalement :

$$d\, a(\mu,\nu) \leq \lim_{M'\to\infty} (\sup \rho(\mathcal{A},\mu,\nu) + \frac{\log M'}{M'}) \leq \sup \rho(\mathcal{A}, \mu,\nu)$$

ce qui achève la démonstration de la propriété ii).

Le théorème 2.2 suit alors de 2.4 et 2.5 par un argument de compacité. Appellons en effet $M(\mathcal{A},\mu)$ l'ensemble des mesures ν de I_μ telles que :

$$\rho(\mathcal{A}, \mu,\nu) \leq \sup \{\rho(\mathcal{A}', \mu, \nu_{a'}) \; ; \; \mathcal{A}'\in \mathcal{R}\} \; .$$

D'après 2.4 l'ensemble $M(\mathcal{A},\mu)$ est un fermé non vide de I_μ. De plus si $\mathcal{A}'$ raffine $\mathcal{A}$, $M(\mathcal{A}',\mu)$ est contenu dans $M(\mathcal{A}, \mu)$. L'intersection M_0 des $M(\mathcal{A},\mu)$ pour tous les $\mathcal{A}$ de $\mathcal{R}$ n'est donc pas vide et si ν_0 est une mesure de M_0, nous avons :

$$d\alpha(\mu) \leq d\, a(\mu,\nu_0) \text{ par définition}$$

$$\leq \sup \{\rho(\mathcal{A}, \mu, \nu_0) \mid \mathcal{A}\in \mathcal{R}\} \quad \text{par (2.5 ii))}$$

$$\leq \sup \{\rho(\mathcal{A}', \mu,\nu_{a'}), \mathcal{A}'\in \mathcal{R}\} \quad \text{car } \nu_0 \text{ appartient à } M(\mathcal{A},\mu) \text{ pour}$$

tout $\mathcal{A}$,

$$\leq \sup_{\mathcal{A}} \inf_{I_\mu} \rho(\mathcal{A}', \mu, \nu) \text{ par définition de } \nu_{\mathcal{A}'}$$

$$\leq \inf_{I_\mu} \sup_{\mathcal{A}'} \rho(\mathcal{A}', \mu, \nu) = d\, \alpha(\mu).$$

Ces inégalités sont donc des égalités, ce qui montre à la fois que $\alpha(\mu) = a(\mu,\nu_0)$ pour ν_0 dans M_0 et que $\alpha(\mu) = \frac{1}{d} \sup \{\rho(\mathcal{A}', \mu,\nu_{a'}), \mathcal{A}'\in\mathcal{R}\}$.

D'après 2.4 la fonction $\mu \longrightarrow \alpha(\mu)$ est donc donnée par la plus grande valeur d'une famille de fonctions semi-continues inférieurement. C'est encore une fonction semi-continue inférieurement, et ceci achève la démonstration du théorème 2.2

Références :

Pour les lemmes 2.3 et 2.5, nous avons adaptés à nos conditions des résultats connus de théorie de l'information (théorème de Gel'fand, Yaglom et Perez):

I M GEL'FAND AND A.M. YAGLOM : Calculation of the amount of information about a random function contained in another such function. American Math. Soc. Translation Série 2 12 (1959), p. 199-246.

M.S. PINSKER : Information and information stability of Random variables and processes. Holden day (1964). Voir en particulierles chapitres 2 et 3 et les notes du traducteur.

A. PEREZ : Notions généralisées d'incertitude, d'entropie et d'information du point de vue de la théorie des martingales.
Transactions of the first Prague conference on Information theory, statistical theory, statistical Decision functions, Random processes. Prague (1957) p. 183-208.

3. Critère de Fürstenberg

Le but de ce paragraphe est d'établir :

3.1 - Théorème

Soient μ une mesure de M , $\alpha(\mu)$ son entropie, $\lambda_1(\mu)$ le plus grand exposant caractéristique $\alpha(\mu) \leq \lambda_1(\mu)$.

Démonstration :

Pour démontrer 3.1 nous considérons d'abord le cas particulier où il existe ν dans I_μ équivalente à la mesure de Lebesgue sur $\mathbb{P}^{d-1}$. Dans ce cas la formule se déduit facilement de 1.1. Puis nous montrerons que ce cas particulier est en fait dense dans M et nous utiliserons les continuités 1.2 et 2.2.

Soit λ la mesure de probabilité sur $\mathbb{P}^{d-1}$ invariante par l'action des matrices orthogonales. Remarquons tout d'abord que la fonction $F(g,\dot{t})$

$F(g,\dot{t}) = \log \frac{\|gt\|}{\|t\|}$ qui intervient dans 1.1 constitue pour tout g une version continue de $-\frac{1}{d} \log \frac{dg^{-1}\lambda}{d\lambda}(\dot{t})$.

3.2 - Lemme

Soit μ une mesure de M telle que il existe ν dans I_μ de la forme $\nu = h_\lambda$ avec $|\log h|$ borné. Alors $a(\mu,\nu) = \lambda_1(\mu)$.

Démonstration de 3.2 :

Nous avons en effet :

$$a(\mu,\nu) = -\frac{1}{d}\int \log \frac{dg^{-1}\nu}{d\nu}(t)\, \nu(dt)\, \mu(dg)$$

$$= -\frac{1}{d}\int \log \left(\frac{h(gt)}{h(t)} \frac{dg^{-1}\lambda}{d\lambda}(t)\right) \nu(dt)\, \mu(dg) .$$

Comme, par hypothèse, $|\log h|$ est borné, nous pouvons séparer les termes de l'intégrale et écrire :

$$a(\mu,\nu) = -\frac{1}{d}\int \log h(gt)\, \nu(dt)\, \mu(dg)$$

$$-\frac{1}{d}\int \log \frac{dg^{-1}\lambda}{d\lambda}(t) \quad \nu(dt)\, \mu(dg)$$

$$+\frac{1}{d}\int \log h(t)\, \nu(dt).$$

Comme $\mu * \nu = \nu$ les deux termes extrêmes s'annulent et il reste, d'après la remarque et 1.1 :

$$a(\mu,\nu) = \int \log F(g,t)\, \nu(dt)\, \mu(dg) = \lambda_1(\mu) \quad \blacksquare$$

Montrons que le lemme 3.2 s'applique à une famille dense de mesures dans M . Notons dg la mesure de Haar sur $SL(d,\mathbb{R})$, dk la mesure de Haar sur le sous-groupe orthogonal K de $SL(d, \mathbb{R})$.

3.3 - Lemme

Si la mesure de probabilité μ est de la forme pdg, où p est une fonction bornée à support compact sur $SL(d,\mathbb{R})$, alors μ appartient à M et si la mesure ν sur $\mathbb{P}^{d-1}$ est invariante par μ, ν est absolument continue par rapport à λ , de densité bornée.

Démonstration :

Il est d'abord clair que la mesure μ appartient à M : la mesure μ intègre $\log||g||$ car p est à support compact et la mesure μ est irréductible car l'ensemble des g qui laissent fixe une réunion finie de sous-espaces propres donnée quelconque est dg-négligeable.

Posons $P(g) = \sup_{k\in K} p(gk)$.

La fonction P est également bornée à support compact et la relation d'invariance s'écrit :

$$\int f(t)\, \nu(dt) = \int f(gt)\, p(g)\, dg\, \nu(dt)$$

$$= \int \left[\int f(gt)\, p(g)\, dg \right] \nu(dt).$$

Par invariance de dg par l'action de K, nous avons encore :

$$\int f\, d\nu = \int f(gkt)\, p(gk)\, dg\, dk\ \ \nu(dt)$$

$$\leq \int f(gkt)\, P(g)\, dg\, dk\, \nu(dt),$$

en intégrant en k, comme $dk * \nu = \lambda$ pourtoute mesure ν sur $\mathbb{P}^{d-1}$, nous avons :

$$\int f\, d\nu \leq \int f(gy)\, P(g)\, dg\ \ \lambda(dy)$$

$$= \int f(y)\, P(g)\, \frac{dg\lambda}{d\lambda}(y)\quad dg\ \lambda(dy).$$

Sur l'ensemble compact $P > 0$, la fonction $\frac{dg\lambda}{d\lambda}(y)$ est bornée par une constante C, et nous avons :

$$\int f\, d\nu \leq C \sup \mathbb{P} \int f(y)\, \lambda(dy) \int_{\text{supp } P} dg \qquad \text{q.e.d.}$$

3.4 - Lemme

> Si la mesure μ sur $SL(d,\mathbb{R})$ est de la forme pdg, où p est une fonction continue à support compact, strictement positive sur un voisinage de K, pour toute mesure ν de I_μ, la mesure λ est absolument continue par rapport à ν, de densité bornée.

Démonstration :

Remarquons tout d'abord que μ appartient à M pour les mêmes raisons que dans 3.3. Posons $P(g) = \inf_{k \in K} p(g\,k)$. La fonction P(g) est strictement positive sur un voisinage de l'identité de mesure positive.

En écrivant l'équation d'invariance par μ et en effectuant les mêmes transformations que pour le lemme précédent, nous obtenons $\int f d\nu \geq \frac{1}{D} \int f d\lambda$ pour une certaine constante D, q.e.d.

Soit alors μ une mesure quelconque de M. Prenons une fonction p_0 positive, continue à support compact sur $SL(d,\mathbb{R})$, strictement positive sur un voisinage de K, et p_n une suite de fonctions positives à support compact telles que les mesures $p_n\,dg$ convergent vaguement vers δ_e ; nous posons pour n,M entiers, μ_M la restriction de μ à l'ensemble $||g|| \leq M$,

$$\mu_{n,M} = (1 - \frac{1}{M})\ (\mu_M * p_n dg) + a_M\, p_0\, dg,$$

où $$a_M = 1 - (1 - \frac{1}{M}).\ \mu_M(SL(d,\mathbb{R})).$$

Chaque mesure $\mu_{n,M}$ est une mesure de probabilité sur $SL(d,\mathbb{R})$, et d'après les lemmes 3.3 et 3.4, $\mu_{n,M}$ appartient à M et les mesures ν de $I_{\mu_{n,M}}$ sont équivalentes à λ , de densités bornées inférieurement et supérieurement. D'après le lemme 3.2, nous avons $\alpha(\mu_{n,M}) = \lambda_1(\mu_{n,M})$.

Quand n et M tendent vers l'infini, les mesures $\mu_{n,M}$ tendent vers μ dans M et d'après 1.2 et 2.2 nous avons à la limite :

$$\alpha(\mu) \leq \liminf_{n,M} \alpha(\mu_{n,M}) = \liminf_{n,M} \lambda_1(\mu_{n,M}) = \lambda_1(\mu)$$

Le théorème 3.1 a deux corollaires importants :

3.5 - Corollaire. Critère de Fürstenberg

Soit μ une mesure de M . S'il existe pas de mesure ν telle que $g\nu = \nu$ μ presque partout, alors $\lambda_1(\mu) > 0$.

En effet si λ_1 est nul, nous avons $\alpha(\mu) = 0$ et d'après 2.2 cela signifie qu'il existe ν_0 probabilité sur $\mathbb{P}^{d-1}$ telle que :

$$- \int \log \frac{dg^{-1}\nu_0}{d\nu_0}(t) \quad \nu_0(dt)\,\mu(dg) = 0.$$

Ceci n'est possible que si pour presque tout g $\frac{dg^{-1}\nu_0}{d\nu_0}(t) = 1$ ν_0 presque partout, c'est-à-dire $g^{-1}\nu_0 = \nu_0$, ou encore $\nu_0 = g\,\nu_0$.

3.6 - Corollaire

Soit μ une mesure de M . Il existe une mesure invariante ν_0 telle que pour μ-presque tout g la mesure $g\,\nu_0$ est absolument continue par rapport à la mesure ν_0.

En effet d'après 3.1 et 2.2, il existe une mesure ν_0 de I_μ telle que :

$$- \int \log \frac{dg^{-1}\nu_0}{d\nu_0}(t)\ \nu_0(dt)\,\mu(dg) \leq d\lambda_1(\mu) < +\infty \quad .$$

En particulier, l'intégrale $-\int \log \frac{dg^{-1}\nu_0}{d\nu_0}(t)\ \nu_0(dt)$ est finie pour presque tout g.

En particulier, nous avons pour ces mêmes g la mesure ν_0 absolument continue par rapport à la mesure $g^{-1}\nu_0$. Nous avons bien pour presque tout g l'absolue continuité de $g\,\nu_0$ par rapport à ν_0.

Références :

La démonstration que nous donnons ici est celle originale de Fürstenberg (op. cit. paragraphe 8), à ceci près que nous avons explicité le rôle de l'entropie $\alpha(\mu)$. L'avantage est de nous permettre d'obtenir le corollaire 3.6 et il semble que l'on peut retrouver ainsi les extensions de 3.5 en particulier les résultats de :

A.D. VIRTSER : On products of random matrices and operators. Th. Prob. Appl. 24 (1979) 367-377.

G. ROYER : Croissance exponentielle de produits markoviens de matrices aléatoires Ann. IHP 16 (1980) p. 49-62.

Y. GUIVARC'H : Marches aléatoires à pas markovien. C.R.A.S. Paris 289 (1979) 211-213.

F. LEDRAPPIER - G. ROYER : Croissance exponentielle de certains produits aléatoires de matrices. C.R.A.S. Paris 290 (1980) 513-514.

4. Majoration de la dimension de la mesure invariante

Nous nous plaçons dans le cas où d=2 et nous notons λ la valeur commune de λ_1 et $-\lambda_2$. Les mesures invariantes de I_μ sont alors des mesures sur $\mathbb{P}^1$. Nous considérons la distance sur $\mathbb{P}^1$ définie par le plus petit angle de deux directions de $\mathbb{R}^2$. Nous cherchons à obtenir des relations analogues à celles des paragraphes II.4 et II.5. Nous obtiendrons seulement des convergences en probabilité. Nous avons d'abord la majoration :

4.1 - Théorème

Soit μ une mesure de M_2 avec $\lambda(\mu) > 0$. Il existe une unique mesure invariante ν et pour tout $\chi > 0$, il existe $\varepsilon_0 > 0$ tel que si $\varepsilon \leq \varepsilon_0$

$$\nu(\{t \; ; \; \frac{\log \nu(B(t,\varepsilon))}{\log \varepsilon} \leq \frac{\alpha(\mu)}{\lambda(\mu)} + \chi\}) \geq 1-\chi \; .$$

Démonstration :

La première étape consiste à identifier la mesure ν. Considérons le système (X', m', θ', x_0) où (X', m') est l'espace des suites bilataires de matrices de déterminant 1, $(X',m') = (SL(2,R), \mu) \otimes \mathbb{Z}$. Si $\lambda(\mu) > 0$, d'après le théorème I.4.2, il existe pour presque tout x' deux directions $E^+_{x'}$ et $E^-_{x'}$ telles que :

$$v \in E^+_{x'}, \; v \neq 0 \; , \text{ si et seulement si :}$$

$$\frac{1}{n} \log \|x'_{n-1} \ldots, x'_0 \, v\| \to \lambda \quad \text{et} \quad \frac{1}{n} \log \|x'^{-1}_{-n} \ldots . \, x'^{-1}_{-1} \, v\| \to -\lambda$$

quand n tend vers l'infini et $v \in E^-_{x'}$, $v \neq 0$, si et seulement si les mêmes limites sont échangées.

4.2 - Lemme

La direction de $E^+_{x'}$ dans $\mathbb{P}^1$ est indépendante des coordonnées $\{x'_s, \; s \geq 0\}$ et a pour loi l'unique mesure invariante ν de I_μ.

Démonstration de 4.2 :

Considérons sur $X' \times \mathbb{P}^1$ la transformation $\hat\theta$ $\hat\theta(x',t) = (\theta x', x'_0 \, t)$ et les mesures invariantes par $\hat\theta$ qui se projettent sur X' en m'. D'après I.5.1 les mesures ergodiques sont nécessairement l'une ou l'autre de :

$$m'_+(dx',d\dot{t}) = \delta_{E^+_{x'}}(d\dot{t})\, m'(dx') \text{ et}$$

$$m'_-(dx',d\dot{t}) = \delta_{E^-_{x'}}(d\dot{t})\, m'(dx') ,$$

avec $\int F(x',t)\, m'_\pm(dx',dt) = \pm\lambda$.

Considérons alors ν une mesure de I_μ et sur $X \times \mathbb{P}^1$ la mesure $m \otimes \nu$. C'est une mesure $\hat\theta$ invariante qui se projette sur X en m et il est possible de la prolonger de manière unique en une mesure $\hat\theta$ invariante sur $X' \times \mathbb{P}^1$ qui se projette sur X' en m'.

En effet, considérons dans $X' \otimes \mathbb{P}^1$ la σ-algèbre $\mathcal{G}$ engendrée par la projection sur $X \times \mathbb{P}^1$.

Il existe un unique prolongement à la σ-algèbre $\hat\theta^k \mathcal{G} = \mathcal{G}_k$ de la mesure $m \times \nu$ compatible avec l'invariance et c'est une mesure de la forme :

$$m^k_{x'}(dt)\, \mu(dx'_{-k})\, \mu(dx'_{-k+1}) \ldots\ldots$$

Les relations de compatibilité montrent que les mesures $m^k_{x'}(dt)$ forment une martingale par rapport à la suite $\mathcal{G}_k$ de σ-algèbres.

Cette martingale converge presque sûrement vers une mesure $m_{x'}(dt)$ et l'unique prolongement de la mesure $m \times \nu$ à $X' \times \mathbb{P}^1$ qui est $\hat\theta$ invariant est la mesure $m_{x'}(dt)\, m'(dx')$.

C'est une mesure $\hat\theta$ invariante de marginale m', c'est donc une combinaison des mesures m'_+ et m'_-.

D'après 1.1, de plus, l'intégrale de la fonction $F(x',t)$ pour cette mesure est donnée par $\int F(x_0,t)\, \mu(dx_0)\, \nu(dt) = \lambda$.

La mesure $m_{x'}(dt)\, m'(dx')$ coïncide donc avec m'_+, autrement dit la mesure m'_+ se projette sur $X \times P^1$ en la mesure $m \times \nu$, ce qui veut exactement dire l'énoncé du lemme.

En particulier la mesure ν est unique puisque c'est la loi de $E^+_{x'}$.

De la démonstration de 4.2 retenons encore :

4.3 - Corollaire

La suite $x'_{-1} \dots x'_{-k}\nu$ converge presque partout vers la mesure $\delta_{E^+_{x'}}$.

On trouve en effet cette mesure si on explicite $m^k_{x'}$.

4.4 - Corollaire

La mesure $m \times \nu$ sur $X \times \mathbb{P}^1$ est $\widehat{\theta}$ ergodique et est la projection de la mesure m'_+ .

Revenons à la démonstration de 4.1. D'après le théorème 2.1 nous pouvons calculer $\alpha(\mu)$ comme $a(\mu,\nu)$ pour l'unique mesure invariante

$$\alpha(\mu) = -\frac{1}{2}\int \log \frac{dg^{-1}\nu}{d\nu}(t)\nu(dt)\mu(dg) = -\frac{1}{2}\int \log \frac{dx_0^{-1}\nu}{d\nu}(t)\nu(dt)\, m(dx_0).$$

D'après 3.1 et 4.4 et le théorème ergodique appliqué à la fonction $-\frac{1}{2}\log \frac{dx_0^{-1}\nu}{d\nu}(t)$, nous avons, $(m \times \nu)$ presque partout et en moyenne :

$$\alpha(\mu) = \lim_n -\frac{1}{2n}\sum_{i=0}^{n-1} \log \frac{dx_i^{-1}\nu}{d\nu}(x_{i-1} \dots x_0\, t)$$

$$= \lim_n -\frac{1}{2n}\log \frac{dx_0^{-1}\dots x_{n-1}^{-1}\nu}{d\nu}(\dot{t}).$$

Nous fixons x' dans l'ensemble de m'mesure 1 où la convergence

$$\alpha(\mu) = \lim_n -\frac{1}{2n}\log \frac{dx_0'^{-1}\dots x_{n-1}'^{-1}\nu}{d\nu}(\dot{t})$$

a lieu ν presque partout et en moyenne dans $L^1(\nu)$. Choissons $C_{x'}$, un sous-ensemble de P^1 non négligeable pour ν .

Nous avons encore :

$$\alpha(\mu) = \lim_n -\frac{1}{2n\nu(C_{x'})}\int_{C_{x'}} \log \frac{dx_0'^{-1}\dots x_{n-1}'^{-1}\nu}{d\nu}(\dot{t})\ \nu(d\dot{t})$$

et (1) $$\alpha(\mu) \geq \limsup_n -\frac{1}{2n}\log(\nu(x'_{n-1}\dots x'_0\, C_{x'}))$$

par l'inégalité de Jensen.

Le choix de $C_{x'}$ sera fait par le lemme suivant :

4.5 - Lemme

Pour tout $\varepsilon > 0$ et pour presque tout x', il existe un ensemble $C_{x'}$ et un entier N tels que :

$\nu(C_{x'}) > 0$ et si $n \geq N$,

$$x'_{n-1} \dots x'_0\, C_{x'} \subset B(E^+_{\theta^n x'}, e^{-2n(\lambda-\varepsilon)}) .$$

Démonstration :

Nous écrivons la matrice $x'_{n-1} \dots x'_0$ dans les bases orthogonales dont le premier vecteur est $E^+_{x'}$, pour l'espace de départ et $E^+_{\theta^n x'}$ pour l'espace d'arrivée

$$x'_{n-1} \cdot \dots \cdot x'_0 = \begin{pmatrix} a_n(x') & b_n(x') \\ 0 & a_n^{-1}(x') \end{pmatrix}$$

avec $\frac{1}{n} \log|a_n(x')| \to \lambda$ quand $n \to +\infty$.

Pour estimer b_n écrivons la propriété des vecteurs de l'espace

$$E^-_{x'} = \left\{ \begin{matrix} r \cos\alpha' \\ r \sin\alpha' \end{matrix} \right\} r \in \mathbb{R} .$$

Il vient :

$$b_n(x') = -\frac{a_n(x')}{\mathrm{tg}\,\alpha'} + C_n(x') \text{ avec}$$

$$\limsup_n \frac{1}{n} \log|C_n(x')| \leq -\lambda .$$

Nous posons alors :

$$C_{x'} = \{t \in \mathbb{P}^1, \ |\mathrm{tg}\,t| \leq \min(\tfrac{1}{3}|\mathrm{tg}\,\alpha'|, \tfrac{1}{2})\} .$$

Nous avons $\nu(C_{x'}) > 0$ m'-presque partout car par 4.2 , $E^+_{x'}$ appartient au support de ν m'presque partout.

D'autre part si t appartient à $C_{x'}$, nous avons :

$$(x'_{n-1} \dots x'_0\, t) = \begin{pmatrix} r \cos u_n \\ r \sin u_n \end{pmatrix} \quad \text{avec}$$

$$\mathrm{tg}\,u_n = a_n^{-2} \frac{\mathrm{tg}\,t}{1 + b_n a_n^{-1} \mathrm{tg}\,t} = a_n^{-2} \frac{\mathrm{tg}\,t}{1 - \frac{\mathrm{tg}\,t}{\mathrm{tg}\,\alpha'} + C_n a_n^{-1} \mathrm{tg}\,t}$$

Nous avons donc $\limsup_n \frac{1}{n} \log|\mathrm{tg}\,u_n| \leq -2\lambda$ q.e.d ∎

Soit alors $\varepsilon = \frac{\lambda^2\chi}{\alpha + \lambda\chi}$.

Si nous choisissons x' dans l'ensemble de mesure 1 où (1) est vrai et où nous pouvons choisir $C_{x'}$, par 4.5 nous avons :

$$\alpha(\mu) \geq \limsup_n - \frac{1}{2n} \log \nu(B(E^+_{\theta^n x'}, e^{-2n(\lambda-\varepsilon)})) .$$

Comme la loi de la direction $E^+_{\theta^n x'}$ est également la mesure ν , nous avons sur des ensembles de mesure de plus en plus grande quand n tend vers l'infini :

$$\frac{1}{2n(\lambda-\varepsilon)} \log \nu(B(t, e^{-2n(\lambda-\varepsilon)})) \leq \frac{\alpha}{\lambda - \varepsilon} = \frac{\alpha}{\lambda} + \chi,$$

ce qui montre le théorème 4.1.

Il est possible de traduire le résultat en termes de dimension (cf. notes du paragraphe II.4). Soit (X,d,m) un espace métrique muni d'une mesure de probabilité . Posons pour $\delta > 0$, $\varepsilon > 0$ $N(\delta,\varepsilon)$ le cardinal minimal d'un ensemble I tel que $m(\bigcup B(x,\varepsilon) ; x \in I) \geq 1-\delta$.

Nous avons alors les notions de dimension à δ près et de dimension essentielle suivantes :

$$\begin{cases} \dim \sup_\delta (X,m) = \limsup_{\varepsilon\to 0} \frac{1}{\log 1/\varepsilon} \log N(\delta,\varepsilon) \\ \dim \sup (X,m) = \lim_{\delta\to 0} \dim \sup (X,m). \end{cases}$$

4.6 - <u>Corollaire</u>

Soit μ une mesure de M_2 avec $\lambda(\mu) > 0$. Il existe une unique mesure invariante ν et $\dim \sup (P^1,\nu) \leq \frac{\alpha}{\lambda}$.

En effet, choisissons $\delta > \chi > 0$. Nous pouvons recouvrir l'ensemble A_χ du théorème 4.1 :

$$A_\chi = \{t ; \nu(B(t,\varepsilon)) \geq \varepsilon^{\frac{\alpha}{\lambda} + \chi}\}$$

par des intervalles de largeur 2ε centrés sur A_χ de manière à ce qu'au plus deux intervalles recouvrent chaque point. La somme de leurs mesures est alors plus petite que 2 et donc $N(\delta,\varepsilon) \leq 2\, \varepsilon^{-(\frac{\alpha}{\lambda} + \chi)}$. Nous avons donc $\dim \sup_\delta (P^1,\nu) \leq \frac{\alpha}{\lambda} + \chi$, pour tout $0 < \chi < \delta$. q.e.d ■

5. Exposant, entropie et dimension en dimension 2

Dans ce paragraphe nous établissons la convergence en probabilité :

5.1 - Théorème

Soit μ une mesure de M_2, $\lambda(\mu) > 0$, et soit ν l'unique mesure invariante. Quand ε tend vers 0, les fonctions $\frac{\log\nu(B(t,\varepsilon))}{\log\varepsilon}$ convergent en probabilité vers $\frac{\alpha(\mu)}{\lambda(\mu)}$.

Démonstration :

Si nous orientons $\mathbb{P}^1$ et que nous considérons $s < \frac{\Pi}{2}$, l'intervalle $(t,t+s)$ est défini sans ambiguïté. Pour s_1, $s_2 < \frac{\Pi}{4}$, nous notons $B(t,s_1,s_2)$ l'intervalle $(t-s_1,\ t+s_2)$.

5.2 - Lemme

Pour tout $\varepsilon > 0$ il existe une fonction mesurable positive $\mu \otimes \nu$ presque partout $h(g,t,\varepsilon)$ telle que si s_1, $s_2 < h(g,t,\varepsilon)$

$$\frac{\nu(g\ B(t,s_1,s_2))}{\nu(B(t,s_1,s_2))} \leq e^{\varepsilon}\ \frac{dg^{-1}\nu}{d\nu}\ (t) \qquad \mu \otimes \nu \text{ presque partout .}$$

En effet les expressions $\frac{\nu(g\ B(t,0,s))}{\nu(B(t,0,s))} = \frac{1}{\nu(B(t,0,s))} \int_{B(t,0,s)} \frac{dg^{-1}\nu}{d\nu}\, d\nu$ convergent presque partout vers $\frac{dg^{-1}\nu}{d\nu}$ (t) car ν est diffuse. De même pour $\frac{\nu(g\ B(t,s,0))}{\nu(B(t,s,0))}$ ∎

Nous avons également la propriété d'intégrabilité suivante :

5.3 - Lemme

Posons $f^*(g,t) = \sup\limits_{s,u<\frac{\Pi}{4}} \frac{\nu(g\ B(t,s,u))}{\nu(B(t,s,u))}$.

La fonction $\log^+ f^*$ est intégrable.

Démonstration :

D'après l'inégalité maximale (par exemple, en se ramenant à des intervalles de (R,dt) par changement de variable, on peut utiliser l'inégalité classique), nous avons $\mu\ \times\ \nu(f^* > \beta) \leq \frac{1}{\beta}$ et donc :

$$\int \log^+ f^* \, d\mu \times \nu = \lim_{M\to\infty} \int \inf(\log^+ f^*, M)\, d\mu \times \nu$$

$$= \lim_{M\to\infty} \int_1^M \int_{X \times P^1} \chi_{(f^* > \beta)} \frac{1}{\beta}\, d(\mu \times \nu) d\beta$$

$$\leq \lim_{M\to\infty} \int_1^M \frac{1}{\beta^2}\, d\beta = 1.$$

La première étape de la démonstration consiste à approcher α par une moyenne sur des intervalles de plus en plus petits.

5.4 - Lemme

Pour tout ε il existe ρ_0 tel que si $\rho \geq \rho_0$, m' presque partout :

$$\lim_n \inf - \frac{1}{2n} \log \frac{\nu(x'_{n-1} \dots x'_0 B(t, e^{-n\rho}))}{\nu(B(t, e^{-n\rho}))} \geq \alpha - \varepsilon .$$

Démonstration de 5.4

Nous avons en effet :

$$- \log \frac{\nu(x'_{n-1} \dots x'_0 B(t, e^{-n\rho}))}{\nu(B(t, e^{-n\rho}))} = - \sum_{i=1}^{n-1} \log \frac{\nu(x'_i \dots x'_0 B(t, e^{-n\rho}))}{\nu(X'_{i-1} \dots X'_0 B(t, e^{-n\rho}))}$$

et remarquons que l'ensemble $x'_{i-1} \dots x'_0 B(t, e^{-n\rho})$ est de la forme $B(x'_{i-1} \dots x'_0 t, s_1 s_2)$ avec les s_j majorés par $2C||x'_{i-1} \dots x'_0||^2 e^{-n\rho}$.

(Rappelons que $\frac{dg^{-1}\lambda}{d\lambda}(t) = ||gt||^{-2} \leq ||g^{-1}||^2 \leq C||g||^2$).

D'après 5.2 nous avons donc :

$$- \log \frac{\nu(x'_i \dots x'_0 B(t, e^{-n\rho}))}{\nu(x'_{i-1} \dots x'_0 B(t, e^{-n\rho}))} \geq - \varepsilon - \log \frac{dx_0'^{-1} \dots x_i'^{-1} \nu}{dx_0'^{-1} \dots x_{i-1}'^{-1} \nu}(t)$$

dès que $2C||x'_{i-1} \dots x'_0||^2 e^{-n\rho} \leq h(x'_i, x'_{i-1} \dots x'_0 t, \varepsilon)$.

Posons $H(x', \rho) = \sup_{n \geq i \geq 1} 2C||x'_{-1} \dots x'_{-i}||^2 e^{-n\rho}$; nous avons $H(x', \rho) < +\infty$ presque partout dès que $\rho > \int \log||g||\, \mu(dg)$ et nous pouvons écrire :

$$- \frac{1}{2n} \log \frac{\nu(x'_{n-1} \dots x'_0 B(t, e^{-n\rho}))}{\nu(B(t, e^{-n\rho}))} \geq - \frac{\varepsilon}{2} - \frac{1}{2n} \log \frac{x_0'^{-1} \dots x_{n-1}'^{-1} \nu}{d\nu}(t)$$

$$- \frac{1}{2n} \sum_{i=1}^{n-1} \chi . (\log^+ f^*(\hat{\theta}^i(x', t)) + |\log \frac{dx_0^{-1}\nu}{d\nu}(\hat{\theta}^i(x', t))|)$$

$$\{H(\hat{\theta}^i(x', t), \rho) > h(\hat{\theta}^i(x', t), \varepsilon)\} .$$

Nous avons minoré notre fonction par des moyennes de Birkhoff de fonctions intégrables. La limite inférieure cherchée est donc plus grande que :

$$-\frac{\varepsilon}{2}+\alpha-\frac{1}{2}\int_{\{H(x',\rho)>h(x'_0,t,\varepsilon)\}} \log^+ f^* + \left|\log \frac{dx_0'^{-1}\nu}{d\nu}(t)\right| \quad m'_+(dx',dt) .$$

(Rappelons que d'après 4.4, m'_+ est ergodique pour $\hat{\theta}$ et que d'après 5.3 et 3.1 la fonction $\log^+ f^* + \left|\log \frac{dx_0^{-1}\nu}{d\nu}(t)\right|$ est intégrable).

D'après la définition de $H(x',\rho)$, quand ρ tend vers l'infini, $H(x',\rho)$ tend presque sûrement vers zéro.

Une fois $h(x'_0,t,\varepsilon)$ choisi, comme $h(x'_0, t,\varepsilon) > 0$ presque partout, nous pouvons trouver ρ_0 assez grand pour que dès que $\rho \geq \rho_0$, l'ensemble $H(x',\rho) > h(x'_0, t,\varepsilon)$ est de mesure assez petite pour que la dernière intégrale soit plus petite que ε q.e.d ■

Soit $\chi > 0$. Appelons $\beta(\chi)$ la plus grande valeur telle que pour ε assez petit, on ait $\nu\left(\frac{\log\nu(B(t,\varepsilon))}{\log \varepsilon} > \beta(\chi)\right) \geq 1-\chi$.

Compte tenu de 4.1, il suffit de montrer que $\lim_{\chi\to 0} \beta(\chi) \geq \frac{\alpha(\mu)}{\lambda(\mu)}$ pour établir la convergence en probabilité. Fixons donc ε et choisissons $\rho \geq \rho_0(\varepsilon)$ donné par 5.4. Comme la mesure m'_+ est la mesure $\delta_{E^+_{x'}}(dt)\, m'(dx')$, nous avons à considérer l'ensemble $x'_{n-1}\dots x'_0\, B(E^+_{x'}, e^{-n\rho})$. Reprenons les notations de la démonstration du lemme 4.4 et les mêmes calculs. Si n est assez grand, $e^{-n\rho} \leq \frac{1}{2}\,|\mathrm{tg}\ \alpha'|$ et nous avons pour n très grand,

$$x'_{n-1}\dots x'_0\, B(E^+_{x'}, e^{-n\rho}) \supset B(E^+_{\theta^n x'}, e^{-n(\rho+2\lambda+\varepsilon)}).$$

En reportant dans le résultat de 5.4, nous pouvons trouver pour tout $\chi > 0$ et n assez grand un certain ensemble A_n, $m'(A_n) \geq 1-\chi$ tel que pour X' dans A_n :

$$-\frac{1}{2n}\log\ \nu(B(E^+_{\theta^n x'}, e^{-n(\rho+2\lambda+\varepsilon)}) \geq \alpha-\varepsilon-\frac{1}{2n}\log \nu(B(E^+_{x'}, e^{-n\rho})).$$

Appelons $B_{\chi,n}$ l'ensemble où $-\frac{1}{2n}\log B(E^+_{x'}, e^{-n\rho}) > \frac{\rho}{2}\beta(\chi)$.

Par définition de $\beta(\chi)$ et comme la loi de $E^+_{x'}$ est la mesure ν (lemme 4.2), nous avons pour n assez grand $m'(B_{\chi,n}) \geq 1-\chi$.

Nous avons donc alors si n est assez grand et si x' appartient à

$\theta^n(A_n \cap B_{\chi,n})$:

$$-\frac{1}{2n}\log \nu(B(E^+_{x'}, e^{-n(\rho+2\lambda+\varepsilon)}) \geq \alpha - \varepsilon + \frac{\rho}{2}\beta(\chi).$$

Par invariance de m', $m'(\theta^n(A_n \cap B_{\chi,n})) \geq 1-2\chi$ et par définition de β, il vient :

$$\left(\frac{\rho}{2}+\lambda+\frac{\varepsilon}{2}\right)\beta(2\chi) \geq \alpha - \varepsilon + \frac{\rho}{2}\beta(\chi).$$

En faisant tendre χ vers 0, il vient :

$$\left(\frac{\rho}{2}+\lambda+\frac{\varepsilon}{2}\right)\lim_{\chi\to 0}\beta(\chi) \geq \alpha - \varepsilon + \frac{\rho}{2}\lim_{\chi\to 0}\beta(\chi).$$

D'où $\lim_{\chi\to 0}\beta(\chi) \geq \frac{\alpha-\varepsilon}{\lambda+\frac{\varepsilon}{2}}$.

Le théorème 5.1 suit alors de l'arbitraire de ε ■

Avec les notations du paragraphe 4, posons :

$$\dim\inf_\delta(X,m) = \liminf_{\varepsilon\to 0}\frac{1}{\log 1/\varepsilon}\log N(\delta,\varepsilon) \text{ et}$$

$$\dim\inf(X,m) = \lim_{\delta\to 0}\dim\inf_\delta(X,m).$$

On note dim(X,m) la valeur commune, si elle existe de dim inf(X,m) et dim sup(X,m).

5.5 - <u>Corollaire</u>

Soit μ une mesure de M_2 avec $\lambda(\mu) > 0$. Il existe une unique mesure invariante ν et $\dim(P^1,\nu) = \frac{\alpha(\mu)}{\lambda(\mu)}$.

En effet un recouvrement à δ près en intervalles de largeur 2ε rencontre l'ensemble où $\nu(B(t,2\varepsilon)) \leq (2\varepsilon)^{\frac{\alpha}{\lambda}-\chi}$ sur au moins une mesure $(1-\delta-\chi)$. Il doit donc posséder au moins $(1-\delta-\chi)(2\varepsilon)^{-\frac{\alpha}{\lambda}+\chi}$ éléments. Nous avons donc $\dim\inf_\delta(P^1,\nu) \geq \frac{\alpha}{\lambda}-\chi$ pour tout $\chi > 0$. Nous avons donc déjà $\dim\inf_\delta(P^1,\nu) \geq \frac{\alpha}{\lambda}$. Le corollaire suit en comparant avec 4.6.

Remarquons encore que l'ensemble des mesures μ telles que $\alpha(\mu) = \lambda(\mu) > 0$ est d'après 1.2, 2.2 et la démonstration du théorème 3.1 une intersection dénombrable d'ouverts denses dans M_2. Nous avons donc pour beaucoup de mesures μ une unique mesure ν dans I_μ et $\dim(P^1,\nu) = 1$. Pour obtenir un exemple de mesures de M_2 avec $\dim(P^1,\nu) < 1$, considérons l'action de $SL(2,\mathbb{R})$ sur le demi-plan de Poincaré Im $z > 0$ par les applications homographiques réelles

$$g = \begin{pmatrix} a & b \\ c & d \end{pmatrix} \qquad \varphi_g(z) = \frac{az+b}{cz+d}.$$

L'action à l'infini sur $R \cup \{\infty\}$ s'identifie alors à l'action naturelle de $SL(2,\mathbb{R})$ sur $\mathbb{P}^1$ par l'application qui à X associe la droite de pente Arc tgX. Si une mesure μ est portée par un sous groupe discret Γ de $SL(2,R)$ la mesure invariante ν doit être portée par l'ensemble limite Λ_Γ de Γ . Il est facile de construire des sous-groupes discrets où Λ_Γ ne peut pas porter de mesure de dimension supérieure à un certain $\delta(\Gamma) < 1$.
(cf. par exemple A.F. Beardon : The Hausdorff dimension of singular sets of properly discontinuous groups. American J of Maths. 88(1966) 722-736).

Par analogie avec le paragraphe II.5, il est naturel de demander si on a convergence presque sûre dans le théorème 5.1.

D'autre part Guivarc'h et Raugi ont montré (communication personnelle)

$$\liminf_{\varepsilon\to 0} \frac{\log\nu(B(t,\varepsilon))}{\log \varepsilon} > 0 \text{ en tout point } t.$$

IV - EXEMPLES

1. Spectre singulier des opérateurs de Schrödinger aléatoires. L'argument de Pastour

On considère le modèle suivant : soient (X,m,θ) un système ergodique et V une fonction réelle mesurable bornée sur X.

Pour chaque x de X, nous définissons un opérateur H_x auto adjoint borné de $\ell^2(\mathbb{Z})$ dans lui-même par $H_x \varphi(n) = \varphi(n-1) + \varphi(n+1) - V(\theta^n x)\, \varphi(n)$.

Par le théorème spectral, il existe un isomorphisme unitaire entre $\ell^2(\mathbb{Z})$ et $\bigoplus_{k \geq 1} L^2(\mathbb{R}, \mu_k)$ où la mesure μ_{k+1} est absolument continue par rapport à μ_k, qui transporte l'opérateur H_x en la multiplication par l'identité.

On appelle la classe de μ_1 le type spectral maximal et le nombre minimal de mesures non nulles dans la décomposition la multiplicité. Ce sont des invariants de H_x dont on veut déterminer les propriétés.

Soient δ_j la fonction de $\ell^2(Z)$ définie par $\delta_j(n) = 0$ sauf si $n = j$ et $\delta_j(j) = 1$, $\sigma_j(x,d\lambda)$ la mesure spectrale de δ_j définie par les relations

$$\int_{\mathbb{R}} \lambda^n \, \sigma_j(x,d\lambda) = H_x^n \, \delta_j(j).$$

Posons :

$$\sigma(x,d\lambda) = \frac{1}{2} (\sigma_0(x,d\lambda) + \sigma_1(x,d\lambda)).$$

1.1 - Proposition

Le type spectral maximal de H_x est celui de $\sigma(x,d\lambda)$ et la multiplicité est inférieure à 2.

Ces deux propriétés suivent du fait que les fonctions $H_x^m \delta_0$, $H_x^n\, \delta_1$, $n, m \in \mathbb{Z}$ engendrent $\ell^2(\mathbb{Z})$.

L'ensemble des types spectraux maximaux est décrit par une mesure M sur $X \times \mathbb{R}$:

$$M(dx,d\lambda) = m(dx)\, \sigma(x,d\lambda).$$

Sa projection sur $\mathbb{R}$ est appelée densité d'états k :

$$k(d\lambda) = \int \sigma(x,d\lambda)\, m(dx).$$

1.2 - Théorème

Considérons pour tout e réel la matrice aléatoire $A_e(x) = \begin{pmatrix} e+V(x) & -1 \\ 1 & 0 \end{pmatrix}$

et l'exposant $\lambda(e)$ de la famille $(X,m\ ;\ \theta;\ A_e)$.

Si $\lambda(e)$ est positif k presque partout, la mesure M n'a pas de partie absolument continue par rapport à une mesure produit.

En particulier si $\lambda(e) > 0$ k presque partout, l'opérateur H_x a pour m presque tout x, un spectre singulier ; pour tout λ, l'ensemble des x tels que la mesure de Dirac δ_λ appartienne au type spectral de H_x est aussi négligeable.

Soient e réel, x dans X. Un vecteur ϕ de $\mathbb{R}^2$ diverge si ou bien $||A_e^{(n)}(x)\phi||$, ou bien $||A_e^{(-n)}(x)\phi||$ croît exponentiellement vite quand n tend l'infini.

Si pour k presque tout e, $\lambda(e) > 0$, pour chacune de ces valeurs de e, nous avons que pour m-presque tout x, tout vecteur non nul de $\mathbb{R}^2$ diverge. L'ensemble des (e,x) tels que tout vecteur non nul de $\mathbb{R}^2$ diverge est donc de mesure 1 pour k x m. Au contraire le lemme 1.3 montre que pour M presque tout (e,x) il existe un vecteur ϕ de $\mathbb{R}^2$ qui ne diverge pas.

1.3 - Lemme

Pour $\sigma(x,d\lambda)$ presque tout λ, il existe une fonction réelle sur $\mathbb{Z}$, $\Psi(n)$ avec :

$$H_x\Psi = \lambda\Psi \text{ et } \sum_{n\in\mathbb{Z}} \frac{|\Psi(n)|^2}{n^2} < +\infty .$$

Démonstration :

D'après la proposition 1.1 nous pouvons choisir deux éléments $e_1(\lambda)$ et $e_2(\lambda)$ et représenter $\ell^2(\mathbb{Z})$ comme l'ensemble des fonctions

$\lambda \to p_1(\lambda)\ e_1(\lambda) + p_2(\lambda)\ e_2(\lambda)$ avec la norme

$$||(p_1,p_2)||^2 = \int (p_1^2\ (\lambda) + p_2^2(\lambda)\ )\ \sigma(x,d\lambda).$$

L'opérateur H_x se représente alors par la multiplication par λ. Notons $(p_{1,n}(\lambda),\ p_{2,n}(\lambda))$ un représentant de la fonction δ_n, $n\in\mathbb{Z}$. En identifiant la définition de $H_x\ \delta_n$, il vient que nous avons $\sigma(x,.)$ presque partout :

$$p_{i,n+1}(\lambda) + p_{i,n-1}(\lambda) - V_n(\theta^n x)\ p_{i,n}(\lambda) = \lambda\ p_{i,n}(\lambda).$$
$$i = 1,2.$$

Autrement dit en notant $\Psi_\lambda(n) = p_{1,n}(\lambda)$ si $p_{1,n}(\lambda)$ n'est pas identiquement nul, $\Psi_\lambda(n) = p_{2,n}(\lambda)$ sinon, nous obtenons pour $\sigma(x,.)$ presque tout λ une fonction Ψ_λ sur $\mathbb{Z}$ satisfaisant $H_x \Psi_\lambda = \lambda\ \Psi_\lambda$.

De plus $\sum_{n\in\mathbb{Z}} \frac{|\Psi_\lambda(n)|^2}{n^2} \leq \sum_{n\in\mathbb{Z}} \frac{p_{1,n}^2(\lambda) + p_{2,n}^2(\lambda)}{n^2}$ et l'intégrale de $\sum_{n\in\mathbb{Z}} \frac{|\Psi_\lambda(n)|^2}{n^2}$ en λ est majorée par $\sum_{n\in\mathbb{Z}} \frac{\|\delta_n\|^2}{n^2} < +\infty$.

La fonction $\sum_{n\in\mathbb{Z}} \frac{|\Psi_\lambda(n)|^2}{n^2}$ est donc $\sigma(x,.)$ presque partout finie ce qui montre le lemme 1.3.

Donc pour M presque tout (x,λ), nous pouvons trouver Ψ_λ grâce au lemme 1.3. En posant $\phi = \begin{pmatrix} \Psi_\lambda(1) \\ \Psi_\lambda(0) \end{pmatrix}$ nous avons $A_e^{(n)}(x)\ \phi = \begin{pmatrix} \Psi_\lambda(n+1) \\ \Psi_\lambda(n) \end{pmatrix}$ et le vecteur ϕ ne diverge pas. Les mesures M et m x k sont étrangères, q.e.d.

Nous citons maintenant deux exemples classiques d'applications du critère de Pastour 1.2 :

1.4 - Corollaire

Considérons l'équation de Schrödinger à coëfficients indépendants. Alors pour presque tout x, le spectre de M_x est singulier dès que V prend deux valeurs distinctes.

Il suffit en effet d'appliquer le critère de Fürstenberg III.3.5 aux matrices $\begin{pmatrix} e+V & -1 \\ 1 & 0 \end{pmatrix}$ pour montrer que $\lambda(e) > 0$ partout.

Or il est clair que si a est différent de b il n'y a pas de mesure sur P^1 invariante à la fois par l'action des deux matrices $\begin{pmatrix} a & -1 \\ 1 & 0 \end{pmatrix}$ et $\begin{pmatrix} b & -1 \\ 1 & 0 \end{pmatrix}$.

1.5 - Corollaire

Soient $X = R/Z$, $m = dx$, $\theta x = x + \alpha \pmod 1$ pour un nombre α irrationnel et $V(x) = \varepsilon \cos 2\Pi x$. Le spectre de l'opérateur de Schödinger H_x est singulier dès que $|\varepsilon| > 2$.

Il suffit encore d'appliquer le critère de Pastour 1.2 et le lemme 1.6.

1.6 - Lemme

Considérons la famille $(X, m, \theta; A_e)$ où $X = R/Z$, $m = dx$, $\theta x = x + \alpha$ mod 1, α irrationnel et $A_e(x) = \begin{pmatrix} e + \varepsilon \cos 2\Pi x & -1 \\ 1 & 0 \end{pmatrix}$. Alors $\lambda \geq \log \frac{|\varepsilon|}{2}$.

Démonstration de 1.6 (M. Herman) :

Le système (X, m, θ) est isomorphe à la restriction au cercle unité de la multiplication dans C par β, $\beta = e^{2i\Pi\alpha}$.

$$\text{Si } |Z| = 1 \quad A_e(Z) = \begin{pmatrix} e + \dfrac{\varepsilon Z + \varepsilon Z^{-1}}{2} & -1 \\ 1 & 0 \end{pmatrix} = \frac{1}{Z} B_e(Z) \text{ avec}$$

$$\text{avec } B_e(Z) = \begin{pmatrix} \frac{\varepsilon}{2} + eZ + \frac{\varepsilon}{2} Z^2 & -Z \\ Z & 0 \end{pmatrix}.$$

Nous avons à calculer :

$$\lambda = \inf_n \frac{1}{n} \log \int_{|Z|=1} ||A_e(\beta^{n-1} Z) \ldots A_e(Z)||$$

$$= \inf_n \frac{1}{n} \log \int_{|Z|=1} ||B_e(\beta^{n-1} Z) \ldots B_e(Z)||$$

Pour tout $n \geq 1$, la fonction $Z \to B_e(\beta^{n-1}Z) \ldots B_e(Z)$ est holomorphe de C dans $\mathcal{L}(C^2, C^2)$ et donc le logarithme de sa norme est une fonction sous-harmonique. L'intégrale de cette fonction sur le cercle est minorée par sa valeur au centre :

$$\lambda \geq \lim_n \inf \frac{1}{n} ||B_e^n(0)|| = \lim \inf \frac{1}{n} \log ||\begin{pmatrix} (\varepsilon/2)^n & 0 \\ 0 & 0 \end{pmatrix}|| = \log \frac{|\varepsilon|}{2}.$$

Références :

Ce paragraphe fait seulement une brève incursion dans un domaine passionnant, aussi bien pour la signification des résultats que par les techniques employées.

Nous citons simplement nos sources : l'argument de Pastour est explicite dans :

L.A. PASTUR : Spectral properties of disordered systems in the One. Body Approximation. Commun. Math. Phys. 75(1980), p. 179-186.

Le résultat 1.4 semble remonter à :

A. CASHER et J. LEBOWITZ : J. Math. Phys. 12(1970) 1701 - 1711

H. MATSUDA et K ISHII : Progress Theor. Phys. Suppl. 45 (1970) 56-89.

Le corollaire 1.5 a été conjecturé par G. André et S. Aubry ; la démonstration de 1.6 que nous donnons est due à M. Herman et se trouve (avec bien d'autres résultats) dans :

M. HERMAN : "Une méthode pour minorer les exposants de Lyapunov et quelques exemples montrant le caractère local d'un théorème d'Arnold et de Moser sur le tore en dimension 2". Prétirage Ecole Polytechnique (1982).

Pour en savoir plus ..., on trouvera une bibliographie pour le cas quasi-périodique (c'est-à-dire $X = R^d/ Z^d$, $\theta x = x + \alpha$) dans :

B. SIMON : Almost-periodic Schödinger operators : A review. Adv. in Applied math. (1983)

et pour le cas indépendant ou Markovien dans :

G. ROYER : Etude des opérateurs de Schrödinger à potentiel aléatoire en dimension 1. Bull. Soc. Math. France 110 (1982) p. 27-48.

H. KUNZ et B. SOUILLARD : Commun math. Phys. 78 (1980) 201-246.

J. LACROIX : Localisation pour l'opérateur de Schrödinger aléatoire dans un ruban. Prétirage.

Dans le théorème 1.2 on se trouve en présence d'une famille de matrices aléatoires (X, m, θ, A_e) dépendant d'un paramètre e. Déjà sur cet exemple on peut voir qu'on ne peut pas s'en tirer par un éventuel théorème ergodique sous-additif à valeurs dans un Banach (faux d'après Y. Derriennic et U. Krengel) et que les dépendances en e des objets liés au théorème d'Osseledets ne sont pas simplettes.
Voir à ce sujet I. Golsheid.

Y. DERRIENNIC et U. KRENGEL : Subadditive mean ergodic theorem. Ergod. Th. & Dynam. Syst. I(1981) p. 33-48.

I. GOLSHEID : Asymptotic properties of the product of random matrices depending on a parameter. Advances in Proba. and related topics 6. Multicomponents random systems (1980), p. 239-284.

2. Marche aléatoire sur $\mathbb{Z}$ dans un milieu aléatoire

Soit (X,m,θ) un système ergodique et $p(x,.)$ une application mesurable de X dans l'espace des probabilités sur Z.

On appelle marche aléatoire en milieu aléatoire le processus qui pour chaque x de X est la chaîne de Markov homogène sur Z de probabilité de transition données par :

$$P^x(X^{i+1} = k \mid X^i = \ell) = p(\theta^\ell x, k-\ell).$$

La loi de la marche est obtenue en intégrant en x les lois P^x des chaînes : globalement le processus n'est plus de Markov car le passé donne des informations sur le point x qui dirige la chaîne. Nous supposerons que pour presque tout x la chaîne est irréductible. Nous supposerons également qu'il existe L et R entiers positifs tels que pour presque tout x :

$$p(x, [-L, +R]) = 1 \ , \ p(x, [1,R]) > 0 \text{ et } p(x, [-L, -1]) > 0.$$

La marche a alors en un sens le même comportement asymptotique en presque tout point :

2.1 - Proposition

Sous les hypothèses précédentes :

i) Ou bien pour m presque tout x et tout n > 0,

$$\mathbb{P}_0^x \ (\underline{X}(.) \text{ entre dans } [n, +\infty\ )) = 1$$

ii) Ou bien, pour m presque tout x, on a :

$$\limsup_{n\to+\infty} \frac{1}{n} \log (P_0^x \ (\underline{X}(.) \text{ entre dans } [n,+\infty\ )) \leq \lambda < 0.$$

On a également l'alternative symétrique pour l'entrée dans $(-\infty, -n]$.

Nous avons naturellement noté $\mathbb{P}_0^x$ la loi de la chaîne de probabilité de transition $\mathbb{P}^x(k,\ell)$ partant de $X_0 = 0$. Nous démontrerons cette proposition en appendice (paragraphe 3).

2.2 - Corollaire

Les propriétés i) et ii) de 2.1 sont respectivement équivalentes aux propriétés i)' et ii)' :

i)' pour presque tout x, $\limsup_t X_t = +\infty$ $\mathbb{P}_0^x$ p.s.

ii)' pour presque tout x, $\lim_t X_t = -\infty$ $\mathbb{P}_0^x$ p.s.

Comme i) ou ii) sont les seuls cas possibles il suffit de montrer que i) $\Rightarrow$ i)' et ii) $\Rightarrow$ ii)'.

Si i) est réalisé, nous avons par définition pour tout n,

$$\mathbb{P}_0^x\ (\limsup_t X_t \geq n) = 1 \qquad \text{q.e.d.}$$

Si ii) est réalisé, nous avons à partir d'un certain rang $\mathbb{P}_0^x(\underline{X}(.)$ entre dans $[n,+\infty\ ) \leq e^{-n\ (\lambda+\varepsilon)}$ et donc $\limsup_t X_t < +\infty$ $\mathbb{P}_0^x$ p.s. par Borel Cantelli.

Mais par irréductibilité la loi de la variable $\limsup_t X_t$ sur $\mathbb{Z} \cup \{-\infty\} \cup \{+\infty\}$ sous $\mathbb{P}_0^x$ vérifie, pour tout n de $\mathbb{Z}$:

$$\begin{aligned} \mathbb{P}_0^x\ (\limsup X_t \leq n) &= \mathbb{P}_1^x\ (\limsup X_t \leq n) \\ &= \mathbb{P}_0^{\theta x}\ (\limsup X_t \leq n-1) \\ &\leq \mathbb{P}_0^{\theta x}\ (\limsup X_t \leq n). \end{aligned}$$

Par invariance de m, on a égalité dans l'inégalité ci-dessus ce qui montre à la fois que la loi de $\limsup_t X_t$ est constante et ne charge pas $\{n\}$. Si donc $\limsup_t X_t < +\infty$ $\mathbb{P}_0^x$ p.s., cette loi est portée par $\{-\infty\}$ q.e.d.

2.3 - <u>Théorème</u>

On suppose que $\int \log p(x,-L)\ m(dx) > -\infty$ et $\int \log\ p(x,R)\ m(dx) > -\infty$.
Soient A(x) la matrice (R+L) x (R+L) donnée par :

$$A(x) = \begin{pmatrix} a_{R-1}(x) & \cdots & \cdots & a_{-L}(x) \\ 1 & 0 & \cdots & 0 \\ 0 & 1 & 0 \cdots & 0 \\ \vdots & \vdots & \ddots & \vdots \\ 0 & 0 & \cdots & 1 \end{pmatrix}$$

avec $a_j(x) = \dfrac{\delta_{0,j} - p(x,j)}{p(x,R)}$ et $\lambda_1 \geq \lambda_2 \geq \ldots \geq \lambda_{R+L}$ les exposants caractéristiques de la famille $(X,m,\theta\ ;\ A)$.
Nous avons alors les trois cas suivants :

i) $\lambda_R + \lambda_{R+1} < 0$ si et seulement si $X_t \longrightarrow +\infty$ p.s.

ii) $\lambda_R + \lambda_{R+1} = 0$ si et seulement si $-\infty = \liminf_t X_t < \limsup_t X_t = +\infty$ p.s.

iii) $\lambda_R + \lambda_{R+1} > 0$ si et seulement si $X_t \longrightarrow -\infty$ p.s.

Dans le cas où R = L = 1, ce résultat s'écrit :

2.4 - Corollaire

> Soient p(x), q(x) deux fonctions positives :
>
> $p+q \leq 1 \quad -\int \log p < \infty$ et $-\int \log q < \infty$.
>
> Posons $p(x,+1) = p(x) \quad p(x,-1) = q(x)$
>
> $p(x,0) = 1-p(x)-q(x)$.
>
> Le comportement de la marche est déterminé par le signe de $\int \log \frac{p}{q}$.

En effet la matrice s'écrit $A = \begin{pmatrix} \frac{p+q}{p} & -\frac{q}{p} \\ 1 & 0 \end{pmatrix}$ et dans ce cas :

$$\lambda_1 + \lambda_2 = \int \log \ |\det A| \ = \ \int \log \frac{q}{p} = - \int \log \frac{p}{q} .$$

Démonstration de 2.3 :

Il est clair d'abord que les hypothèses assurent $\int \log^+ ||A(x)|| \ m(dx) < +\infty$. Soit $m \geq 0$. Considérons la loi $\mathbb{P}_m^x$ de la chaîne de Markov partant de $x_0 = m$ et de probabilités de transition $\mathbb{P}^x(k,l)$. Nous pouvons regarder sous $\mathbb{P}_m^x$ le premier instant τ où la trajectoire est négative. Nous dirons que $\underline{X}$ entre dans $(-\infty,0)$ en j si $X_\infty = j$. Les valeurs possibles de j sont -L,...-1.

Posons $F_j(x,m) = \mathbb{P}_m^x$ ($\underline{X}$ entre dans $(-\infty,0)$ en j) si $m \geq 0$

$F_j(x,m) = \delta_{j,m}$ si $m < 0$.

Nous avons clairement si $m \geq 0$

(1) $$F_j(x,m) = \sum_{n \in \mathbb{Z}} p(\theta^m x, n-m) \, F_j(x,n)$$

par la propriété de Markov et la définition de $\mathbb{P}^x$. Nous appelons $\tilde{F}_j(x,m)$ le (R+L) vecteur des coordonnées :

$$\tilde{F}_j(x,m) = \begin{pmatrix} F_j(x,m+R-1) \\ \vdots \\ F_j(x,m-L) \end{pmatrix}$$

La relation (1) s'écrit alors :

$$\tilde{F}_j(x,m+1) = A(\theta^m x) \, \tilde{F}_j(x,m) \quad \text{pour } m \geq 0.$$

Montrons d'abord que nous avons $\lambda_{R+1} \leq 0$. Les vecteurs $\tilde{F}_j(x,0)$ sont en effet linéairement indépendants et nous avons :

$$A^{(n)}(x)\ \tilde{F}_j(x,0) = \tilde{F}_j(x,n).$$

Les coefficients de $\tilde{F}_j(x,n)$ étant compris entre 0 et 1, il vient :

$$\limsup_n \frac{1}{n} \log ||A^{(n)}(x)\ \tilde{F}_j(x,0)|| \leq 0.$$

Nous avons ceci pour L vecteurs indépendants, il y a donc au moins L exposants négatifs ou nuls. Nous avons donc bien $\lambda_{R+1} \leq 0$.

Examinons maintenant les deux cas de la proposition 2.1 pour l'entrée dans $(-\infty,-n]$.

2.5 - <u>Lemme</u>

Si $\limsup_n \frac{1}{n} \log \mathbb{P}_0^x(\underline{X}$ entre dans $(-\infty,-n]) \leq \lambda < 0$, alors $\lambda_{R+1} < 0$.

<u>Démonstration de 2.5</u> :

Pour chacun des L vecteurs indépendants $\tilde{F}_j(x,0)$, d'après I,3.1, $\lim_n \frac{1}{n} \log ||A^{(n)}(x)\ \tilde{F}_j(x,0)||$ existe presque sûrement.

D'autre part, par définition de $\tilde{F}_j(x,m)$,

$$\frac{1}{n} \log ||\tilde{F}_j(x,n)|| \leq \sup_{-L\leq j\leq R} \frac{1}{n} \log (P_{n+j}^x(\underline{X} \text{ entre dans } (-\infty,0)))$$

$$\leq \sup_{-L\leq j\leq R} \frac{1}{n} \log (P_0^{\theta_x^{n+j}}(\underline{X} \text{ entre dans } (-\infty,-n-j)))$$

par définition des chaînes $\mathbb{P}^x$ et $\mathbb{P}^{\theta_x^{n+j}}$.

Par invariance de la mesure m les fonctions $\frac{1}{n}\log(P_0^{\theta_x^{n+j}}(\underline{X}$ entre dans $(-\infty,-n-j)))$ ont même loi que $\frac{1}{n}\log(P_0^x(\underline{X}$ entre dans $(-\infty,-n-j)))$ et donc sont majorées par $\lambda < 0$ sur des ensembles de mesure arbitrairement proches de 1. Les limites presque sûres de $\frac{1}{n}\log||A^{(n)}(x)\ \tilde{F}_j(x,0)||$ sont donc strictement négatives. Il y a donc au moins L exposants négatifs. q.e.d.

Dans l'autre cas, nous avons d'abord un lemme :

2.6 - <u>Lemme</u>

Soient x tel que pour la mesure P_z^x partant de $X_0 = z$, quelque soit $z \geq 0$, la chaîne $\underline{X}$ atteint $(-\infty,0)$ et g une fonction bornée vérifiant :

$$g(m) = \sum_{n\in\mathbb{Z}} p(\theta^n x, n-m)\, g(n) \text{ pour } m \geq 0.$$

$$\text{Alors } g(m) = \sum_{j=-1}^{-L} g(j)\, F_j(x,m).$$

Démonstration de 2.6 :

$$\text{Posons } Q^x(k,\ell) = P^x(k,\ell) \quad \text{si } k \geq 0$$
$$= \delta_{k,\ell} \quad \text{si } k < 0.$$

Nous avons alors pour tout m de Z, $g(m) = \sum_{n\in Z} Q^x(m,n)\, g(n)$.

En notant Q^x_m la loi du processus de Markov de probabilité de transition $Q^x(k,l)$ partant de $Y_0 = m$, d'après l'hypothèse, sous Q^x_m, Y_t finit par atteindre une valeur Y_∞, $-L \leq Y_\infty \leq -1$, et y rester.

Nous avons $Y_\infty = j$ avec la probabilité $F_j(x,m)$. Comme $g(Y_t)$ est une Q^x_m martingale bornée, nous avons :

$$g(m) = \int g(Y_\infty)\, dQ^x_m = \sum_{j=-1}^{-L} g(j)\, F_j(x,m) \quad \text{q.e.d.}$$

2.7 - Lemme

Si pour presque tout x et tout n, P^x_0 (X entre dans $(-\infty, -n]$) = 1, alors $\lambda_{R+1} = 0$.

Démonstration de 2.7 :

Considérons une application mesurable $\tilde{G}$ de X dans $\mathbb{R}^{R+1}$ correspondant à un exposant strictement négatif m presque partout, c'est à dire tel que $\limsup_n \frac{1}{n} \log ||A^{(n)} \tilde{G}(x)|| < 0$ m.p.s. Nous allons montrer que $\tilde{G}(x)$ appartient au sous-espace engendrée par les vecteurs $\tilde{F}_j(x)$. En effet en posant g(x,m) égal à la R-ième composante du vecteur $A^{(m)}(x)\,\tilde{G}(x)$, nous obtenons pour tout x une suite g(x,m) satisfaisant $g(x,m) = \sum_{n\in\mathbb{Z}} p(\theta^m x, n-m)\, g(x,m)$ pour $m \geq 0$, et qui est bornée pour presque tout x.

D'autre part, par invariance les hypothèses de 2.6 sont satisfaites en presque tout x. Nous avons donc par 2.6, en presque tout x,

$$g(x,m) = \sum_{j=-1}^{-L} g(x,j)\, F_j(x,n),$$

ou encore :

$$\tilde{G}(x) = \sum_{j=-1}^{-L} g(x,j)\, \tilde{F}_j(x).$$

D'après I.3.1 nous venons de voir que tous les exposants strictement négatifs sont obtenus en considérant l'espace de dimension L engendré par les vecteurs $\tilde{F}_i$. Dans cet espace, par hypothèse, il y a aussi le vecteur $\tilde{1}$ de coordonnées constantes toutes égales à 1, $\tilde{1} = \sum_i \tilde{F}_j$.

Ce vecteur $\tilde{1}$ est invariant et donne donc l'exposant 0. Il ne peut y avoir au plus que L-1 exposants strictement négatifs. Nous avons donc $\lambda_{R+1} = 0$, q.e.d.

Revenons à la démonstration du théorème. D'après 2.2, 2.5 et 2.7 nous avons :

$$\lambda_{R+1} = 0 \quad \text{si et seulement si } \lim_t \inf X_t = -\infty \qquad \text{p.s.}$$

$$\lambda_{R+1} < 0 \quad \text{si et seulement si } \lim_t X_t = +\infty \qquad \text{p.s.}$$

Pour examiner le comportement à droite, nous avons à faire le même raisonnement partant de m < 0 et en regardant l'entrée du processus $\underline{X}$ dans $[0,+\infty)$.

Les fonctions $F'_{j,m}$ sont définies par :

$$F'_j(x,m) = P^x_m(\underline{X} \text{ entre dans } [0,+\infty) \text{ en } j) \qquad \text{si } m < 0$$

$$F'_j(x,m) = \delta_{j,m} \qquad \text{si } m \geq 0$$

pour j=0,..., R-1.

Nous avons encore l'équation (1) mais désormais nous voulons exprimer $F'_j(x,m-L)$ en fonction de $F'_j(x, m+s)$ s = - L + 1, ... R. Comme c'est la même équation, nous obtenons une matrice A'(x) et nous devons avoir en posant :

$$\mathcal{F}'_j(x,m) = \begin{pmatrix} F_j(x,m-L) \\ \\ F_j(x,m+R-1) \end{pmatrix}$$

$\mathcal{F}'_j(x,m) = A'(\theta^{-m}x)\, \mathcal{F}'_j(x,m+1)$, $m < -1$ et $J\,A'(x)J = A^{-1}(x)$ si J désigne l'involution qui renverse l'ordre des coordonnées. Les exposants de la famille $(X, \mathcal{a}, m, \theta^{-1}, A')$ sont les mêmes que ceux de $(X, \mathcal{a}, m, \theta^{-1}, A^{-1})$ et d'après I.4.1 sont donc les nombres $-\lambda_{R+L} \geq \ldots\ldots \geq -\lambda_1$.

La même démonstration nous dit alors de considérer le $(L+1)^{\text{ième}}$ exposant c'est-à-dire $-\lambda_R$, et que :

$-\lambda_R = 0$ si et seulement si $\lim_t \sup X_t = +\infty$ p.s.

$-\lambda_R < 0$ si et seulement si $\lim_t X_t = -\infty$ p.s.

En comparant ces résultats et en examinant les différentes possibilités, $\lambda_R > 0$ et $\lambda_{R+1} < 0$ est impossible et les autres cas correspondent à l'énoncé du théorème 2.3 .

Remarques et références :

Les résultats de ce paragraphe (et du suivant) sont empruntés à la thèse d'Eric Key (Cornell University - Janvier 83). Le corollaire 2.4 est dû à Solomon.

F. SOLOMON : Random walks in a random environment . Ann. Proba 3(1975) p. 1-31.

Le lemme 2.6 est classique. Voir par exemple :

J.M. MERTENS. E. SAMUEL-COHN, S.YAMIR : J. Appl. Proba. 15 (1978) 848-851.

La discussion, en général, de l'irréductibilité est assez délicate. Voir E. Key pour le cas indépendant.

3. Appendice. Démonstration de la proposition 2.1

Fixons x dans X et posons Q(x,j) la probabilité sur $\{1,2,\ldots, R, \Delta\}$ définie par :

$$Q(x,j) = P_0^x(\underline{X}_\tau = j \text{ si } \tau \text{ est le premier instant où } \underline{X}. > 0$$

$$Q(x,\Delta) = 1 - \sum_{j=1}^{R} Q(x,j) .$$

La chaîne de Markov $\underline{Y}$ sur $N \cup \Delta$ de probabilités de transition :

$$Q^x(Y_{k+1} = m + n \mid Y_k = n) = Q(\theta^n x, m)$$

si n appartient à N, m = 1,2,... R, Δ

et $\quad Q^x(Y_{k+1} = \Delta | Y_k = \Delta) = 1$

est le processus des avancées successives du processus X dirigé par le même point x de X.

Si $\int Q(x,\Delta)\, m(dx) = 0$, pour presque tout x, le processus $\underline{Y}$ reste dans N et progresse de au moins un à chaque instant.

Le processus $\underline{X}$ entre donc dans chaque demi-droite $[n,+\infty)$ et on est dans le cas i).

Supposons au contraire que $\int Q(x,\Delta)\, m(dx) > 0$. Si le point x est tel que le processus $\underline{X}$ dirigé par x entre dans $[n,+\infty)$, il a dû faire au moins $\left[\frac{n}{R}\right] -1$ avancées. Le processus $\underline{Y}$ n'est donc pas tombé en Δ avant le $(\left[\frac{n}{R}\right] - 1)^{\text{ème}}$ instant. Ceci montre que :

(1) $$P_0^x\ (\underline{X} \text{ entre dans } [n,+\infty\ )) \leq Q_0^x\ (Y_{\left[\frac{n}{R}\right]-1} \neq \Delta\ ) .$$

Pour estimer le deuxième membre de (1) remarquons d'abord que pour tout x et tout j=1,...,R, il existe une trajectoire du processus $\underline{X}$, finie et de probabilité positive telle que :

$$X_0 = 0 \quad X_i < 0 \quad 0 < i < s_x \qquad X_{s_x} = - j,$$

et que nous avons

$$Q(x,\Delta) \geq \tau_j(x) \text{ en posant :}$$

$$\tau_j(x) = P_0^x(X_i = x_i \quad 0 \leq i \leq s_x)\ \ Q(\theta^{-j}\, x, \Delta) .$$

Remarquons alors que si $\sigma(x) = \inf_{0 \leq j \leq R} \tau_j(\theta^j x)$ nous avons $\sigma(x) > 0$ sur un ensemble de mesure positive et que $Q(x,\Delta) \geq \sup_{0 \leq j \leq R} \sigma(\theta^{-j} x)$.

Soit B_ρ le sous-ensemble de X où $\sigma(x) > \rho$ et choisissons $\rho > 0$ tel que $m(B_\rho) > 0$ et x dans l'ensemble de mesure 1 où les moyennes $\frac{1}{n} \sum_{i=0}^{n-1} \chi_{B_\rho}(\theta^i x)$ convergent vers $m(B_\rho)$.

Nous découpons alors N en intervalles disjoints $[n_t, n_t + R]$ $t = 0,1,\ldots$ tels que $\theta x^{n_t} \in B_\rho$ et $\theta x^j \notin B_\rho$ sur le complémentaire.

De chaque suite strictement croissante de pas inférieur à R $y_1, y_2, \ldots y_j$, nous extrayons la sous-suite y'_t où y'_t est le premier élément de la suite y_i dans l'intervalle $[n_{2t}, n_{2t} + R]$ $n_{2t} \leq j$, $t = 1,\ldots$

Il est facile de voir que la sous-suite y'_t a au moins $\frac{1}{2(R+2)} \sum_{i=0}^{j} \chi_{B_\rho}(\theta^i x)$ éléments et que par construction $Q(\theta^{y'_t} x,\Delta) \geq \rho$ pour tout t.

Nous pouvons alors estimer le deuxième membre de (1) par :

$$\sum_{y_1,\ldots y_{[\frac{n}{R}]-1}} Q(x,y_1)\, Q(\theta^{y_1} x, y_2) \ldots Q(\theta^{y_{[\frac{n}{R}]-2}} x, y_{[\frac{n}{R}]-1})$$

où la somme porte sur toutes les suites croissantes de pas $\leq R$.

En découpant selon la sous-suite y'_t associée à y_j , en majorant pour toutes les suites où $y'_t = y_{a(t)}$ la somme $\sum_{y_{a(t)+1}} Q(\theta^{y_{a(t)}} x, y_{a(t)+1})$ par $1-Q(\theta^{y'_t} x,\Delta) \leq 1 - \rho$, nous obtenons :

$$Q_0^x\,(y_{[\frac{n}{R}]-1} \neq \Delta) \leq (1-\rho)^{\frac{1}{2(R+2)} \sum_{i=0}^{[\frac{n}{R}]-1} \chi_{B_\rho}(\theta^i x)}.$$

Nous obtenons donc d'après (1) pour presque tout x :

$$\limsup_n \frac{1}{n} \log P_0^x (\underline{X} \text{ entre dans } [n, +\infty))$$

$$\leq \limsup_n \Big(\frac{1}{2n(R+2)} \sum_{i=0}^{[\frac{n}{R}]-1} \chi_{B_\rho}(\theta^i x)\Big) \times \log (1-\rho)$$

$$\leq \frac{m(B_\rho)}{2R(R+2)} \log(1-\rho) < 0.$$

Ceci achève la démonstration de 2.1.

REFERENCES GENERALES

- P. BILLINGSLEY : Ergodic theory and information. Wiley and Sons (1965).

- H. FÜRSTENBERG : Non-commuting random products Transactions Amer. Math. Soc. 108 (1963) p. 377-428.

- H. FURSTENBERG and H. KESTEN : Products of random matrices. Ann. Math. Stat. 31 (1960) p. 457-469.

- Y. GUIVARC'H : Quelques propriétés asymptotiques des produits de matrices aléatoires. Ecole d'été de probabilités de Saint-Flour VIII 1978. Springer Verlag in maths. 774 (1980).

- G. LETAC V. SESHADRI : Z für W. 62 (1983) p. 485-489.

- R. MANE : A proof of Pesin's formula. Ergod th. and Dynam. Sys 1(1981) 77-93.

- V.I. OSELEDEC : A multiplicative ergodic theorem. Lyapunov characteristic numbers for dynamical systems. Trans. Mocow Math. Soc. 19 (1968)197-231.

- Ya. B. PESIN : Lyapunov characteristic exponents and Smooth ergodic theory Russ. Math. Surveys 32;4 (1977) 55-114.

- D. RUELLE : Ergodic theory of differentiable dynamical systems. Publ. Math. IHES 50 (1979) 27-58.

- L.S. YOUNG : Dimension, entropy and Lyapunov exponents. Ergod. th. and Dynam. syst. II (1982) p. 109-124.

Vol. 952: Combinatorial Mathematics IX. Proceedings, 1981. Edited by E. Billington, S. Oates-Williams, and A.P. Street. XI, 443 pages. 1982.

Vol. 953: Iterative Solution of Nonlinear Systems of Equations. Proceedings, 1982. Edited by R. Ansorge, Th. Meis, and W. Törnig. VII, 202 pages. 1982.

Vol. 954: S.G. Pandit, S.G. Deo, Differential Systems Involving Impulses. VII, 102 pages. 1982.

Vol. 955: G. Gierz, Bundles of Topological Vector Spaces and Their Duality. IV, 296 pages. 1982.

Vol. 956: Group Actions and Vector Fields. Proceedings, 1981. Edited by J.B. Carrell. V, 144 pages. 1982.

Vol. 957: Differential Equations. Proceedings, 1981. Edited by D.G. de Figueiredo. VIII, 301 pages. 1982.

Vol. 958: F.R. Beyl, J. Tappe, Group Extensions, Representations, and the Schur Multiplicator. IV, 278 pages. 1982.

Vol. 959: Géométrie Algébrique Réelle et Formes Quadratiques, Proceedings, 1981. Edité par J.-L. Colliot-Thélène, M. Coste, L. Mahé, et M.-F. Roy. X, 458 pages. 1982.

Vol. 960: Multigrid Methods. Proceedings, 1981. Edited by W. Hackbusch and U. Trottenberg. VII, 652 pages. 1982.

Vol. 961: Algebraic Geometry. Proceedings, 1981. Edited by J.M. Aroca, R. Buchweitz, M. Giusti, and M. Merle. X, 500 pages. 1982.

Vol. 962: Category Theory. Proceedings, 1981. Edited by K.H. Kamps D. Pumplün, and W. Tholen, XV, 322 pages. 1982.

Vol. 963: R. Nottrot, Optimal Processes on Manifolds. VI, 124 pages. 1982.

Vol. 964: Ordinary and Partial Differential Equations. Proceedings, 1982. Edited by W.N. Everitt and B.D. Sleeman. XVIII, 726 pages. 1982.

Vol. 965: Topics in Numerical Analysis. Proceedings, 1981. Edited by P.R. Turner. IX, 202 pages. 1982.

Vol. 966: Algebraic K-Theory. Proceedings, 1980, Part I. Edited by R.K. Dennis. VIII, 407 pages. 1982.

Vol. 967: Algebraic K-Theory. Proceedings, 1980. Part II. VIII, 409 pages. 1982.

Vol. 968: Numerical Integration of Differential Equations and Large Linear Systems. Proceedings, 1980. Edited by J. Hinze. VI, 412 pages. 1982.

Vol. 969: Combinatorial Theory. Proceedings, 1982. Edited by D. Jungnickel and K. Vedder. V, 326 pages. 1982.

Vol. 970: Twistor Geometry and Non-Linear Systems. Proceedings, 1980. Edited by H.-D. Doebner and T.D. Palev. V, 216 pages. 1982.

Vol. 971: Kleinian Groups and Related Topics. Proceedings, 1981. Edited by D.M. Gallo and R.M. Porter. V, 117 pages. 1983.

Vol. 972: Nonlinear Filtering and Stochastic Control. Proceedings, 1981. Edited by S.K. Mitter and A. Moro. VIII, 297 pages. 1983.

Vol. 973: Matrix Pencils. Proceedings, 1982. Edited by B. Kågström and A. Ruhe. XI, 293 pages. 1983.

Vol. 974: A. Draux, Polynômes Orthogonaux Formels – Applications. VI, 625 pages. 1983.

Vol. 975: Radical Banach Algebras and Automatic Continuity. Proceedings, 1981. Edited by J.M. Bachar, W.G. Bade, P.C. Curtis Jr., H.G. Dales and M.P. Thomas. VIII, 470 pages. 1983.

Vol. 976: X. Fernique, P.W. Millar, D.W. Stroock, M. Weber, Ecole d'Eté de Probabilités de Saint-Flour XI – 1981. Edited by P.L. Hennequin. XI, 465 pages. 1983.

Vol. 977: T. Parthasarathy, On Global Univalence Theorems. VIII, 106 pages. 1983.

Vol. 978: J. Ławrynowicz, J. Krzyż, Quasiconformal Mappings in the Plane. VI, 177 pages. 1983.

Vol. 979: Mathematical Theories of Optimization. Proceedings, 1981. Edited by J.P. Cecconi and T. Zolezzi. V, 268 pages. 1983.

Vol. 980: L. Breen. Fonctions thêta et théorème du cube. XIII, 115 pages. 1983.

Vol. 981: Value Distribution Theory. Proceedings, 1981. Edited by I. Laine and S. Rickman. VIII, 245 pages. 1983.

Vol. 982: Stability Problems for Stochastic Models. Proceedings, 1982. Edited by V.V. Kalashnikov and V.M. Zolotarev. XVII, 295 pages. 1983.

Vol. 983: Nonstandard Analysis-Recent Developments. Edited by A.E. Hurd. V, 213 pages. 1983.

Vol. 984: A. Bove, J.E. Lewis, C. Parenti, Propagation of Singularities for Fuchsian Operators. IV, 161 pages. 1983.

Vol. 985: Asymptotic Analysis II. Edited by F. Verhulst. VI, 497 pages. 1983.

Vol. 986: Séminaire de Probabilités XVII 1981/82. Proceedings. Edited by J. Azéma and M. Yor. V, 512 pages. 1983.

Vol. 987: C.J. Bushnell, A. Fröhlich, Gauss Sums and p-adic Division Algebras. XI, 187 pages. 1983.

Vol. 988: J. Schwermer, Kohomologie arithmetisch definierter Gruppen und Eisensteinreihen. III, 170 pages. 1983.

Vol. 989: A.B. Mingarelli, Volterra-Stieltjes Integral Equations and Generalized Ordinary Differential Expressions. XIV, 318 pages. 1983.

Vol. 990: Probability in Banach Spaces IV. Proceedings, 1982. Edited by A. Beck and K. Jacobs. V, 234 pages. 1983.

Vol. 991: Banach Space Theory and its Applications. Proceedings, 1981. Edited by A. Pietsch, N. Popa and I. Singer. X, 302 pages. 1983.

Vol. 992: Harmonic Analysis, Proceedings, 1982. Edited by G. Mauceri, F. Ricci and G. Weiss. X, 449 pages. 1983.

Vol. 993: R.D. Bourgin, Geometric Aspects of Convex Sets with the Radon-Nikodým Property. XII, 474 pages. 1983.

Vol. 994: J.-L. Journé, Calderón-Zygmund Operators, Pseudo-Differential Operators and the Cauchy Integral of Calderón. VI, 129 pages. 1983.

Vol. 995: Banach Spaces, Harmonic Analysis, and Probability Theory. Proceedings, 1980–1981. Edited by R.C. Blei and S.J. Sidney. V, 173 pages. 1983.

Vol. 996: Invariant Theory. Proceedings, 1982. Edited by F. Gherardelli. V, 159 pages. 1983.

Vol. 997: Algebraic Geometry – Open Problems. Edited by C. Ciliberto, F. Ghione and F. Orecchia. VIII, 411 pages. 1983.

Vol. 998: Recent Developments in the Algebraic, Analytical, and Topological Theory of Semigroups. Proceedings, 1981. Edited by K.H. Hofmann, H. Jürgensen and H.J. Weinert. VI, 486 pages. 1983.

Vol. 999: C. Preston, Iterates of Maps on an Interval. VII, 205 pages. 1983.

Vol. 1000: H. Hopf, Differential Geometry in the Large. VII, 184 pages. 1983.

Vol. 1001: D.A. Hejhal, The Selberg Trace Formula for PSL(2, IR). Volume 2. VIII, 806 pages. 1983.

Vol. 1002: A. Edrei, E.B. Saff, R.S. Varga, Zeros of Sections of Power Series. VIII, 115 pages. 1983.

Vol. 1003: J. Schmets, Spaces of Vector-Valued Continuous Functions. VI, 117 pages. 1983.

Vol. 1004: Universal Algebra and Lattice Theory. Proceedings, 1982. Edited by R.S. Freese and O.C. Garcia. VI, 308 pages. 1983.

Vol. 1005: Numerical Methods. Proceedings, 1982. Edited by V. Pereyra and A. Reinoza. V, 296 pages. 1983.

Vol. 1006: Abelian Group Theory. Proceedings, 1982/83. Edited by R. Göbel, L. Lady and A. Mader. XVI, 771 pages. 1983.

Vol. 1007: Geometric Dynamics. Proceedings, 1981. Edited by J. Palis Jr. IX, 827 pages. 1983.

Vol. 1008: Algebraic Geometry. Proceedings, 1981. Edited by J. Dolgachev. V, 138 pages. 1983.

Vol. 1009: T. A. Chapman, Controlled Simple Homotopy Theory and Applications. III, 94 pages. 1983.

Vol. 1010: J.-E. Dies, Chaînes de Markov sur les permutations. IX, 226 pages. 1983.

Vol. 1011: J.M. Sigal. Scattering Theory for Many-Body Quantum Mechanical Systems. IV, 132 pages. 1983.

Vol. 1012: S. Kantorovitz, Spectral Theory of Banach Space Operators. V, 179 pages. 1983.

Vol. 1013: Complex Analysis – Fifth Romanian-Finnish Seminar. Part 1. Proceedings, 1981. Edited by C. Andreian Cazacu, N. Boboc, M. Jurchescu and I. Suciu. XX, 393 pages. 1983.

Vol. 1014: Complex Analysis – Fifth Romanian-Finnish Seminar. Part 2. Proceedings, 1981. Edited by C. Andreian Cazacu, N. Boboc, M. Jurchescu and I. Suciu. XX, 334 pages. 1983.

Vol. 1015: Equations différentielles et systèmes de Pfaff dans le champ complexe – II. Seminar. Edited by R. Gérard et J.P. Ramis. V, 411 pages. 1983.

Vol. 1016: Algebraic Geometry. Proceedings, 1982. Edited by M. Raynaud and T. Shioda. VIII, 528 pages. 1983.

Vol. 1017: Equadiff 82. Proceedings, 1982. Edited by H.W. Knobloch and K. Schmitt. XXIII, 666 pages. 1983.

Vol. 1018: Graph Theory, Łagów 1981. Proceedings, 1981. Edited by M. Borowiecki, J.W. Kennedy and M.M. Sysło. X, 289 pages. 1983.

Vol. 1019: Cabal Seminar 79–81. Proceedings, 1979–81. Edited by A.S. Kechris, D.A. Martin and Y.N. Moschovakis. V, 284 pages. 1983.

Vol. 1020: Non Commutative Harmonic Analysis and Lie Groups. Proceedings, 1982. Edited by J. Carmona and M. Vergne. V, 187 pages. 1983.

Vol. 1021: Probability Theory and Mathematical Statistics. Proceedings, 1982. Edited by K. Itô and J.V. Prokhorov. VIII, 747 pages. 1983.

Vol. 1022: G. Gentili, S. Salamon and J.-P. Vigué. Geometry Seminar "Luigi Bianchi", 1982. Edited by E. Vesentini. VI, 177 pages. 1983.

Vol. 1023: S. McAdam, Asymptotic Prime Divisors. IX, 118 pages. 1983.

Vol. 1024: Lie Group Representations I. Proceedings, 1982–1983. Edited by R. Herb, R. Lipsman and J. Rosenberg. IX, 369 pages. 1983.

Vol. 1025: D. Tanré, Homotopie Rationnelle: Modèles de Chen, Quillen, Sullivan. X, 211 pages. 1983.

Vol. 1026: W. Plesken, Group Rings of Finite Groups Over p-adic Integers. V, 151 pages. 1983.

Vol. 1027: M. Hasumi, Hardy Classes on Infinitely Connected Riemann Surfaces. XII, 280 pages. 1983.

Vol. 1028: Séminaire d'Analyse P. Lelong – P. Dolbeault – H. Skoda. Années 1981/1983. Edité par P. Lelong, P. Dolbeault et H. Skoda. VIII, 328 pages. 1983.

Vol. 1029: Séminaire d'Algèbre Paul Dubreil et Marie-Paule Malliavin. Proceedings, 1982. Edité par M.-P. Malliavin. V, 339 pages. 1983.

Vol. 1030: U. Christian, Selberg's Zeta-, L-, and Eisensteinseries. XII, 196 pages. 1983.

Vol. 1031: Dynamics and Processes. Proceedings, 1981. Edited by Ph. Blanchard and L. Streit. IX, 213 pages. 1983.

Vol. 1032: Ordinary Differential Equations and Operators. Proceedings, 1982. Edited by W.N. Everitt and R.T. Lewis. XV, 521 pages. 1983.

Vol. 1033: Measure Theory and its Applications. Proceedings, 1982. Edited by J.M. Belley, J. Dubois and P. Morales. XV, 317 pages. 1983.

Vol. 1034: J. Musielak, Orlicz Spaces and Modular Spaces. V, 222 pages. 1983.

Vol. 1035: The Mathematics and Physics of Disordered Media. Proceedings, 1983. Edited by B.D. Hughes and B.W. Ninham. VII, 432 pages. 1983.

Vol. 1036: Combinatorial Mathematics X. Proceedings, 1982. Edited by L.R.A. Casse. XI, 419 pages. 1983.

Vol. 1037: Non-linear Partial Differential Operators and Quantization Procedures. Proceedings, 1981. Edited by S.I. Andersson and H.-D. Doebner. VII, 334 pages. 1983.

Vol. 1038: F. Borceux, G. Van den Bossche, Algebra in a Localic Topos with Applications to Ring Theory. IX, 240 pages. 1983.

Vol. 1039: Analytic Functions, Błażejewko 1982. Proceedings. Edited by J. Ławrynowicz. X, 494 pages. 1983

Vol. 1040: A. Good, Local Analysis of Selberg's Trace Formula. III, 128 pages. 1983.

Vol. 1041: Lie Group Representations II. Proceedings 1982–1983. Edited by R. Herb, S. Kudla, R. Lipsman and J. Rosenberg. IX, 340 pages. 1984.

Vol. 1042: A. Gut, K.D. Schmidt, Amarts and Set Function Processes. III, 258 pages. 1983.

Vol. 1043: Linear and Complex Analysis Problem Book. Edited by V.P. Havin, S.V. Hruščëv and N.K. Nikol'skii. XVIII, 721 pages. 1984.

Vol. 1044: E. Gekeler, Discretization Methods for Stable Initial Value Problems. VIII, 201 pages. 1984.

Vol. 1045: Differential Geometry. Proceedings, 1982. Edited by A.M. Naveira. VIII, 194 pages. 1984.

Vol. 1046: Algebraic K-Theory, Number Theory, Geometry and Analysis. Proceedings, 1982. Edited by A. Bak. IX, 464 pages. 1984.

Vol. 1047: Fluid Dynamics. Seminar, 1982. Edited by H. Beirão da Veiga. VII, 193 pages. 1984.

Vol. 1048: Kinetic Theories and the Boltzmann Equation. Seminar, 1981. Edited by C. Cercignani. VII, 248 pages. 1984.

Vol. 1049: B. Iochum, Cônes autopolaires et algèbres de Jordan. VI, 247 pages. 1984.

Vol. 1050: A. Prestel, P. Roquette, Formally p-adic Fields. V, 167 pages. 1984.

Vol. 1051: Algebraic Topology, Aarhus 1982. Proceedings. Edited by I. Madsen and B. Oliver. X, 665 pages. 1984.

Vol. 1052: Number Theory. Seminar, 1982. Edited by D.V. Chudnovsky, G.V. Chudnovsky, H. Cohn and M.B. Nathanson. V, 309 pages. 1984.

Vol. 1053: P. Hilton, Nilpotente Gruppen und nilpotente Räume. V, 221 pages. 1984.

Vol. 1054: V. Thomée, Galerkin Finite Element Methods for Parabolic Problems. VII, 237 pages. 1984.

Vol. 1055: Quantum Probability and Applications to the Quantum Theory of Irreversible Processes. Proceedings, 1982. Edited by L. Accardi, A. Frigerio and V. Gorini. VI, 411 pages. 1984.

Vol. 1056: Algebraic Geometry. Bucharest 1982. Proceedings, 1982. Edited by L. Bădescu and D. Popescu. VII, 380 pages. 1984.

Vol. 1057: Bifurcation Theory and Applications. Seminar, 1983. Edited by L. Salvadori. VII, 233 pages. 1984.

Vol. 1058: B. Aulbach, Continuous and Discrete Dynamics near Manifolds of Equilibria. IX, 142 pages. 1984.

Vol. 1059: Séminaire de Probabilités XVIII, 1982/83. Proceedings. Edité par J. Azéma et M. Yor. IV, 518 pages. 1984.

Vol. 1060: Topology. Proceedings, 1982. Edited by L.D. Faddeev and A.A. Mal'cev. VI, 389 pages. 1984.

Vol. 1061: Séminaire de Théorie du Potentiel. Paris, No. 7. Proceedings. Directeurs: M. Brelot, G. Choquet et J. Deny. Rédacteurs: F. Hirsch et G. Mokobodzki. IV, 281 pages. 1984.